Urban Ecology

Urban Ecology

Its Nature and Challenges

Edited by

Pedro Barbosa

Department of Entomology
University of Maryland
College Park, Maryland

CABI

CABI is a trading name of CAB International

CABI
Nosworthy Way
Wallingford
Oxfordshire OX10 8DE
UK

Tel: +44 (0)1491 832111
Fax: +44 (0)1491 833508
E-mail: info@cabi.org
Website: www.cabi.org

CABI
WeWork
One Lincoln St
24th Floor
Boston, MA 02111
USA

T: +1 (617)682-9015
E-mail: cabi-nao@cabi.org

British Library Cataloguing-in-Publication Data
A catalogue record for this book is available from the British Library, London.

Library of Congress Cataloging-in-Publication Data
Names: Barbosa, Pedro, 1944- editor.
Title: Urban ecology : its nature and challenges / edited by Pedro Barbosa, Department of Entomology, University of Maryland, College Park, Maryland.
Description: Wallingford, Oxfordshire ; Boston, MA : CAB International, [2020] | Includes bibliographical references and index. | Summary: "This book contains a series of essays designed for graduate students and researchers on crucial issues in the relatively new field of urban ecology"-- Provided by publisher.
Identifiers: LCCN 2020023623 (print) | LCCN 2020023624 (ebook) | ISBN 9781789242607 (hardback) | ISBN 9781789242614 (ebook) | ISBN 9781789242621 (epub)
Subjects: LCSH: Urban ecology (Biology)
Classification: LCC QH541.5.C6 U718 2020 (print) | LCC QH541.5.C6 (ebook) | DDC 577.5/6--dc23
LC record available at https://lccn.loc.gov/2020023623
LC ebook record available at https://lccn.loc.gov/2020023624

References to internet websites (URLs) were accurate at the time of writing.
ISBN: 978 1 78924 2607 (hardback)
978 1 78924 2614 (ePDF)
978 1 78924 2621 (ePub)

Commissioning Editor: Ward Cooper
Editorial Assistant: Lauren Davies
Production Editor: James Bishop

Typeset by Exeter Premedia Services Pvt Ltd, Chennai, India
Printed and bound in the UK by CPI Group (UK) Ltd, Croydon, CR0 4YY

Contents

Contributors vii

Preface ix

1. **Urban Ecology: What Is It and Why Do We Need It?** 1
Nancy B. Grimm

2. **How Urban Forest Composition Shapes the Structure and Function of Arthropod Communities** 15
Holly M. Martinson, Michael J. Raupp and Steven D. Frank

3. **Wildlife in the City: Human Drivers and Human Consequences** 37
Susannah B. Lerman, Desiree L. Narango, Riley Andrade, Paige S. Warren, Aaron M. Grade and Katherine Straley

4. **Urban Plant Ecophysiology** 67
Nancy Falxa Sonti

5. **Urban Biodiversity: Between Elsewhere-Nature and the Post-Wild World** 85
Lea R. Johnson

6. **Protecting Pollinators in the Urban Environment** 100
Lisa Kuder, Lindsay Miller Barranco and Dennis van Engelsdorp

7. **Urban Ecology as an Integrative Science and Practice** 122
Steward T.A. Pickett, J. Morgan Grove, Shannon L. LaDeau, Emma J. Rosi and Mary L. Cadenasso

8. **Ecology of Urban Climates: The Need for Landscape Biophysics in Cities** 144
Elsa Youngsteadt and Adam J. Terando

9. **The Green Cities Movement: Progress and Future Challenges** 160
John G. Kelcey

10. **Light Pollution: How Urban Conditions Influence Ecological Interactions** 176
Iriana Zuria, Ignacio Castellanos and Pedro Barbosa

11. **Citizen Science in Urban Habitats** 184
Amanda E. Sorensen and Rebecca C. Jordan

12. Ecology of Insects and Other Arthropods in Urban Agroecosystems 193
Heidi Liere and Monika Egerer

13. Future Prospects and Challenges in Urban Ecology 214
Gail A. Langellotto

Index 225

Contributors

Aaron M. Grade (Chapter 3), Organismic and Evolutionary Biology Program, University of Massachusetts, Amherst, Massachusetts

Adam J. Terando (Chapter 8), US Geological Survey, Southeast Climate Adaptation Science Center, Raleigh, North Carolina, and Department of Applied Ecology, North Carolina State University, Raleigh, North Carolina

Amanda E. Sorensen (Chapter 11), Communication and Outreach Specialist, Department of Community Sustainability, College of Agriculture and Natural Resources, Michigan State University, East Lansing, Michigan

Dennis van Engelsdorp (Chapter 6), Department of Entomology, University of Maryland, College Park, Maryland

Desiree L. Narango (Chapter 3), Advanced Science Research Center, City University of New York, New York

Elsa Youngsteadt (Chapter 8), Department of Applied Ecology, North Carolina State University, Raleigh, North Carolina

Emma J. Rosi (Chapter 7), Cary Institute of Ecosystem Studies, Millbrook, New York

Gail A. Langellotto (Chapter 13), Department of Horticulture, College of Agricultural Sciences, Oregon State University, Corvallis, Oregon

Heidi Liere (Chapter 12), Department of Environmental Studies, Seattle University, Seattle Washington

Holly Martinson (Chapter 2), Department of Biology, McDaniel College, Westminster, Maryland

Ignacio Castellanos (Chapter10), Biological Interactions Laboratory, Biological Research Center, Autonomous University of the State of Hidalgo, Mexico

Iriana Zuria (Chapter 10), Institute of Basic Sciences and Engineering, Biological Research Center, Autonomous University of the State of Hidalgo, Mexico

John G. Kelcey (Chapter 9), Independent Consultant

J. Morgan Grove (Chapter 7), USDA Forest Service, Baltimore, Maryland

Katherine Straley (Chapter 3), Organismic and Evolutionary Biology Program, University of Massachusetts, Amherst, Massachusetts

Lea Johnson (Chapter 5), Research and Conservation Division, Longwood Gardens, Kennett Square, Pennsylvania

Lindsay Miller Barranco (Chapter 6), Department of Entomology, University of Maryland, College Park, Maryland

Lisa Kuder (Chapter 6), Department of Entomology, University of Maryland, College Park, Maryland

Mary L. Cadenasso (Chapter 7), Department of Plant Sciences, University of California, Davis, California

Michael J. Raupp (Chapter 2), Department of Entomology, University of Maryland, College Park, Maryland

Monika Egerer (Chapter 12), Environmental Studies Department, University of California, Santa Cruz, California. Department of Ecology and Ecosystem Management, School of Life Sciences Weihenstephan, Technische Universität München, Munich, Germany

Nancy B. Grimm (Chapter 1), School of Life Sciences, Arizona State University, Tempe, Arizona

Nancy Falxa Sonti (Chapter 4), USDA Forest Service, Baltimore, Maryland

Pedro Barbosa (Chapter 10), Department of Entomology, University of Maryland, College Park, Maryland

Paige S. Warren (Chapter 3), Department of Environmental Conservation, University of Massachusetts, Amherst, Massachusetts

Rebecca C. Jordan (Chapter 11), Professor and Chair, Department of Community Sustainability, College of Agriculture and Natural Resources, Michigan State University, East Lansing, Michigan

Riley Andrade (Chapter 3), School of Geographical Sciences and Urban Planning, Arizona State University, Tempe, Arizona

Shannon L. LaDeau (Chapter 7), Cary Institute of Ecosystem Studies, Millbrook, New York

Steven Frank (Chapter 2), Department of Entomology and Plant Pathology, North Carolina State University, Raleigh, North Carolina

Steward T.A. Pickett (Chapter 7), Cary Institute of Ecosystem Studies, Millbrook, New York

Susannah B. Lerman (Chapter 3), USDA Forest Service Northern Research Station, USDA Forest Service Northern Research Station, Amherst, Massachusetts, Amherst, Massachusetts

Preface

Today, 55% of the world's population lives in urban areas. Estimates suggest that the world's population is expected to increase by 68% by 2050 (United Nations, 2005). Further, by 2030, 80–90% of the global population will live in cities (United Nations, 2005; Seto *et al.*, 2012). In the USA, the 2012 census reported that more than 80% of the US population lived in urban areas (Barton and Tobin, 2001). Others estimate that 'more than half the world's population lives in cities and suburbs' (Grimm *et al.*, 2008) and an estimated 80% of the world's population will live in urbanized areas by 2050 (Greenhalgh *et al.*, 2007). Indeed, some have suggested that sometime in the next 20–30 years, developing countries in Asia and Africa are likely to cross a historic threshold, joining Latin America in having a majority of urban residents (Montgomery, 2008). By 2030, according to the projections of the United Nations (United Nations, 2005; United Nations Population Fund, 2007 (Population Division)), each of the major regions of the developing world will hold more urban than rural dwellers. Further, by 2050, two thirds of their inhabitants are likely to live in urban areas. The world's population, as a whole, is estimated to undergo substantial further growth, almost all of which is expected to take place in the cities and towns of poor countries.

More than half of the world's inhabitants currently live in urban environments, whose population size will increase significantly. Further, others predict that, globally, by 2030, it is likely that almost 6 million km^2 of land will be transformed into urban areas and about 1.2 million km^2 will undergo urban expansion. These changes will invariably have important and direct impacts on biodiversity and carbon pools, i.e. reservoirs of carbon that have the capacity to both take in and release carbon (Seto *et al.*, 2012). Although land cover categorized as 'urban' is a relatively small fraction of the total surface of the earth, urban areas are major driving forces in global environmental changes, habitat losses, threats to biodiversity, and the loss of terrestrial carbon stored in vegetation biomass. These and many other factors highlight the need to understand the broad-scale impacts of urban expansion as reflected in emerging threats and unintended consequences of urbanization.

Humans live virtually everywhere on earth. Wherever they settle there are significant transformations of natural habitats (Berry, 1990; Meyer and Turner, 1992; Houghton, 1994; Marzluff and Hamel, 2001). Near the end of the last century, human dwellings occupied 1–6% of the earth's surface; human agriculture covered another 12% (Meyer and Turner, 1992). Virtually all lands have been affected by human settlement or agriculture, or have been used to provide the natural resources or recreational opportunities needed to sustain a burgeoning human population. Models suggest that over the last three centuries forests have declined by 19%, grasslands by 8%, and cropland has increased over 400% (Matthews, 1983; Richards, 1990; Meyer and Turner, 1992; Marzluff and Hamel, 2001). Our 'domination' of earth is manifested in our use of 40% of all terrestrial net primary productivity (Vitousek *et al.*, 1986) and our disruption of natural cycles as illustrated by the extent and nature of illuminated areas, worldwide and visible from space at night (Elvidge *et al.*, 1997; see Fig. 10.1). Human populations are becoming increasingly urban. In 1700, only 14 cities (all in Eurasia) existed, with populations of more than 200,000 people (Berry, 1990). By 1900, 42 cities on four continents (Eurasia, North America, South America and Africa) had such populations, and by 2000, 171 cities on five continents (those above, plus Australia) had populations greater than 200,000. In 1900, only 10% of humans lived in cities; by 2000 nearly 50% did so, and nearly 70% are expected to do so by 2050 (Marzluff, 2001; United Nations, 1996). Thus by 2050, nearly as many humans are expected to live in cities (6.5 billion) as those that occupy the entire earth today (Brown *et al.*, 1998). More than 5% of the total surface area of the USA is urban and other developed land (United States Census Bureau, 2001). Not only is this a lot of land, but it is also more land than is covered by the combined total of national and state parks and areas preserved by the Nature Conservancy. Furthermore, especially foreboding is that growth rate of urban land use is accelerating faster than land preserved as parks or conservation areas by the Nature Conservancy.

The result of human populations increasing and becoming predominantly urban (Marzluff, 2001) is that land cover changes reduce, perforate, isolate and degrade species habitat (such as the habitat of birds) on local and global scales. In some cases, bird density can increase, but richness and evenness decrease, in response to urbanization. As human settlement increases, the associated changes may be favourable for some bird species (such as non-native species), whereas the effects of urbanization on other species (such as hawks, owls and cavity nesters, appear to be less consistent. The factors favouring species in urbanizing areas appear simpler than those reducing species. Decreased habitat availability, reduced patch size, increased edge, increased non-native vegetation, decreased vegetative complexity, and increased nest predation are commonly associated with bird declines in response to human settlement.

Discussions on urban ecology, as provided in this book, provide information on many foundational elements and formative forces in action in urban environments. In addition, the information provided demonstrates the importance and implications of urbanization-induced changes to the interactions between people and nature. However, we suggest that the importance of urban green spaces in enhancing urban areas for people requires a real understanding of the ecological forces in urban areas and the importance of the integration of ecological concepts and values in urban planning and design. Understanding the importance of urban green spaces and the value of ecosystem services in urban areas, in particular the incorporation of those values in urban planning and design, is critical to the future of urban areas. An understanding and appreciation of urban ecology can enhance human social organization, as well as human health and the quality of life. Most importantly, chapters in this book provide the reader with insights that currently are recognized as particularly important, as well as insights which have not received the attention they deserve, such as discussions of the importance of invasive species, protecting pollinators in urban environments, the green cities movement, ecological corridors, and other topics. These and other topics need more attention and study if we are to understand the nature and impact of ecological phenomena in urban environments, and the role played by human inhabitants in these habitats. We respectfully suggest that this book is a 'must-read' for concerned urban dwellers, citizen scientists, undergraduate and graduate students, urban planning practitioners and scholars. Given that urban ecology is an interdisciplinary field, focusing on essential elements of urban planning, the fundamental underpinnings of which are biology, botany and other related fields, it requires broad-scale discussions of many topics, as presented in this book.

Finally, urban ecology is a foundational component of many degree programmes or essential component courses in major universities, such as the Department of Urban Design and Planning, University of Washington, as well as departments such as the Department of Urban Planning and Design at Harvard University. In 1923, the Department of Urban Planning and Design at Harvard University was the first formal North American programme in city and regional planning. Others include the Department of Architecture, Urban Planning and Design, University of Missouri-Kansas City; Hofstra University; The University of Utah, Yale School of Forestry & Environmental Studies; Massachusetts Institute of Technology; Antioch University, and many others. Similarly, there is great interest among individuals who are interested in the topic but not in on academic tracks, including individuals in forestry, geography, landscape design, community planning, or urban resource management and sustainability. Still more efforts are needed.

References

Barton, A.C. and Tobin, K. (2001) Urban science education. *Journal of Research in Scientific Teaching* 38, 843–846.

Berry, B.J.L. (1990) Urbanization. In: Turner II, B.L., Clark, W.C., Kates, R.W., Richards, J.F., Mathews, J.T. and Meyers, W.B. (eds) *The Earth as Transformed by Human Action*. Cambridge University Press, Cambridge, UK, pp. 103–119.

Brown, L.R., Gardner, G. and Halweil, B. (ed. Starke, L.) (1998) Beyond Malthus: sixteen dimensions of the population problem. Worldwatch Institute Paper 143, Washington, DC.

Elvidge, C.D., Baugh, K.E. Hobson, V.R., Kihn, E.A., Kroehl, H.W., Davis, E.R. and Cocero, D. (1997) Satellite inventory of human settlements using nocturnal radiation emissions: a contribution for the global toolchest. *Global Change Biology* 3, 387–395.

Greenhalgh, S., Montgomery, M., Segal, S.J. and Todaro, M.P. (2007) *State of World Population 2007: Unleashing the Potential of Urban Growth*. United Nations Population Fund.

Grimm, N.B., Faeth, S.H., Golubiewski, N.E., Redman, C.L., Wu, J., Bai, X. and Briggs, J.M. (2008) Global change and the ecology of cities. *Science* 319, 756–760.

Houghton, R.A. (1994) The worldwide extent of land-use change. *BioScience* 44, 305–313.

Marzluff, J.M. (2001) Worldwide urbanization and its effects on birds. In: Marzluff, J.M., Bowman, R. and Donnelly, R. (eds) *Avian Ecology and Conservation in an Urbanizing World*. Kluwer Academic Publishers, Norwell, Massachusetts, pp. 19–48.

Marzluff, J.M. and Hamel, N. (2001) Land use issues.. In: Levin, S.A. (ed.) *Encyclopedia of Biodiversity*. Academic Press, San Diego, California, pp. 659–673.

Matthews, E. (1983) Global vegetation and land use: new high-resolution databases for climate studies. *Journal of Applied Meteorology and Climatology* 22, 474–487.

Meyer, W.B. and Turner II, B.L. (1992) Human population growth and global land-use/cover change. *Annual Review of Ecological Systems* 23, 39–61.

Montgomery, M.R. (2008) The urban transformation of the developing world. *Science* 319, 761–764.

Richards, J.F. (1990) Land transformation. In: Turner II, B.L., Clark, W.C., Kates, R.W., Richards, J.F., Mathews, J.T. and Meyers, W.B. (eds) (1990) *The Earth as Transformed by Human Action: Global and Regional Changes in the Biosphere over the Past 300 Years.* Cambridge University Press, Cambridge, UK.

Seto, K.C., Güneralp, B. and Hutyra, L.R. (2012) Global forecasts of urban expansion to 2030 and direct impacts on biodiversity and carbon pools. *PNAS* 109, 16083–16088.

United Nations (1996) *An Urbanizing World: Global Report on Human Settlements*. United Nations Centre for Human Settlements/Oxford University Press, Oxford, UK.

United Nations (2005) *World Urbanization Prospects: The 2005 Revision Population Database.* Department of Economic and Social Affairs, Population Division, United Nations, New York. Available at: www.unpopulation.org or contact Ms Hania Zlotnik, Director, Population Division, DESA, United Nations, New York.

United Nations Population Fund (2007) *State of World Population 2007: Unleashing the Potential of Urban Growth*. United Nations Population Fund, New York.

United States Census Bureau (2001) *Statistical Abstract of the United States: 2001*. Section 6. Geography and Environment, pp. 205–230.

Vitousek, P.M., Ehrlich, P.R., Ehrlich, A.H. and Matson, P.A. (1986) Human appropriation of the products of photosynthesis. *BioScience* 36, 368–373.

1 Urban Ecology: What Is It and Why Do We Need It?

Nancy B. Grimm*
Arizona State University, Tempe, Arizona

The Growth and Rationale for Urban Ecology

Urban ecology has blossomed within a discipline that once shunned cities as unworthy of its attention (Collins *et al.*, 2000), especially in the USA. Hundreds of papers on the topic are now published each year compared to 25 in a five-year period in the 1990s (Collins *et al.*, 2000), and urban ecology sessions at the Ecological Society of America's (ESA) annual meeting have been 'standing room only' in recent years. In the ESA's family of journals, urban ecology papers have increased from just four in the first half of the 1990s to almost 100 between 2015 and 2019 (Fig. 1.1). Accompanying the increased attention to cities has been an expansion of conceptual frameworks guiding urban research (see McPhearson *et al.*, 2016b for a summary). Most of these frameworks build upon the idea of cities as novel ecosystems, rather than seeing cities as disturbances of existing ecosystems. They, by necessity, incorporate social dimensions (Alberti, 2008; Grimm *et al.*, 2000, 2008; Pickett *et al.*, 2001, 2008; Groffman *et al.*, 2017; see also Chapter 7).

The growth in interest in urban ecology is well founded given patterns of human migration in the past century, migration that continues to accelerate along with other drivers of change in the Anthropocene. In the USA, the 2012 census reported that more than 80% of the US population lives in urban areas, the major transition to urban and suburban areas having occurred in the post-World War II era (Grimm *et al.*, 2008). Moreover, the percentage of total surface area in the USA that is developed or built up is projected to increase from 5.2% in 1997 to 9.2% by 2025 (Alig *et al.*, 2004).

The pattern of urbanization in the USA and in Europe is being repeated today in developing countries. Rapid urbanization is occurring in the global south, with the fastest growth in African and Asian cities of less than one million inhabitants (United Nations, 2015). North America, the Caribbean and Europe already are more than 75% urban, and most increases in the urban population are expected to occur in low- to middle-income countries. As in the USA, the rate of urban expansion exceeds the rate of urban population growth in many world regions (Seto *et al.*, 2012). By mid-century, 80–90% of the global population is projected to live in cities (Grimm *et al.*, 2008; Seto *et al.*, 2012). In 1950, 24% of the world's 233 countries were urbanized (i.e. had an urban population greater than the rural population); by 2014, that proportion had increased to 63% and by 2050, over 80% of countries are projected to have more than half of their population living in cities with about half of these countries being more than 75%

*NBGRIMM@asu.edu

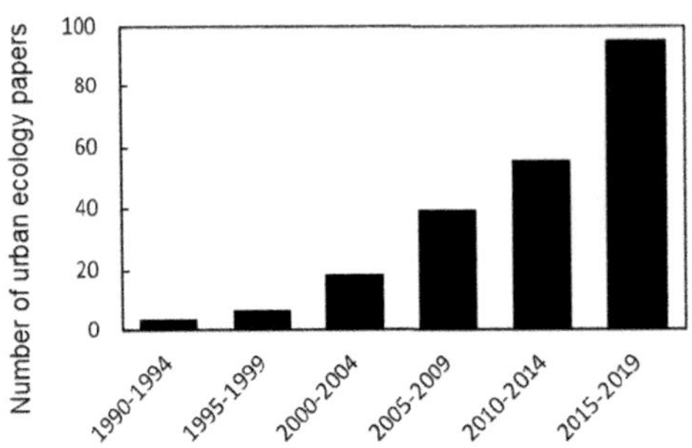

Fig. 1.1. Increase in the number of urban ecology papers published in the Ecological Society of America family of journals, 1990–2019, by half-decade. Search was conducted on the ESA journal website with the following search terms: urban, urbaniz*, city, cities. Journals include *Ecology, Ecological Applications, Ecological Monographs, Ecosphere, Ecosystem Health and Sustainability, Frontiers in Ecology & the Environment,* and *Bulletin of the Ecological Society of America.*

urbanized (United Nations, 2015). Sometime in the next 20–30 years, developing countries in Asia and Africa are likely to cross a historic threshold, joining Latin America in having majority-urban populations. The world's population as a whole is expected to undergo substantial further growth over the period, almost all of which is expected to take place in the cities and towns of poor countries.

Today's cities exhibit a wide range of population sizes and densities. The median urban population density is 5800 people/km^2, equivalent to the population density of Shanghai, China, but the range of densities is huge (Grimm and Schindler, 2018). If the global population rises to 11 billion by the end of this century, an evenly distributed population density would be ~725–1550 people/km^2 – less than today's median (Grimm and Schindler, 2018). But that is an unlikely outcome: in the fast-growing, poor cities of the global south, much of the population growth is occurring in slums and informal settlements, which present huge challenges for meeting infrastructure needs, providing clean water, sanitation and housing, and protecting populations from extreme events.

People live virtually everywhere on earth and significantly transform natural habitats where they settle (Berry, 1990; Meyer and Turner, 1992) and in distant lands they rely on to supply resources. Near the end of the last century, human dwellings occupied 1–6% of the earth's surface; human agriculture covered another 12% (Meyer and Turner, 1992). Virtually all lands have experienced human settlement or agriculture, or have been used to provide the natural resources or recreational opportunities needed to sustain the burgeoning human population. One estimate holds that only 17% of the earth's surface is untouched by human activity (Kareiva *et al.*, 2007). Models suggest that over the last three centuries forests have declined by 19%, grasslands by 8%, and cropland has increased over 400% (Meyer and Turner, 1992; Marzluff and Hamel, 2001). Human domination of planet Earth is evidenced by our use of 40% of all terrestrial net primary productivity (Vitousek *et al.*, 1986) and lights that are visible from space at night (see Fig. 10.1; Elvidge *et al.*, 1997).

We are thus living in an urban century – a part of the epoch of the Anthropocene, which is characterized by the indelible imprint of human impact on the earth's system (Steffen *et al.*, 2018). In this century, we will see the movement of the vast majority of the global human population to cities, accompanied by other accelerating changes in the environment. Changes in human activities, as recorded by exponentially increasing trends in, for example, urban population, foreign investments, vehicle miles and carbon dioxide in the atmosphere, match in scale and acceleration troublesome environmental trends. The earth is getting hotter, extreme events are increasing in frequency and magnitude, water security is increasingly threatened, and species are being lost at astonishing rates. Perhaps most urgent among these are climate change and increases in the frequency and severity of extreme events. The resulting collision course is one that presents opportunities for building better cities or rebuilding existing ones, and in which an ecologist's perspective, along with the perspectives of social scientists, planners, designers, engineers and builders, has potential to move cities along a trajectory toward greater liveability, resilience to extreme events, and sustainability (Childers *et al.*, 2014; McPhearson *et al.*, 2016b).

Social-ecological systems (SES) models enable urban ecologists to describe emergent dynamics among ecosystems, people and institutions, such as how existing social norms influence choices made about landscape vegetation, and thus its appropriateness as habitat for birds (e.g. Cook *et al.*, 2012; Chapter 3). Existing conceptual models, such as the Human Ecosystem Framework (Machlis *et al.*, 1995), the Integrated Social-Ecological System Model (Redman *et al.*, 2004), the Press-Pulse Dynamics Model (Collins *et al.*, 2011) the Long-Term Ecological Research Program, and, most recently, the SES Framework (McGinnis and Ostrom, 2014) have advanced social-ecological systems theory. But to understand cities, we must integrate social, ecological and built infrastructure (including roads, buildings, power, transportation systems, and water delivery and removal systems). This built infrastructure and its associated governance, which we refer to as the technological dimension, is often left out of traditional SES research (Ramaswami *et al.*, 2012a; Grimm *et al.*, 2013, 2015; McPhearson *et al.*, 2016b; Advisory Committee for Environmental Research and Education (AC-ERE), 2018; Markolf *et al.*, 2018; Partelow, 2018; Fig. 1.2). Together, the social, ecological, and technological dimensions form the foundation of a truly new urban ecology, an urban systems science. This expanded view is reflected in the conceptual frameworks adopted by the two urban long-term ecological research projects in the USA; the Central Arizona–Phoenix LTER and the Baltimore Ecosystem Study.

The foundations of this new urban ecology are actually old; they can be found in the early writings of Sir Arthur Tansley, who argued that 'The "natural" entities and the *anthropogenic derivates* alike must be analyzed in terms of the most appropriate concepts we can find' (emphasis added). Tansley (1935) made this argument in the same paper in which he defined one of the most enduring concepts in the whole field of ecology, that of the ecosystem. While there are

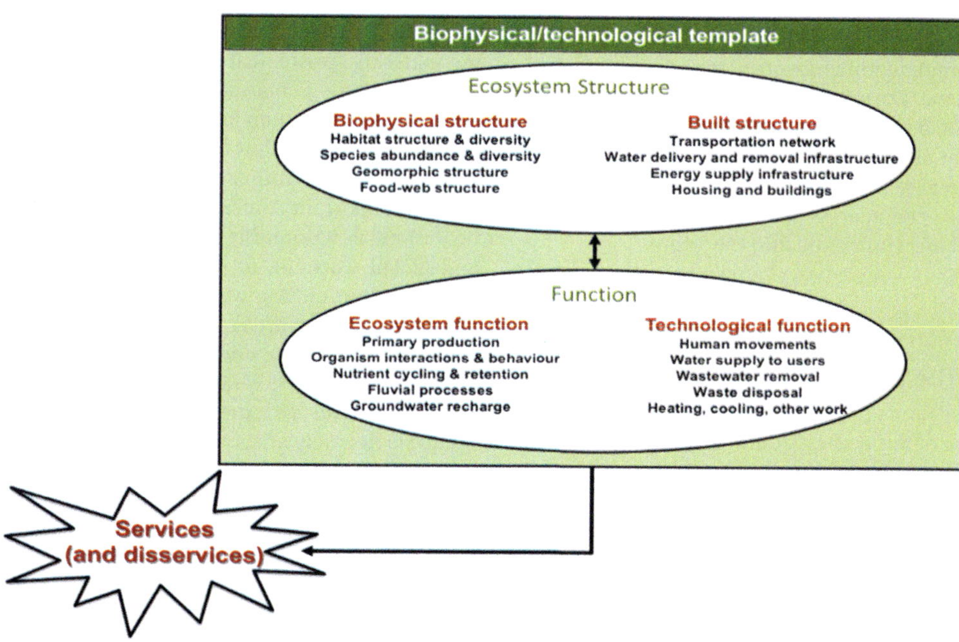

Fig. 1.2. Whereas in the press-pulse dynamics framework for social-ecological systems (Collins *et al.*, 2011) the interaction of ecosystem structure and function within a biophysical template is seen as delivering ecosystem goods and services (and disservices), a SETS (social-ecological-technological systems) framing also identifies the interaction of built structure and technological function as delivering services and disservices to the human population. Note that this diagram depicts only the ecological and technological components of the SETS, which also includes the social dimension as part of the system (see, for example, Grimm *et al.*, 2013).

disparities between ecologists and non-specialists on exactly what constitutes an ecosystem, its utility to scientists, managers and the public's understanding is well established. I write this chapter from the perspective of an ecosystem scientist, asserting that the ecosystem concept is highly appropriate to understanding the structure, dynamics and interactions of ecological, social and technological components in cities, for learning how cities interact with surrounding local and global ecosystems. In addition, it is highly appropriate for predicting how expected changes in landscapes and regions resulting from increased urbanization coupled with other environmental changes will affect the future of the earth system. But as we see from the proliferation of conceptual frameworks to guide ecosystem study of urban areas, ecosystem study, as traditionally applied, is necessary but not sufficient to understand urban ecosystems. Rather, the new urban ecology is an ecology of complex, urban, SETS; it is an interdisciplinary science of the Anthropocene (i.e. the epoch [as yet unofficial] during which human activity has been the dominant influence on climate and the environment). The primary objective of this chapter is to provide an overview of ecosystem study of cities that illustrates the need for integration of SETS, showing how an integrated urban systems science can address the challenges we face in the urban century and into the future.

The Physical Environment of Cities

From the earliest times of established urban centres, beginning some 7500 years ago in the Fertile Crescent (Redman, 1999), urban populations have benefitted from aggregation to solve challenges of living on earth. In many cases, these urban centres have arisen and succeeded where transportation is facilitated, such as along coasts and rivers, and this is true today, with 42% of the US population living in coastal counties (Fleming *et al.*, 2018). Other cities have grown up in proximity to railroads (Cronon, 1991) or in inland, arid regions (e.g. Phoenix (Gober, 2011), Albuquerque and Denver) where life outside a concentrated urban centre would be difficult.

The most obvious feature of a city is its built or engineered elements. Indeed, when one thinks of a city, it is likely that a skyline of tall buildings, bridges, or rows of brownstones or apartment buildings come to mind. Infrastructure that supports human well-being and livelihoods includes road networks, water and power delivery systems, stormwater and wastewater systems, and buildings for home and work activities. Built infrastructure, thus, is a basic component of the structure of a city (Pickett and Grove, 2009) and its physical environment that has a strong influence on climate and hydrology. The built environment also presents habitat, stresses such as noise and light pollution, or barriers to movement (and direct mortality) for organisms (see also discussion in Chapter 3).

Urban climate and the urban heat island, a phenomenon wherein temperature in the city exceeds temperature outside the city (Oke, 1973), provides an example of modulation of local climate by built environment and human activity (see also Chapter 3). Contributing factors include the high heat absorption by building materials, waste heat from urban activities (air conditioning, manufacturing etc. (Chow *et al.*, 2014)), reduction in vegetative cover, and changes in the wind flow owing to urban geometry (Oke, 1973). Younstead *et al.* (Chapter 8) draw an important contrast between the urban heat island as a primarily surface phenomenon and global warming as an atmospheric phenomenon, but outline ways in which similarities among the two drivers of urban heat can be exploited for a better understanding of evolutionary and adaptive responses to heat. Urban heat island and extreme heat in cities often disproportionately affect the poor and minority communities who may lack access to air conditioning and/or the cooling benefits of an urban tree canopy (Jenerette *et al.*, 2011; Harlan *et al.*, 2013). The urban heat island also has substantial impacts on urban plant and animal populations, as discussed in Chapters 2, 3, 4, 6 and 8.

The built environment and human manipulation alter urban hydrology. Streams are buried or paved over (Elmore and Kaushal, 2008), rivers are dammed or diverted, and the properties of urban surfaces reduce infiltration and heighten peak storm flows (Walsh *et al.*, 2012), with implications for recipient stream ecosystems (Walsh *et al.*, 2005) as well as property and livelihoods exposed to harmful flooding.

Urban Ecosystem Structure

Traditional elements of ecosystem structure are soils, vegetation, water bodies, animals and microbes. An architecture of ecosystems is often considered as part of its infrastructure; for example, the canopy, understorey and ground cover of a forest ecosystem. Such elements can also be seen in cities, where built infrastructure adds an additional dimension. Canopy may be conferred by tall buildings and ground cover by pavement; yet soils, vegetation, animals and microbes do persist in urban SETS, albeit with some important modifications. For example, Nancy Sonti (see Chapter 4) points out that little is known of below-ground processes in cities because they often are hidden beneath built infrastructure or pavement. Organismal populations must exist in cities alongside the most dominant population of all, the human population. As an element of ecosystem structure, the human population dominates, achieving population density of tens of thousands of individuals per square kilometre in some world cities to less than 1000/km^2 in most USA cities (Grimm and Schindler, 2018). But it is the design of cities, i.e. the configuration of built structures, unseen infrastructure, 'natural' elements, governance institutions, and social, cultural and economic entities, rather than the bodies of humans themselves, that makes up what is familiar to us as a city.

Urban green space comprises a network (sometimes very fragmented) of parks, open space and vacant parcels that are managed to varying extents and may support species and ecological processes that are little altered from the surrounding environment. Much of this book describes the dynamics of populations, ecophysiology, species interactions, and other ecological topics in urban green space, including urban agriculture (Chapter 12). However, in public spaces as well as in residential landscapes, choice of species to plant, whether to use chemicals to prevent unwanted species from colonizing, and mechanisms to attract pollinators and other desirable species are the dominant controls on structure (Cook *et al.*, 2012; Avolio *et al.*, 2015, 2018). Indeed, the choices and preferences of human actors in urban landscape are often so strong that they converge in cities located in very different biomes (Wheeler *et al.*, 2017), although there are larger climate-related limitations to the full range of possible tree species (Jenerette *et al.*, 2016).

Much has been written about urban biodiversity, both decrying its loss under urbanization as well as expressing hope that urban habitats can be used as species refuges (Lerman and Warren, 2011; Lerman *et al.*, 2012). Communities of greatest interest are usually plants and birds because of the value that people place upon these organisms (Lerman and Warren, 2011); there is less concern, or even negative opinions, about insect pollinators or mammalian or herpetological populations in cities (but see detailed discussion about protecting bees in urban habitats in Chapter 6). The general consensus is that diversity of urban habitats is lower than corresponding ex-urban habitats, although in warm climates where many species can thrive, plant diversity may actually be higher owing to people's preferences for diverse landscapes (Jenerette *et al.*, 2016). Long-term studies in central Arizona have suggested that bird diversity is declining in both urban and desert riparian sites, with the latter communities becoming more similar to those of engineered urban sites (Banville *et al.*, 2017). Mechanisms that explain patterns of diversity in urban areas are under increased scrutiny (Faeth *et al.*, 2005; Shochat *et al.*, 2006; Bang *et al.*, 2012, see also Chapters 3 and 12), with findings that species interactions may play a greater role in reducing diversity than was previously thought.

People occupy urban SETS at varying densities and with differential access to the benefits of urban life, including biodiversity (Lerman and Warren, 2011). Socio-spatial heterogeneity in distributions of urban amenities or disamenities is a common feature of cities in the USA, many of which have a history of environmental racism (Bullard, 1996; Mielke *et al.*, 1999; Morgan Grove *et al.*, 2006; Boone *et al.*, 2009; Bolin *et al.*, 2013; Schwarz *et al.*, 2015). A resulting legacy is that wealthy, white populations have access to urban forest cover and quality housing, while environmental disamenities like toxic release sites and polluted soils and water disproportionately affect poor, minority populations. The Baltimore

Ecosystem Study has led the way in developing an understanding of socio-spatial heterogeneity, which is discussed in some detail in Chapter 7. This heterogeneity is one way in which social-ecological interactions have not worked to the benefit of all urban residents. The provision of ecosystem services (the benefits that people derive from ecosystems) has been uneven in many cities.

The arrangements and types of built structure and green space comprise a city's urban form. Urban form has implications for how ecosystem processes play out across the landscape. Movements of water, materials and organisms are interrupted by unfavourable barriers (e.g. highways). Concentration of impervious surfaces in highly built-up urban centres exacerbates the urban heat island effect. Built structure replaces vegetation and covers soils, thus reducing primary production. Generation of air pollutants by traffic concentrates pollution near roadways but may also extend far from the city in air movements. Unique types of 'pollution', including noise (Katti and Warren, 2004) and light (Chapter 10) characterize cities and alter organismal life cycles, physiological responses and, potentially, interspecific interactions.

Urban Ecosystem Function

Ecosystem processes in cities are affected by urban form, species that are selected by people or able to survive in cities, and ways in which water flows are altered, curtailed or enhanced. Ecosystem functions underlie the ecosystem services that have potential to benefit people (Gómez-Baggethun and Barton, 2013). However, they may be undermined when overstressed with pollutant loads, overuse, and loss of biodiversity.

Whereas most ecosystems have a productive base that supports energy flow and food webs, metabolism of most urban systems demands massive imports from external, productive ecosystems. Of course, the supplier of the imported energy and materials is ultimately nature, but it is nature external to the city – natural capital derived from the extraction of minerals, rock and fossil fuels from the earth, the extensive planting of agricultural lands, and feeding operations that raise food for the urban population. Thus, urban energy flow is dominated by imported energy and consumption of that energy through food webs and, most importantly, the burning of fossil fuels (Odum and Odum, 1980). Primary production is usually much reduced in cities owing to development, but the primary production that does occur supports grazing and detrital food webs just as in non-urban ecosystems.

Nutrient flows in cities are similarly dominated by imports (Baker *et al.*, 2001; Groffman *et al.*, 2004; Kaye *et al.*, 2006; Fissore *et al.*, 2011; Metson *et al.*, 2012), with variable levels of nutrient retention depending upon the element and structure of the system examined. Human activities in cities influence biogeochemical cycles through alterations of hydrology, additions (intentional, i.e. fertilizer, and inadvertent, i.e. by-products of fossil fuel combustion), changes in land use and land cover that drive changes in soil processes or vegetation–soil interactions, and local climate changes that influence process rates. Pollutants that are unique to cities, such as pharmaceuticals, present an entirely new challenge for microbial communities (Rosi *et al.*, 2018).

Impacts of human activities in urban systems on biogeochemical cycles and metabolism are profound and extend to scales far beyond those of the city itself, both through demand for materials and energy and production of wastes that can influence regional and even global ecosystems (Kaye *et al.*, 2006; Grimm *et al.*, 2008; Ramaswami *et al.*, 2012b). Although the surface area of cities accounts for only 2–4% of the earth's land surface, their ecological footprint, which is the productive land area required to supply all resources and assimilate all waste of a population, can exceed city area by orders of magnitude (Rees and Wackernagel, 1996; Luck *et al.*, 2001; see also Chapter 13). Cities produce waste (including carbon dioxide) that is transported by air and affects global biogeochemical cycles and climate, and accounts for up to 80% of greenhouse gas emissions in the USA (Maxwell *et al.*, 2018). Concentrated human demand for food, water and materials drives changes in land cover and hydrological systems at least regionally; these changes may have profound influences on ecosystem function and biodiversity at some distance from the city. Demand for 'luxury items' from wealthy urban areas in the USA has a much farther reach in terms of impact. Impacts such as these drive local, regional and global environmental change.

Urban Ecosystems and Global Environmental Change: Why We Need Urban Ecology

The Anthropocene represents an age of compounded challenges of global urban growth and climate change that threaten the earth system's sustainability. Cities are the places where 80% of the world population will live by the end of this century; thus, the problem of sustainability, at least for the human population, will be solved (or not) in cities. Cities and urban areas are complex, and this complexity is further compounded by long-term futures that are uncertain, subject to non-stationarity, and difficult to prepare for. Many of our greatest environmental and societal challenges, including climate change, will be experienced in cities. The international community recognized this challenge in identifying 'Sustainable Cities and Communities' as one of 17 United Nations Sustainable Development Goals for 2030. The 'wicked problems' of the urban century, including increased frequency and magnitude of extreme events affecting cities, inadequate infrastructure in rapidly growing cities, and ageing infrastructure in existing cities, require a transdisciplinary approach. Transdisciplinary work features multiple perspectives and brings together researchers and practitioners to co-produce the needed knowledge and move toward solutions (Muñoz-Erickson *et al.*, 2017). Urban ecology has much to offer in this arena, especially in its capacity to integrate across the social, ecological and technological domains. Pickett *et al.* (Chapter 7) discuss some of the insights that their long-term study in Baltimore has yielded; among them, they make a strong case for place-based research, welcoming multiple perspectives, linking social and environmental factors as both drivers and responses, issues of social equity, and that our basic research can be use-inspired; all of which are needed perspectives for the new urban systems science.

Urban ecology investigates how urban SETS drive and respond to environmental change at all scales. The interplay between driver and responder is subject to change as global environmental changes accelerate. Five major categories of global change have effects at various scales (Grimm *et al.*, 2008): land use and land-cover change (LULCC), altered biogeochemical cycles, loss of biodiversity, climate change, and altered hydrological systems. LULCC is pervasive and crosses all scales, whereas biodiversity changes in cities have primarily local effects. On the other hand, altered biogeochemical cycles reach the global scale, such as through greenhouse gas emissions. Hydrological systems are severely altered on a local scale, but large-scale diversions and inter-basin water transfers can also reach regional and even continental scales. In terms of responses, for urban dwellers, the top-down effects of many global environmental changes are often swamped by even more dramatic changes in the local environment, including the urban heat island, depauperate species pools of birds and pollinators, socio-spatial inequities, and local pollution. In these cases, the interactions of urbanization and global environmental change are asymmetrical.

Although this asymmetry has been the rule for past decades, climate change impacts are beginning to be felt much more in cities. Extreme climate events are on the rise (Munich RE, 2015) and cities are especially vulnerable, given their concentration of people and infrastructure that is either ageing (ASCE, 2013) or inadequate, coupled with the fact that many are located along rivers and coasts or in drought-prone drylands. Rising sea levels, flooding, drought and heatwaves pose significant risks to human settlements, communities and infrastructure – risks that are increasing in every part of the world. Thus there is an urgent need for urban ecologists to understand how cities will respond, and to help build resilience in the face of these risks (Royal Society, 2014). SETS is a useful framework to organize the concepts of vulnerability and resilience of the social, ecological and infrastructural components of the urban system (Markolf *et al.*, 2018).

Resilience concepts from ecology have been adopted in social-ecological systems research (Romero-Lankao *et al.*, 2016), where resilience is defined as the ability of a system to maintain its characteristic composition, organization and function over time while remaining adaptive and economically viable, and sustaining human communities (Carpenter *et al.*, 2001; Folke *et al.*,

2010). Resilience is a system characteristic that governs its response to stresses, shocks or disturbances, which can arise from biophysical or social drivers (Grimm *et al.*, 2017; Elmqvist *et al.*, 2019). The capacity of a system to self-organize, cope and transform from its current state to an alternative, desirable state in the face of change, i.e. its transformability (Schlüter and Pahl-Wostl, 2007) has also been seen as a component of resilience. In order to more fully incorporate the technological/ infrastructural components of urban SETS into this understanding of resilience, a more flexible, systems-based concept of infrastructure is needed (Pandit *et al.*, 2017; Chester and Allenby, 2018).

Urban SETS: Cities Provide Solutions

Complex sustainability challenges face urban areas as they continue to expand and are exposed to greater threats from global environmental change. Resilient solutions should provide ecosystem services, improve social well-being, and exploit new technologies in ways that benefit all segments of urban populations; in other words, they should attend to all three SETS domains. In fact, many cities are leaders in implementing climate-change adaptation and mitigation strategies even while state and national entities are lagging in such efforts. The Rockefeller Foundation's 100 Resilient Cities programme was meant to rapidly develop resilience plans for select world cities. Other entities like ICLEI and the Urban Sustainability Directors Network in the USA are organizing efforts to prepare for climate change.

Many cities are considering or implementing nature-based solutions, also referred to as green infrastructure, low-impact development, or ecosystem-based adaptations, to restore or use natural hydrologic and ecological processes to provide ecosystem services (Nesshöver *et al.*, 2017; Depietri and McPhearson, 2017; Kabisch *et al.*, 2017; Hobbie and Grimm, 2020; see also Chapter 4). In the USA, investment in green infrastructure saw a rapid increase following the release of a memorandum supporting its use by the Environmental Protection Agency in 2007 (Hopkins *et al.*, 2018). Many city practitioners are developing sustainability and resilience plans, in which nature-based solutions are often featured, and have adopted resilience as a goal for urban transformation and dealing with the uncertainty of future climate conditions (Moser *et al.*, 2019). However, despite the investment in nature-based solutions and the embracing of the resilience concept, the relationship between these strategies and resilience is still poorly known (Munroe *et al.*, 2012).

Urban nature has the potential to improve air and water quality, mitigate flooding, enhance physical and mental health, and promote social and cultural well-being. These benefits are often described as urban ecosystem services, defined as the benefits humans derive from urban nature (Gómez-Baggethun and Barton, 2013; Elmqvist *et al.*, 2013). Several chapters in this book touch on ecosystem services. Nature-based solutions are a subset of urban ecosystem services (Grimm *et al.*, 2015; Kabisch *et al.*, 2017; Grimm and Schindler, 2018; Hobbie and Grimm, 2020) that may provide air-pollution absorption, stormwater retention, coastal flood protection, water purification or climate modulation, all examples of regulating services that can reduce the impacts of climate change. Certain cities are investing heavily in nature-based solutions. For example, New York City has a US$1 billion fund for green infrastructure as a stormwater solution (New York City Green Infrastructure Plan: A Sustainable Strategy for Clean Waterways, 2010), and four other cities were identified as 'green leaders' by Hopkins *et al.* (2018) for investments of more than 20% of their plan funds in green infrastructure. Investment among 19 US cities was highly variable and often seemed to reflect different objectives of stormwater management (i.e. volume control or pollution control) or the dedication of a single individual, as was the case for Syracuse, New York (Hopkins *et al.*, 2018).

Besides implementing nature-based solutions, what else can city practitioners do and how can urban ecology help? In the USA and Latin America, a team of urban ecologists, social scientists and engineers has been working with practitioners in nine cities to explore alternative, positive visions for their cities under conditions of future (2080) climate. There is a need for cities to revolutionize the planning process to create positive, transformative visions, given the urgency of the challenge (McPhearson *et al.*, 2016a). Scenario co-development is the process of articulating and exploring alternative future pathways based on stakeholders' knowledge, values, preferences and underlying worldviews to envision plausible futures (Wiek and Iwaniec, 2014). A scientific basis is provided by input data, but the process of developing scenarios brings together multiple disciplines, perspectives and sectors to create visions that can be used to explore trade-offs and express shared values (Iwaniec *et al.*, 2020).

New Directions for Urban Ecology

Although I have not attempted to cover the now very broad range of topics addressed by urban ecologists, instead choosing to focus primarily on urban ecosystems or, more correctly, urban SETS, it is clear from the remaining chapters of this book that the knowledge base in urban ecology has come a long way in a few decades. No longer is the urban realm shunned as an inappropriate area for ecologists to study. Indeed, I hope I have made the case that urban ecologists need to be 'in the trenches' along with other scientists and practitioners of cities because cities are at the nexus of environmental change and human population dynamics. With a new urban systems science continuing to develop the multi-scaled, transdisciplinary frameworks for understanding cities and their interacting ecological, social and technological components in a rapidly changing environment, we can begin to put this extensive knowledge to use in solving the challenges we face.

Several knowledge gaps and barriers to achieving this vision are evident, however, and have been pointed out in the chapters in this book. For organisms inhabiting urban environments, we have only begun to scratch the surface in our understanding of how functional traits, species interactions, physiology, and life history traits are altered by urban environments and the stresses they impose (i.e. 'filters' – see Chapter 3). Stresses emphasized in the chapters of this book include: non-native species; pollutant loads, including novel pollutants; altered urban climate and hydrology; and direct human-caused mortality. An exciting new direction in evolutionary ecology asks to what extent responses of urban species represents evolutionary change, and the feedbacks between urban ecology and urban evolution. New research is beginning on impacts of light on ecological dynamics (see Chapter 10).

At the ecosystem level, we benefit from many years of research in the two urban LTERs on biogeochemical dynamics (Kaye *et al.*, 2006; Groffman *et al.*, 2004, Pickett *et al.* (Chapter 7), among many others) and budgets. We lack information about below-ground processes and unseen (below-ground) infrastructure, such as pipe networks, sewers, electrical wiring and water delivery networks, as these data often are protected and difficult for ecologists to access. We have only just begun to use the tool of comparative ecosystem science to understand how cities in different contexts differ in terms of their material mass balances (or 'metabolism' as industrial engineers like to call it; but see Metson *et al.*, 2015), or in terms of the kinds of nature-based solutions or other adaptation strategies that are most appropriate to place.

In the Central Arizona–Phoenix LTER, there has always been value placed on the bringing together of perspectives from multiple social sciences, physical science and engineering. Institutional structures at the host university make it simple and practical to interact across disciplinary boundaries. But this remains a difficult challenge for many urban ecologists who either do not have close ties with other disciplines or are in institutions that frown on such integration, either explicitly or implicitly. There is an urgent imperative to break free from traditional thinking, especially for use-inspired research and transdisciplinary work with practitioners. Not only is a new science required, but also a new way of training the next generation of scholars and action-oriented researchers is essential to solving the urgent problems we face.

Acknowledgement

Support from the NSF (CAP LTER, #1,832,016, UREx SRN, #1444755, NATURA, #1927468 and SETS Convergence, #1934933) and the Virginia M. Ullman Professorship at Arizona State University is gratefully acknowledged.

References

ASCE (2013) 2013 report card for America's infrastructure. American Society of Civil Engineers. Available at: http://www.infrastructurereportcard.org/a/#p/home(accessed 25 June 2020).

Advisory Committee for Environmental Research and Education (AC-ERE) (2018) *Sustainable Urban Systems: Articulating a Long-term Convergence Research Agenda*. A report from the NSF Advisory Committee for Environmental Research and Education. Prepared by the Sustainable Urban Systems Subcommittee. Available at: http://www.nsf.gov/ere/ereweb/ac-ere/sustainable-urban-systems.pdf (accessed 1 February 2019).

Alberti, M. (2008) *Advances in Urban Ecology: Integrating Humans and Ecological Processes in Urban Ecosystems*. Springer.

Alig, R.J., Kline, J.D. and Lichtenstein, M. (2004) Urbanization on the US landscape: looking ahead in the 21st century. *Landscape and Urban Planning* 69(2-3), 219–234. DOI: 10.1016/j.landurbplan.2003.07.004.

Avolio, M., Pataki, D.E., Gillespie, T.W., Jenerette, G.D., McCarthy, H.R. *et al*. (2015) Tree diversity in southern California's urban forest: the interacting roles of social and environmental variables. *Ecology and Evolution* 3, 73–87.

Avolio, M.L., Pataki, D.E., Trammell, T.L.E. and Endter-Wada, J. (2018) Biodiverse cities: the nursery industry, homeowners, and neighborhood differences drive urban tree composition. *Ecological Monographs* 88(2), 259–276. DOI: 10.1002/ecm.1290.

Baker, L.A., Hope, D., Xu, Y., Edmonds, J. and Lauver, L. (2001) Nitrogen balance for the central Arizona-Phoenix (CAP) ecosystem. *Ecosystems* 4(6), 582–602. DOI: 10.1007/s10021-001-0031-2.

Bang, C., Faeth, S.H. and Sabo, J.L. (2012) Control of arthropod abundance, richness, and composition in a heterogeneous desert city. *Ecological Monographs* 82(1), 85–100. DOI: 10.1890/11-0828.1.

Banville, M.J., Bateman, H.L., Earl, S.R. and Warren, P.S. (2017) Decadal declines in bird abundance and diversity in urban riparian zones. *Landscape and Urban Planning* 159, 48–61. DOI: 10.1016/j.landurbplan.2016.09.026.

Berry, B.J.L. (1990) Urbanization. In: Turner II, B.J.L. (ed.) *The Earth as Transformed by Human Action: Global and Regional Changes in the Biosphere over the Past 300 Years*. Cambridge University Press with Clark University, Cambridge, New York, pp. 103–119.

Bolin, B., Barreto, J., Hegmon, M., Meierotto, L. and York, A. (2013) Double exposure in the Sunbelt: the sociospatial distribution of vulnerability in Phoenix, Arizona. In: Boone, C.G. and Fragkias, M. (eds) *Urbanization and Sustainability*. Springer, Netherlands, pp. 159–178.

Boone, C.G., Buckley, G.L., Grove, J.M. and Sister, C. (2009) Parks and people: an environmental justice inquiry in Baltimore, Maryland. *Annals of the Association of American Geographers* 99(4), 767–787. DOI: 10.1080/00045600903102949.

Bullard, R.D. (1996) Environmental justice: It's more than waste facility siting. *Social Science Quarterly* 77, 493–499.

Carpenter, S., Walker, B., Anderies, J.M. and Abel, N. (2001) From metaphor to measurement: resilience of what to what? *Ecosystems* 4(8), 765–781. DOI: 10.1007/s10021-001-0045-9.

Chester, M.V. and Allenby, B.R. (2018) Toward adaptive infrastructure: flexibility and agility in a non-stationarity age. *Sustainable and Resilient Infrastructure* 3, 1–19.

Childers, D.L., Pickett, S.T.A., Grove, J.M., Ogden, L. and Whitmer, A. (2014) Advancing urban sustainability theory and action: challenges and opportunities. *Landscape and Urban Planning* 125, 320–328. DOI: 10.1016/j.landurbplan.2014.01.022.

Chow, W.T.L., Salamanca, F., Georgescu, M., Mahalov, A., Milne, J.M. *et al*. (2014) A multi-method and multi-scale approach for estimating city-wide anthropogenic heat fluxes. *Atmospheric Environment* 99, 64–76. DOI: 10.1016/j.atmosenv.2014.09.053.

Collins, J.P., Kinzig, A., Grimm, N.B., Fagan, W.F., Hope, D. *et al*. (2000) A new urban ecology. *American Scientist* 88, 416–425.

Collins, S.L., Carpenter, S.R., Swinton, S.M., Orenstein, D.E., Childers, D.L. *et al*. (2011) An integrated conceptual framework for long-term social–ecological research. *Frontiers in Ecology and the Environment* 9(6), 351–357. DOI: 10.1890/100068.

Cook, E.M., Hall, S.J. and Larson, K.L. (2012) Residential landscapes as social-ecological systems: a synthesis of multi-scalar interactions between people and their home environment. *Urban Ecosystems* 15(1), 19–52. DOI: 10.1007/s11252-011-0197-0.

Cronon, W. (1991) Nature's metropolis. In: Cronon, W. (ed.) *Nature's Metropolis*. W.W. Norton & Company, New York.

Depietri, Y. and McPhearson, T. (2017) Integrating the grey, green, and blue in cities: nature-based solutions for climate change adaptation and risk reduction. In: Kabisch, N., Korn, H., Stadler, J. and Bonn, A. (eds) *Nature-based Solutions to Climate Change Adaptation in Urban Areas*. Springer, pp. 91–109.

Elmore, A.J. and Kaushal, S.S. (2008) Disappearing headwaters: patterns of stream burial due to urbanization. *Frontiers in Ecology and the Environment* 6, 308–312. DOI: 10.1890/070101.

Elmqvist, T., Fragkias, M., Goodness, J., Güneralp, B., Marcotullio, P.J. *et al.* (2013) *Urbanization, Biodiversity and Ecosystem Services: Challenges and Opportunities*. Springer.

Elmqvist, T., Andersson, E., Frantzeskaki, N., McPhearson, T., Olsson, P. *et al.* (2019) Sustainability and resilience for transformation in the urban century. *Nature Sustainability* 2(4), 267–273. DOI: 10.1038/s41893-019-0250-1.

Elvidge, C.D., Baugh, K.E., Kihn, E.A., Kroehl, H.W. and Davis, E.R. (1997) Mapping city lights with nighttime data from the DMSP operational Linescan system. *Photogrammetric Engineering Remote Sensing* 63, 727–734.

Faeth, S.H., Warren, P.S., Shochat, E. and Marussich, W.A. (2005) Trophic dynamics in urban communities. *BioScience* 55(5), 399–407. DOI: 10.1641/0006-3568(2005)055[0399:TDIUC]2.0.CO;2.

Fissore, C., Baker, L.A., Hobbie, S.E., King, J.Y., McFadden, J.P. *et al.* (2011) Carbon, nitrogen, and phosphorus fluxes in household ecosystems in the Minneapolis-Saint Paul, Minnesota, urban region. *Ecological Applications* 21(3), 619–639. DOI: 10.1890/10-0386.1.

Fleming, E., Payne, J., Sweet, W., Craghan, M. and Haines, J. (2018) Coastal effects. In: *Impacts, Risks, and Adaptation in the United States: Fourth National Climate Assessment*. Volume II. Washington, DC.

Folke, C., Carpenter, S.R., Walker, B., Scheffer, M., Chapin, T. *et al.* (2010) Resilience thinking: integrating resilience, adaptability and transformability. *Ecology and Society* 15(4), 20–28. DOI: 10.5751/ES-03610-150420.

Grimm, N.B., Faeth, S.H., Golubiewski, N.E., Redman, C.L., Wu, J. *et al.* (2008) Global change and the ecology of cities. *Science* 319(5864), 756–760. DOI: 10.1126/science.1150195.

Grimm, N.B., Redman, C.L., Boone, C.G., Childers, D.L., Harlan, S.L. *et al.* (2013) Viewing the urban socio-ecological system through a sustainability lens: lessons and prospects from the Central Arizona-Phoenix LTER programme. Chapter 10. In: Singh, S.J., Haberl, H., Chertow, M., Mirtl, M., Schmid, M., *et al.* (eds) *Long Term Socio-Ecological Research: Studies in Society-Nature Interactions across Spatial and Temporal Scales*. Springer.

Grimm, N.B., Cook, E.M., Hale, R.L. and Iwaniec, D.M. (2015) A broader framing of ecosystem services in cities: benefits and challenges of built, natural, or hybrid system function. In: Seto, K.C., Solecki, W.D. and Griffith, V. (eds) *Handbook on Urbanization and Global Environmental Change*. Routledge Handbooks Online, CRC, pp. 203–212.

Grimm, N.B., Pickett, S.T.A., Hale, R.L. and Cadenasso, M.L. (2017) Does the ecological concept of disturbance have utility in urban social–ecological–technological systems? *Ecosystem Health and Sustainability* 3(1), e01255. DOI: 10.1002/ehs2.1255.

Grimm, N.B. and Schindler, S. (2018) Nature of cities and nature in cities: prospects for conservation and design of urban nature in human habitat. In: Lele, S., Brondizio, E.S., Byrne, J., Mace, G.M. and Martinez-Alier, J. (eds) *Rethinking Environmentalism: Linking Justice, Sustainability, and Diversity*. MIT Press, pp. 99–125.

Grimm, N.B., Morgan Grove, J., Pickett, S.T.A. and Redman, C.L. (2000) Integrated approaches to long-term studies of urban ecological systems. *BioScience* 50(7), 571–584. DOI: 10.1641/0006-3568(2000)050[0571:IATLTO]2.0.CO;2.

Groffman, P.M., Cadenasso, M.L., Cavender-Bares, J., Childers, D.L., Grimm, N.B. *et al.* (2017) Moving towards a new urban systems science. *Ecosystems* 20(1), 38–43. DOI: 10.1007/s10021-016-0053-4.

Groffman, P.M., Law, N.L., Belt, K.T., Band, L.E. and Fisher, G.T. (2004) Nitrogen fluxes and retention in urban watershed ecosystems. *Ecosystems* 7(4), 393–403. DOI: 10.1007/s10021-003-0039-x.

Gober, P. (2011) *Metropolitan Phoenix: Place Making and Community Building in the Desert*. University of Pennsylvania Press, Philadelphia.

Gómez-Baggethun, E. and Barton, D.N. (2013) Classifying and valuing ecosystem services for urban planning. *Ecological Economics* 86, 235–245. DOI: 10.1016/j.ecolecon.2012.08.019.

Harlan, S.L., Declet-Barreto, J.H., Stefanov, W.L. and Petitti, D.B. (2013) Neighborhood effects on heat deaths: social and environmental predictors of vulnerability in Maricopa County, Arizona. *Environmental Health Perspectives* 121(2), 197–204. DOI: 10.1289/ehp.1104625.

Hobbie, S.E. and Grimm, N.B. (2020) Nature-based approaches to managing climate change impacts in cities. *Philosophical Transactions of the Royal Society B: Biological Sciences* 375(1794), 20190124.

Iwaniec, D.M., Cook, E.M., Davidson, M.J., Berbés-Blázquez, M., Georgescu, M. *et al*. (2020) The co-production of sustainable future scenarios. *Landscape and Urban Planning* 197, 103744.

Hopkins, K.G., Grimm, N.B. and York, A.M. (2018) Influence of governance structure on green stormwater infrastructure investment. *Environmental Science & Policy* 84, 124–133. DOI: 10.1016/j.envsci.2018.03.008.

Jenerette, G.D., Harlan, S.L., Stefanov, W.L. and Martin, C.A. (2011) Ecosystem services and urban heat riskscape moderation: water, green spaces, and social inequality in Phoenix, USA. *Ecological Applications* 21(7), 2637–2651. DOI: 10.1890/10-1493.1.

Jenerette, G., Clarke, L., Avolio, M., Pataki, D., Gillespie, T. *et al*. (2016) Climate tolerances and trait choices shape continental patterns of urban tree biodiversity. *Global Ecology and Biogeography* 25, 1367–1376.

Kabisch, N., Korn, H., Stadler, J. and Bonn, A. (2017) *Nature-based Solutions to Climate Change Adaptation in Urban Areas: Linkages between Science, Policy, and Practice*. Springer, Cham., Switzerland.

Kareiva, P., Watts, S., McDonald, R. and Boucher, T. (2007) Domesticated nature: shaping landscapes and ecosystems for human welfare. *Science* 316(5833), 1866–1869. DOI: 10.1126/science.1140170.

Katti, M. and Warren, P.S. (2004) Research focus: tits, noise, and urban bioacoustics. *Trends in Ecology & Evolution* 19, 109–110.

Kaye, J.P., Groffman, P.M., Grimm, N.B., Baker, L.A. and Pouyat, R.V. (2006) A distinct urban biogeochemistry? *Trends in Ecology & Evolution* 21(4), 192–199. DOI: 10.1016/j.tree.2005.12.006.

Lerman, S.B., Turner, V.K. and Bang, C. (2012) Homeowner associations as a vehicle for promoting native urban biodiversity. *Ecology and Society* 17(4), 45–57. DOI: 10.5751/ES-05175-170445.

Lerman, S.B. and Warren, P.S. (2011) The conservation value of residential yards: linking birds and people. *Ecological Applications* 21(4), 1327–1339. DOI: 10.1890/10-0423.1.

Luck, M.A., Jenerette, G.D., Wu, J. and Grimm, N.B. (2001) The urban funnel model and the spatially heterogeneous ecological footprint. *Ecosystems* 4(8), 782–796. DOI: 10.1007/s10021-001-0046-8.

Machlis, G.E., Force, J.E. and Burch Jr., W.R. (1995) The human ecosystem as an organizing concept in ecosystem management. *Society & Natural Resources* 10, 347–367.

Markolf, S.A., Chester, M.V., Eisenberg, D.A., Iwaniec, D.M., Davidson, C.I. *et al*. (2018) Interdependent infrastructure as linked social, ecological, and technological systems (SETSs) to address, lock in and enhance resilience. *Earth's Future* 6(12), 1638–1659. DOI: 10.1029/2018EF000926.

Marzluff, J.M. and Hamel, N. (2001) Land use issues. In: Levin, S.A. (ed.) *Encyclopedia of Biodiversity*, pp. 659–673.

Maxwell, K., Grambsch, A., Kosmal, A., Larson, L. and Sonti, N. (2018) Built environment, urban systems, and cities. In: Reidmiller, D.R., Avery, C.W., Easterling, D.R., Kunkel, K.E. and Lewis, K.L.M., *et al*. (eds) *Impacts, Risks, and Adaptation in the United States: Fourth National Climate Assessment*. Vol. II. US Global Change Research Program, Washington, D.C, pp. 438–478.

McGinnis, M.D. and Ostrom, E. (2014) Social-ecological system framework: initial changes and continuing challenges. *Ecology and Society* 19(2), 30. DOI: 10.5751/ES-06387-190230.

McPhearson, T., Iwaniec, D.M. and Bai, X. (2016a) Positive visions for guiding urban transformations toward sustainable futures. *Current Opinion in Environmental Sustainability* 22, 33–40. DOI: 10.1016/j.cosust.2017.04.004.

McPhearson, T., Pickett, S.T.A., Grimm, N.B., Niemelä, J., Alberti, M. *et al*. (2016b) Advancing urban ecology toward a science of cities. *BioScience* 66(3), 198–212. DOI: 10.1093/biosci/biw002.

Metson, G.S., Hale, R.L., Iwaniec, D.M., Cook, E.M., Corman, J.R. *et al*. (2012) Phosphorus in Phoenix: a budget and spatial representation of phosphorus in an urban ecosystem. *Ecological Applications* 22(2), 705–721. DOI: 10.1890/11-0865.1.

Metson, G.S., Iwaniec, D.M., Baker, L.A., Bennett, E.M., Childers, D.L. *et al*. (2015) Urban phosphorus sustainability: systemically incorporating social, ecological, and technological factors into phosphorus flow analysis. *Environmental Science & Policy* 47, 1–11. DOI: 10.1016/j.envsci.2014.10.005.

Meyer, W.B. and Turner, B.L. (1992) Human population growth and global land-use/cover change. *Annual Review of Ecology and Systematics* 23(1), 39–61. DOI: 10.1146/annurev.es.23.110192.000351.

Mielke, H.W., Gonzales, C.R., Smith, M.K. and Mielke, P.W. (1999) The urban environment and children's health: soils as an integrator of lead, zinc, and cadmium in New Orleans, Louisiana, USA. *Environmental Research* 81(2), 117–129. DOI: 10.1006/enrs.1999.3966.

Morgan Grove, J., Cadenasso, M.L., Burch, W.R., Pickett, S.T.A., Schwarz, K. *et al.* (2006) Data and methods comparing social structure and vegetation structure of urban neighborhoods in Baltimore, Maryland. *Society & Natural Resources* 19(2), 117–136. DOI: 10.1080/08941920500394501.

Moser, S., Meerow, S., Arnott, J. and Jack-Scott, E. (2019) The turbulent world of resilience: interpretations and themes for transdisciplinary dialogue. *Climatic Change* 153(1-2), 21–40. DOI: 10.1007/s10584-018-2358-0.

Munich RE (2015) Natural catastrophes 2014: Analyses, assessments, positions. Available at: https://www.munichre.com/site/corporate/get/documents_E1018449711/mr/assetpool.shared/Documents/5_Touch/_Publications/302-08606_en.pdf (accessed accessed 25 June 2020).

Munroe, R., Roe, D., Doswald, N., Spencer, T., Möller, I. *et al.* (2012) Review of the evidence base for ecosystem-based approaches for adaptation to climate change. *Environmental Evidence* 1, 13–23. DOI: 10.1186/2047-2382-1-13.

Muñoz-Erickson, T., Miller, C. and Miller, T. (2017) How cities think: knowledge co-production for urban sustainability and resilience. *Forests* 8(6), 203–219. DOI: 10.3390/f8060203.

Nesshöver, C., Assmuth, T., Irvine, K.N., Rusch, G.M., Waylen, K.A. *et al.* (2017) The science, policy and practice of nature-based solutions: an interdisciplinary perspective. *Science of The Total Environment* 579, 1215–1227. DOI: 10.1016/j.scitotenv.2016.11.106.

New York City Green Infrastructure Plan: A Sustainable Strategy for Clean Waterways (2010). Available at: https://www.adaptationclearinghouse.org/resources/new-york-city-green-infrastructure-plan-a-sustainable-strategy-for-clean-waterways.html (accessed 25 June 2020).

Odum, H.T. and Odum, E.P. (1980) *Energy Basis for Man and Nature*. McGraw-Hill, New York.

Oke, T.R. (1973) City size and the urban heat island. *Atmospheric Environment* 7(8), 769–779. DOI: 10.1016/0004-6981(73)90140-6.

Pandit, A., Minné, E.A., Li, F., Brown, H., Jeong, H. *et al.* (2017) Infrastructure ecology: an evolving paradigm for sustainable urban development. *Journal of Cleaner Production* 163, S19–S27. DOI: 10.1016/j.jclepro.2015.09.010.

Partelow, S. (2018) A review of the social-ecological systems framework: applications, methods, modifications, and challenges. *Ecology and Society* 23(4), 36–60. DOI: 10.5751/ES-10594-230436.

Pickett, S.T.A., Cadenasso, M.L., Grove, J.M., Groffman, P.M., Band, L.E. *et al.* (2008) Beyond urban legends: an emerging framework of urban ecology, as illustrated by the Baltimore ecosystem study. *BioScience* 58(2), 139–150. DOI: 10.1641/B580208.

Pickett, S.T.A., Cadenasso, M.L., Grove, J.M., Nilon, C.H., Pouyat, R.V. *et al.* (2001) Urban ecological systems: linking terrestrial ecological, physical, and socioeconomic components of metropolitan areas. *Annual Review of Ecology and Systematics* 32(1), 127–157. DOI: 10.1146/annurev.ecolsys.32.081501.114012.

Pickett, S.T.A. and Grove, J.M. (2009) Urban ecosystems: what would Tansley do? *Urban Ecosystems* 12(1), 1–8. DOI: 10.1007/s11252-008-0079-2.

Ramaswami, A., Chavez, A. and Chertow, M. (2012a) Carbon footprinting of cities and implications for analysis of urban material and energy flows. *Journal of Industrial Ecology* 16(6), 783–785. DOI: 10.1111/j.1530-9290.2012.00569.x.

Ramaswami, A., Weible, C., Main, D., Heikkila, T., Siddiki, S. *et al.* (2012b) A social-ecological-infrastructural systems framework for interdisciplinary study of sustainable City systems. *Journal of Industrial Ecology* 16(6), 801–813. DOI: 10.1111/j.1530-9290.2012.00566.x.

Redman, C.L. (1999) *Human Impact on Ancient Environments*. University of Arizona Press, Tucson.

Redman, C.L., Grove, J.M. and Kuby, L.H. (2004) Integrating social science into the long-term ecological research (LTER) network: social dimensions of ecological change and ecological dimensions of social change. *Ecosystems* 7(2), 161–171. DOI: 10.1007/s10021-003-0215-z.

Rees, W. and Wackernagel, M. (1996) Urban ecological footprints: Why cities cannot be sustainable—And why they are a key to sustainability. *Environmental Impact Assessment Review* 16(4-6), 223–248. DOI: 10.1016/S0195-9255(96)00022-4.

Romero-Lankao, P., Gnatz, D., Wilhelmi, O. and Hayden, M. (2016) Urban sustainability and resilience: from theory to practice. *Sustainability* 8(12), 1224. DOI: 10.3390/su8121224.

Rosi, E.J., Bechtold, H.A., Snow, D., Rojas, M., Reisinger, A.J. *et al.* (2018) Urban stream microbial communities show resistance to pharmaceutical exposure. *Ecosphere* 9(1), 1–16. DOI: 10.1002/ecs2.2041.

Royal Society (2014) *Resilience to Extreme Weather*. London.

Schlüter, M. and Pahl-Wostl, C. (2007) Mechanisms of resilience in common-pool resource management systems: an agent-based model of water use in a river basin. *Ecology and Society* 12(2), 4–25. DOI: 10.5751/ES-02069-120204.

Schwarz, K., Fragkias, M., Boone, C.G., Zhou, W., McHale, M. *et al.* (2015) Trees grow on money: urban tree canopy cover and environmental justice. *PLOS ONE* 10(4), e0122051. DOI: 10.1371/journal.pone.0122051.

Seto, K.C., Güneralp, B. and Hutyra, L.R. (2012) Global forecasts of urban expansion to 2030 and direct impacts on biodiversity and carbon pools. *Proceedings of the National Academy of Sciences* 109(40), 16083–16088. DOI: 10.1073/pnas.1211658109.

Shochat, E., Warren, P.S., Faeth, S.H., McIntyre, N.E. and Hope, D. (2006) From patterns to emerging processes in mechanistic urban ecology. *Trends in Ecology & Evolution* 21(4), 186–191. DOI: 10.1016/j.tree.2005.11.019.

Steffen, W., Rockström, J., Richardson, K., Lenton, T.M., Folke, C. *et al.* (2018) Trajectories of the earth system in the Anthropocene. *Proceedings of the National Academy of Sciences* 115(33), 8252–8259. DOI: 10.1073/pnas.1810141115.

Tansley, A.G. (1935) The use and abuse of vegetational concepts and terms. *Ecology* 16(3), 284–307. DOI: 10.2307/1930070.

United Nations (2015) *World urbanization prospects: the 2014 revision*. Department of Economic and Social Affairs. United Nations Population Division, Ed. Population Division (ST/ESA/SER.A/366).

Vitousek, P.M., Ehrlich, P.R., Ehrlich, A.H. and Matson, P.A. (1986) Human appropriation of the products of photosynthesis. *BioScience* 36(6), 368–373. DOI: 10.2307/1310258.

Walsh, C.J., Roy, A.H., Feminella, J.W., Cottingham, P.D., Groffman, P.M. *et al.* (2005) The urban stream syndrome: current knowledge and the search for a cure. *Journal of the North American Benthological Society* 24(3), 706–723. DOI: 10.1899/04-028.1.

Walsh, C.J., Fletcher, T.D. and Burns, M.J. (2012) Urban stormwater runoff: a new class of environmental flow problem. *PLOS ONE* 7(9), e45814. DOI: 10.1371/journal.pone.0045814.

Wheeler, M.M., Neill, C., Groffman, P.M., Avolio, M., Bettez, N. *et al.* (2017) Continental-scale homogenization of residential lawn plant communities. *Landscape and Urban Planning* 165, 54–63. DOI: 10.1016/j.landurbplan.2017.05.004.

Wiek, A. and Iwaniec, D. (2014) Quality criteria for visions and visioning in sustainability science. *Sustainability Science* 9(4), 497–512. DOI: 10.1007/s11625-013-0208-6.

2 How Urban Forest Composition Shapes the Structure and Function of Arthropod Communities

Holly M. Martinson[1]*, Michael J. Raupp[2] and Steven D. Frank[3]

[1]Department of Biology, McDaniel College, Westminster, Maryland; [2]Department of Entomology, University of Maryland, College Park, Maryland; [3]Department of Entomology and Plant Pathology, NC State University, Raleigh, North Carolina

Urban forests and greenspace include the woody and herbaceous plants found in gardens, parks, forest remnants, and along streets in urbanized areas (McBride and Jacobs, 1976; Konijnendijk *et al.*, 2006). These forests provide critical ecosystem services, reducing energy use by providing shading and cooling, reducing wind speeds, mitigating soil pollutants, increasing real-estate values, improving human health, and providing habitat for plants and wildlife. These urban forests also provide recreational opportunities and foster community well-being (Moll and Ebenreck, 1989; McPherson and Nowak, 1993; Abdollahi and Ning, 1996; Nowak *et al.*, 2000, 2002; Aukema *et al.*, 2011; Donovan *et al.*, 2013). Over half the world's human population now lives in cities, putting increasing pressure on urban forests and the services they provide.

In this chapter, we describe the composition and historical development of urban forests, how urban development shapes tree density and diversity, and how herbivores and herbivory respond to these changes in urban vegetation. We then examine the pattern of the increased prevalence of exotic plants in cities, the novel plant–insect interactions they create, and the consequences for urban forest health. In doing so, we highlight how the density, diversity and geographic origin of trees in urban forests can lead to the simplification of plant and arthropod communities and greater potential for herbivore outbreaks.

The Composition of Urban Forests

Historical development of urban forests

The use of woody plants as sources of food, for providing shade and for beautification of landscapes dates back several thousand years in urban centres throughout Europe, the Middle East and China (Hauer *et al.*, 2017). The composition of modern urban forests depends, at least in part, on the biome in which the city occurs. In densely forested regions, land was cleared for agriculture, settlement and the growth of cities, which dramatically reduced the density of

*Corresponding author: hmartinson@mcdaniel.edu

woody plants and changed their species composition. The extant urban forests in these regions consist of native trees and their progeny, with the addition of exotic trees grown deliberately for fruit, shade or amenity value or which have become naturalized; urban forest remnants and secondary growth forests are common in these regions. In contrast, savannas, arid regions and Mediterranean biomes are generally devoid of dense native forests. As cities developed in these regions, native and exotic trees were added to cities so that both the density and diversity of trees were higher in urban forests compared to surrounding areas (McBride and Jacobs, 1976; Dreistadt *et al.*, 1990). Overall, the trees present in urban forests represent a combination of native trees retained during urbanization, new trees deliberately planted, and spontaneous reproduction of trees in unmanaged locations (McBride and Jacobs, 1976; Dreistadt *et al.*, 1990; Williams *et al.*, 2009). Thus, the composition of urban forests varies among biomes and with the land use and cultural history of cities, factors which will govern how arthropods respond to changes in the density, diversity and composition of urban trees.

Patterns of tree density in urban forests

The density of trees typically declines as impervious surfaces and buildings replace vegetation in the built-up areas of cities, at least in previously forested regions. In city centres, where street trees may comprise the majority of woody plants, the abundance of trees is extremely low; individual trees become surrounded by impervious surfaces and, in places, trees may be absent altogether. An analysis of street trees in more than 100 cities, villages and towns in the state of New York, USA, revealed that street tree densities can be surprisingly low, ranging from 2.9 to 7.1 trees per 100 metres of street (Cowett and Bassuk, 2014). Urbanization associated with Minnesota's Twin Cities Metropolitan Area revealed a clear inverse relationship between the percentage of impervious surface and the percentage of tree canopy cover in 4000 sample locations (Berland, 2012). A similar study of seven major metropolitan areas in eastern North America revealed strong negative correlations between tree density and the percentage of urban area covered by impervious surfaces and buildings (Nock *et al.*, 2013).

Unsurprisingly, urban forest remnants or biological preserves within cities often support the highest density of trees (Blair and Launer, 1997; Ortega-Álvarez *et al.*, 2011). However, at larger spatial scales, tree canopy cover is often greatest at intermediate sites along rural-to-urban gradients. This pattern has been attributed to three factors along the gradient: the paucity of trees in agricultural and newly developed residential areas surrounding the city, the growth and maturation of abundant trees in low-density residential neighbourhoods at intermediate levels of urbanization, and intense urbanization in the city centre where trees are few (Berland, 2012).

Patterns of tree diversity in urban forests

Several studies have addressed changes in woody plant diversity along urbanization gradients, and, in general, a negative relationship could be predicted between the intensity of urbanization and woody plant diversity. A survey of 26 forests in the urban region of Basel, Switzerland, found that species richness and the Shannon Diversity Index declined with increasing levels of urbanization (Melliger *et al.*, 2018). A comparison of the Shannon Diversity Index, functional diversity (the variety of plant traits such as leaf nitrogen per unit mass, seed mass, wood density, maximum plant height) and ecosystem services provided by plants (see Cardinale *et al.*, 2011) can provide ways to compare urban and non-urban habitats. Indeed, in seven major North American cities, these variables were inversely correlated with the percentage of buildings and hard surfaces (Nock *et al.*, 2013). Overall, extra-urban sites (commuter towns for urban areas) supported the greatest plant diversity. Sites with low or moderate amounts of human influence had significantly greater diversity and functional diversity than sites with the highest amounts of human influence, supporting the notion that urbanization contributed to the loss of valuable functional diversity. At smaller spatial scales, however, several more nuanced patterns of tree richness and diversity emerge. Often, parks, residences and vacant lots support some of the highest levels of tree diversity within cities. Among ten Nordic cities, the species richness of trees in parks was significantly higher than that of street trees (Sjöman *et al.*, 2012). In Mexico City, Mexico, residential sites supported the highest tree species richness, even when compared to green areas in the city (Ortega-Álvarez *et al.*, 2011). In addition, urban vacant lots can sometimes support substantial tree diversity. For example, in Cleveland, Ohio, vacant lots are minimally managed, but they can support a large proportion of the city's urban forest. In fact, inner-city vacant lots support three times as many trees and greater tree diversity than trees in suburban residential lots, thus providing valuable ecosystem services (Riley *et al.*, 2018). Whether tree species are native or exotic can also be an important variable. In the New York metropolitan region, native plants are negatively associated with urbanization, but exotic plants are positively associated. The combined effect of these opposing trends results in a lack of relationship between diversity and urbanization (Aronson *et al.*, 2014).

Simplification and homogenization of urban forests

The replacement of native plants and animals with fewer, geographically widespread exotic biota may lead to the simplification and homogenization (greater similarity between different regions) of biota in urban areas (McKinney and Lockwood, 1999). The effect of exotic species on homogenization, however, may depend on exotic species richness. Where exotic species are present at low levels and different exotic species colonize and establish in different regions, they may serve to differentiate localities (McKinney, 2006). On the other hand, as exotic species richness increases, the homogenizing effects may become increasingly pronounced. Overall patterns of increased homogenization with increasing levels of urbanization are found in urban forests in North America (Nock *et al.*, 2013) and Germany (Kühn and Klotz, 2006). However, not all species respond to urbanization similarly, and important differences among exotic plant groups are evident. In Germany, urban sites dominated by native and long-established exotic tree species are more similar to each other than are rural sites; that is, they reflect homogenization. In contrast, urban sites containing exotic species introduced after the 1500s are more differentiated relative to rural sites (Kühn and Klotz, 2006). Similarly, the urban flora examined in the metropolitan area of New York is not more homogeneous than plant communities at rural sites (Aronson *et al.*, 2015). In New York, a decline in native plant species with urbanization is offset by an increase in exotic species, which vary enough among sites that beta diversity, and therefore community differentiation, is higher in urban compared to rural locations. At this intra-city scale, homogenization may be less likely to occur than at the larger, inter-city scale at which homogenization has been evaluated by others (McKinney, 2004, 2006; Kühn and Klotz, 2006).

Regardless of plant geographic origin, large-scale homogenization of street trees in cities remains a major threat to urban forest canopies in North America and Europe (Nowak *et al.*, 2001; Raupp *et al.*, 2006; Sjöman and Östberg, 2019) in the face of invasive pathogens and insects. Disproportionate planting of elm trees in European and North American cities predisposed them to catastrophic loss with the introduction of Dutch elm disease, *Ophiostoma ulmi* and *Ophiostoma novo-ulmi* (Raupp *et al.*, 2012). In North America, elms were replaced with a small palette of street trees (Raupp *et al.*, 2006), thus predisposing them to catastrophic loss with the introduction of emerald ash borer, *Agrilus planipennis*, and Asian long-horned beetle, *Anoplophora glabripennis*. This has resulted in losses of hundreds of millions of trees in North America (http://www.emeraldashborer.info/), many of which are in urban centres.

In summary, at spatial scales encompassing the rural–urban gradient, the abundance and diversity of woody plants often decline with increasing levels of urbanization, especially in temperate regions and when impervious surfaces replace vegetation in city centres. Within cities, urban sites can encompass considerable variation in tree density and diversity, and in many cities, residential areas and parks with their attendant human-contrived landscapes may support relatively high levels of plant diversity and abundance. Next, we describe how these patterns of urban tree abundance, diversity and composition shape communities of herbivorous arthropods, with consequences for biodiversity, tree health, and ecosystem services.

Response of Herbivorous Arthropods to Vegetation Changes in Urban Forests

The urban heat island effect directly and indirectly affects herbivores

Changes in the density of trees and other vegetation in urban landscapes affect the biotic and abiotic conditions in which herbivores exist. In densely urbanized areas, the high concentration of impervious surface cover and the low density of trees increases temperature (the urban heat island effect, Oke, 1973), which directly affects herbivorous arthropods. Increasing temperature typically leads to higher overwinter survival, faster development time, and more generations per year in herbivorous insects (Bale *et al.*, 1988; Yamamura and Kiritani, 1998; Bale *et al.*, 2002; Menéndez, 2007; Trân *et al.*, 2007; Netherer and Schopf, 2010; Mitton and Ferrenberg, 2012).

The effects of urban warming in particular have been studied in several herbivorous arthropods. Cicadas in Korea, for example, reach higher densities in warmer urban compared to cooler urban or rural areas (Moriyama and Numata, 2008; Nguyen *et al.*, 2018). Higher temperatures increase overwinter survival of mimosa webworm pupae sheltering near buildings compared to those in less urban locations (Hart *et al.*, 1986; Miller and Hart, 1987). Likewise, the gloomy scale, *Melanaspis tenebricosa*, has higher survival over winter and during the growing season as tree canopy temperature increases just 2.5°C (Dale and Frank, 2014a, b). The soft-scale complex *Parthenolecanium* spp. have higher survival during the growing season as tree canopy temperature increases, as documented in urban field observations and growth chamber experiments (Meineke *et al.*, 2013; Meineke and Frank, 2018).

In addition to direct effects on herbivores, impervious surface cover and the urban heat island effect also have indirect effects on herbivores by changing the quality of their host plants. Water stress can be induced by a combination of elevated urban temperatures and limited access to water because of impervious surfaces over roots. In some cases, water stress can improve host plant quality by increasing leaf nitrogen availability (McClure, 1980; Koricheva *et al.*, 1998; Huberty and Denno, 2004). On the other hand, continuous water stress can decrease nutrient availability for some herbivores or even lead to increased plant defences (Koricheva *et al.*, 1998; Huberty and Denno, 2004). Thus, the indirect effects of urban warming on arthropods via their host plants are inconsistent, or, at a minimum, highly variable. For example, sucking insects were found to be more abundant on water-stressed oaks in Lincoln, Nebraska, but less abundant on water-stressed mulberries in College Park, Maryland (Hanks and Denno, 1993; Cregg and Dix, 2001). Heat and drought have additive benefits for gloomy scale. Fecundity of gloomy scale increases in response to both factors, such that scales on the hottest, driest trees produce 50% more embryos than watered trees just 2.5°C cooler. Likewise, the abundance of *Parthenolecanium* spp. and oak spider mites, *Oligonychus bicolor*, increase in response to heat and drought (Meineke *et al.*, 2013; Meineke and Frank, 2018). Unfortunately, the nutritional or defensive mechanisms responsible for these changes are not known for these or most other interactions.

Herbivores respond to patterns of plant density, species richness and diversity

Herbivorous insects often respond to variation in plant abundance, density and diversity, especially for those that depend on particular host plants. Whereas the abundance of many herbivores increases where host plants are abundant or spatially aggregated (Root, 1973; Connor *et al.*, 2000), variation in how herbivores sense and move toward host plants or respond to the palatability of adjacent plants can lead to neutral or negative responses to plant density and patch size (Hambäck and Englund, 2005; Barbosa *et al.*, 2009). To investigate how urban herbivores respond to patterns of abundance and diversity of host plants, Raupp *et al.* (2001) sampled herbivorous arthropods on woody plants in 212 residential landscapes bi-weekly over the course of two consecutive growing seasons. Guilds of insects and mites included defoliators, sap-suckers, leaf-miners, gall-makers and borers. As expected, more species of herbivores were present in landscapes containing a higher abundance of plants and more plant species, such that plant abundance and species richness each explained 17–18% of the observed variation in herbivore species richness. Overall, the slope of the relationship between herbivore species richness and plant species richness was much steeper than for plant abundance. Thus, adding a new plant species to an urban landscape has a much stronger effect on the number of herbivore species than does simply adding additional plant individuals. This is not surprising, as many herbivores are specialized, and adding new plant species provides opportunities for additional specialist herbivore species to colonize landscapes. However, the slopes of relationships between herbivore richness (as the response variable) and plant abundance and richness (as explanatory variables) were significantly less than one. Many species of plants found in residential landscapes do not support detectable levels of herbivores, thereby suppressing a one-to-one relationship between plant richness and the richness of herbivores (Raupp and Noland, 1984; Raupp *et al.*, 2001).

Butterflies and moths: well-studied urban herbivores

Species of Lepidoptera (butterflies and moths) are particularly well studied globally, including in cities. Studies of urban Lepidoptera communities have permitted comparisons of how herbivore diversity and abundance change through time and with urban intensity, as well as how these responses depend on species' ecological traits. Because of the availability of historical records for Lepidoptera, comparisons of modern and historical records can be made directly. In one of the few studies comparing herbivores over time in urbanizing regions, Pyle (1983) found that 7.6– 40.0% of the butterflies in four US cities (San Francisco, Los Angeles, Staten Island, Aurora) have become extinct or endangered. Examining spatial patterns in one of the first studies of urbanization gradients, Blair and Launer (1997) examined Lepidoptera diversity in what had formerly been forested areas in Palo Alto, California. Butterfly richness and diversity were greatest in open recreational areas, golf courses, residential areas and a forest preserve and were lowest at the most urban locations in an office park and business district. Not surprisingly, butterfly abundance declined monotonically from the least developed sites to the most developed. Notably, the business district was devoid of all butterflies found in rural sites. A similar examination of butterfly assemblages along an urban gradient in Ontario, Canada, demonstrated that several species present at less developed sites were missing from the most urbanized sites (Hogsden and Hutchinson, 2004). Abundance and species richness were higher at moderately disturbed sites compared to minimally disturbed sites. On the other hand, a study of a guild of leaf-mining caterpillars on oaks in the San Francisco Bay area found no association between urban land use and the species richness and abundance of leaf-miners (Rickman and Connor, 2003), suggesting that not all species respond similarly to urbanization.

In many butterfly and moth communities, information about species' ecological traits (i.e. habitat affinity, body size, mobility) can be used to shed light on potential mechanisms underlying variation in species' responses to urbanization and to predict how species in other, more poorly studied herbivore communities might respond. For example, the species richness of both specialist and generalist butterflies decreased with increasing levels of urbanization across 105 study plots in Switzerland (Concepción *et al.*, 2015). However, species' responses depended on the spatial scale at which urbanization was measured and on the level of mobility of the species. Species richness of highly mobile butterflies was negatively affected by urbanization across a broad range of spatial scales. Negative effects of urbanization on the species richness of less mobile species were most evident at smaller spatial scales. When mobility and specialization were considered simultaneously, only highly mobile specialists exhibited negative responses to increases in urban area, especially at the largest spatial scales.

Mobility also plays an important role in determining the community composition of Lepidoptera responding to habitat fragmentation, another fundamental consequence of urbanization (Williams *et al.*, 2009; Beninde *et al.*,

2015). In an English study, moth abundance and species richness were found to be greater on oaks in hedgerows (rows of shrubs or trees) compared to isolated oak trees, and mobile forest specialists tended to be more negatively affected by fragmentation than were habitat generalists (Slade *et al.*, 2013). Moths with larger body size, and hence greater dispersal ability, can be favoured in highly fragmented urban sites, both at the community and the intraspecific levels (Merckx *et al.*, 2018). Such a pattern suggests that mobility may be a key ecological trait selected for in cities, and further work investigating the impacts of urbanization on body size and mobility patterns are needed in the future.

Herbivory in Urban Forests

In a recent review of outbreaks of herbivorous arthropod pests, Raupp *et al.* (2012) provided several examples of insects and mites that attain high densities and cause damaging levels of herbivory in urban forests. However, relatively few studies compare levels of herbivory between cities and ex-urban areas or along gradients of urbanization. Among the first to address this question were Christie and Hochuli (2005), who investigated forest fragments in Sydney, Australia. These authors found that oak trees experienced greater leaf damage in small forest fragments compared to oaks in large forest fragments, at both interior and edge locations. While no mechanistic explanation was determined, the authors suggested nutrient subsidies from the surrounding urban areas or loss of top-down regulation from predators and parasitoids in small remnants may have contributed to higher levels of herbivory by chewing insects. Leaf herbivory can also be greater on trees in urban forest fragments compared to landscape and street trees (Long, 2019), probably due to greater plant and herbivore diversity. In a study in the Cincinnati, Ohio, metropolitan area, leaf herbivory did not differ between woody plants growing in natural areas such as woodlots and wooded parks versus those growing in ornamental plantings in residential gardens and managed recreational areas (Matter *et al.*, 2012). However, in both habitat types, native plants sustained more herbivory than exotics. Native plants sustained more leaf herbivory by cankerworms in simple urban landscapes than did native plants in more structurally complex forest remnants (Frank, 2014). Again, overall herbivory of native plants was greater than exotics in both habitats (Frank, 2014).

Herbivory by non-chewing herbivores like scales and aphids, often measured as the abundance of the insects, shows more consistent responses to urbanization and is often greater on urban trees than trees in woodlots or in complex residential landscapes (reviewed in Raupp *et al.*, 2010, 2012). For example, gloomy scale density was over 400 times greater on red maples along streets than on red maples in urban forest fragments (Long *et al.*, 2019) and rural forests throughout the south-eastern USA (Youngsteadt *et al.*, 2015). A similar pattern was found when comparing *Parthenolecanium* spp. scales along streets and in forests (Long, 2019). Experimental and observational evidence indicate high temperature can be the primary factor increasing gloomy scale and *Parthenolecanium* spp. scale density (Meineke *et al.*, 2013; Youngsteadt *et al.*, 2015; Long *et al.*, 2019). Similarly, crape myrtle aphids can be less abundant on landscape trees surrounded by complex vegetation than on trees surrounded primarily by grass or pavement. This response can be attributed, at least in part, to more robust natural enemy communities in complex habitats (Parsons, 2019). Impervious surface cover, as a measure of urbanization, was also the best predictor of Asian citrus psyllid density on trees in a California study (Thomas *et al.*, 2017) and was a strong predictor of scale density on multiple oak and maple species (Backe, 2019). At a larger spatial scale, gloomy scale density consistently increased with urbanization and resulted in worse tree condition in urban sites across eight US cities (Dale and Frank, 2014a; Just *et al.*, 2018, 2019).

In contrast to the general increases in the abundances of sucking insects found in the above within-city studies, a larger-scale comparison of rural and urban sites across 16 European cities revealed an adverse effect of urbanization on overall herbivory (Kozlov *et al.*, 2017). Damage due to leaf chewers and leaf-miners was significantly lower in urban sites compared to rural sites, and the strongest reductions to overall herbivory were evident in large urban areas. Higher rates of bird attacks on plasticine models of caterpillars in urban sites compared to rural sites revealed increased bird predation as a potential mechanism underlying these patterns (Kozlov *et al.*, 2017). Similarly, in another large study in Europe, leaf herbivory was lower on urban oaks compared to rural oaks (Moreira *et al.*, 2019). As an interesting point of contrast, a recent study in Raleigh, North Carolina, found greater leaf herbivory on willow oaks and greater bird predation in urban forest fragments compared to residential landscapes and downtown locations (Long, 2019). The mechanism leading to these results was suggested to be greater herbivore species diversity in urban forests compared to those urban locations with lower tree density and diversity. This within-city pattern supports observations of an inverse relationship between urbanization and functional diversity (Nock *et al.*, 2013). However, herbivory was not related to city size or leaf defences, so mechanisms for these and other results remain unclear. Reconciling disparate responses of herbivores across cities, feeding guilds, and spatial scales remains an important avenue for future research.

Exotic Plants Are a Major Component of Urban Forests

An increase in the proportion of exotic organisms with increasing levels of urbanization is one of the most consistent patterns in urban ecology (McDonnell and Pickett, 1990; McKinney 2002; Luck and Smallbone, 2010; Kowarik, 2011; Aronson *et al.*, 2014). Thus, although many urban areas retain substantial plant diversity, much of that diversity is derived from the presence of exotic plants. Across the 110 cities for which data were available for synthesis, urban plant communities included an average of 28% exotic species (Aronson *et al.*, 2014). In certain cities, that proportion is substantially greater. For example, 53% of the herbaceous and woody plant species in built-up areas of Beijing, China, are exotic (Zhao *et al.*, 2010), and exotic trees are prevalent in Kumasi, Ghana, making up 65.7% of the street trees along principal roads (Uka and Belford, 2016). In Orléans, France, 11 coniferous tree species are present, and six of these are exotic (Rossi *et al.*, 2016). Of 888 street trees in ten South African towns, 71% are exotic (Gwedla and Shackleton, 2017). In that region, the composition of the street trees varies considerably among towns, such that the proportion of native street trees ranges from 0% to 50%. Similarly, Avolio *et al.* (2018) reported that only 11% of trees planted in neighborhoods in Salt Lake Valley, Nevada, were native. In the few cases in which changes in urban biota have been tracked over time, such as in Adelaide, Australia, losses of native plant and animal species have been dwarfed by nearly five times as many introductions of exotic species (Tait *et al.*, 2005).

The shift toward a higher prevalence of exotic plants with urbanization can be seen across spatial scales, with clear trends across urban gradients and distinct patterns of exotic species prevalence within cities. For example, across the New York metropolitan area, the proportion of exotic woody plant species richness increases with increasing urban land cover (Aronson *et al.*, 2015). The proportion of exotic plant species in cities can be a function of city size, especially for plants introduced since the 1500s (Pyšek, 1998), as well as variables such as total vegetation cover and city age (Aronson *et al.*, 2014). Evidence from studies conducted within cities suggests that particular habitats are more likely to harbour a greater proportion of exotic plants. For example, vacant inner-city lots in Cleveland have a significantly higher abundance of exotic trees compared to native trees, as well as a larger proportion of exotic trees than is found in residential properties (Riley *et al.*, 2018). In Scandinavian cities, there is a greater number of exotic plant species along streets compared to those in parks, although the abundance of native plants is greater than exotics throughout (Sjöman *et al.*, 2012). Residential and residential-commercial sites within Mexico City have a greater number of exotic tree species than green areas (Ortega-Álvarez *et al.*, 2011). Similarly, in Fortaleza, Brazil, exotic trees are common in gardens, streets and squares (Moro *et al.*, 2014). Greater than 70% of the species and greater than 75% of the individuals in those locations are exotic. In contrast, a nearby large vegetation remnant contains only 8.2% exotic species and less than 1% exotic individuals (Moro *et al.*, 2014).

Disturbed sites, such as abandoned properties, waste areas, areas with high impervious surface cover, and forest edges, may generally harbour more exotic species than less disturbed habitats (Godefroid and Koedam, 2003; Kowarik, 2008; Riley *et al.*, 2018), but exotic plants may also be deliberately planted due to species-specific tolerances to the often harsh urban environment. Indeed, cities often are important points of entry, release and trade of exotic plants (Kowarik, 2011). Homeowners may also select exotic trees and shrubs for ornamental value based on current trends (Avolio *et al.*, 2018). In certain situations, exotic trees may be more resistant to herbivores and the harsh abiotic conditions in the city centre than are the available native species (Sjöman *et al.*, 2012; Frank *et al.*, 2019).

Native and exotic arthropods in cities encounter plants from many geographic origins

Globally, plants and insects are among the most commonly introduced taxa, and the rates of introduction of these groups do not appear to be slowing (Seebens *et al.*, 2017). In urban areas, a combination of suitability, availability, cost, aesthetics, consumer demand, as well as spontaneous colonization of exotic plants, have contributed to the high relative abundance of exotic plants (Sjöman *et al.*, 2016; Avolio *et al.*, 2018; Riley *et al.*, 2018). Exotic herbivores also continue to be introduced across habitats, where they can have massive ecological and economic impacts (Liebhold *et al.*, 1995; Pimentel *et al.*, 2005; Holmes *et al.*, 2009; Aukema *et al.*, 2011). Liebhold *et al.* (2018) identified the primary role of plant diversity, both native and exotic, in driving exotic insect diversity globally. Thus, we would expect to find urban communities replete with both native and exotic plants, as well as native and exotic insects. The combinations of native and exotic plant and arthropod species, along with unique abiotic conditions, leads to herbivory and other interactions that are hard to predict and do not always follow ecological theories developed in natural ecosystems (Dale and Frank, 2014b, 2018; Just *et al.*, 2019).

Here, we propose that the higher proportion of exotic trees in cities leads to potential mismatches in coevolutionary history between plants and herbivores, with four outcomes based on the geographic provenance of herbivorous insects and the plants on which they feed (Fig. 2.1). For each combination of plants and herbivores, two outcomes may be possible: either elevated or reduced herbivory, where herbivory is quantified as either herbivore damage or the abundance of herbivores. The outcome of each combination is likely to be governed by the specific details of the abiotic and biotic contexts of the interaction, so caution must be used when making generalities about native and exotic plant and herbivore interactions in the urban forest.

A. Novel interactions between native herbivore and plant communities

In many cities, native trees are retained in remnant forest habitats or planted in street, park or residential settings (Nowak *et al.*, 1996; Cowett and Bassuk, 2014). Native plants have been promoted for the retention of biodiversity and their suitability for the regional abiotic conditions (Tallamy, 2004; Chalker-Scott, 2015), and they are frequently colonized by native insects. However, native plants in cities do not always support arthropod communities equivalent to those found in forests or natural areas, nor are populations of arthropods in cities necessarily stable (McIntyre, 2000; Raupp *et al.*, 2010; Dale and Frank, 2018). Such shifts can be due to changes in biotic conditions, such as plant diversity, structural complexity and biological control by natural enemies, as well as differences in abiotic conditions such as temperature, pollution and water availability (Meineke *et al.*, 2017; McCluney *et al.*, 2018; Long *et al.*, 2019). Across taxa, higher trophic levels may be more sensitive to the harsh abiotic conditions in cities (Van Nuland and Whitlow, 2014; Meineke *et al.*, 2017; Melliger *et al.*, 2018), resulting in relaxed natural enemy control of lower trophic levels and the potential for herbivore outbreaks.

Studies examining native herbivores on native trees in cities document declines of sensitive species, as well as outbreaks of insects that may be facilitated by the warm conditions or altered nutrient regimes. For example, in Charlotte, North Carolina, outbreaks of native cankerworms on the commonly planted native willow oak, *Quercus phellos*, have occurred consistently over the past 20 years (Asaro and Chamberlin, 2015), perhaps due to the abundance of oaks in the urban

Herbivorous insect		Plant: Native	Plant: Exotic
	Native	A Herbivory is higher when abiotic conditions increase plant quality and natural enemies are lacking. Herbivory is lower when native plants are adapted to local conditions and well-defended, herbivore communities are simplified, or natural enemies are active.	B Herbivory is higher when native herbivores experience defence-free space. Herbivory is lower when native herbivores cannot host-switch.
	Exotic	C Herbivory is higher when exotic herbivores experience defence-free space and enemy escape due to absence of specialist natural enemies. Herbivory is lower when herbivores cannot host-switch.	D Herbivory is higher when abiotic conditions increase plant quality and herbivores experience enemy escape. Herbivory is lower when herbivores are regulated by natural enemies.

Fig. 2.1. The higher proportion of exotic trees in cities leads to potential mismatches in coevolutionary history between plants and herbivores or due to the separation of herbivores from natural enemies via enemy escape (enemy-free space or enemy release; Jeffries and Lawton 1984; Keane and Crawley, 2002; Mlynarek *et al.*, 2017). Labels A–D refer to different combinations of native and exotic plants and herbivorous insects, the details of which are described in the text. The coevolutionary matrix depicted herein was first proposed by Raupp *et al.* (2010).

forest and urban conditions (heat, nutrient additions) favourable to the insect. Similarly, repeated outbreaks of orange-striped oak worms, *Anisota senatoria*, in the south-eastern USA can negatively affect tree health, especially for water-stressed urban pin oak, *Quercus palustris* (Mattson *et al.*, 1991; Coffelt *et al.*, 1993). Likewise, native scale insects, mites, aphids, and other sucking herbivores frequently become more abundant on their native hosts due to abiotic conditions or changes in host plant quality (Meineke *et al.*, 2013; Dale and Frank, 2014b, 2017; Meineke and Frank, 2018). Due to evolutionary relationships and abiotic conditions, native and exotic herbivores often become damaging herbivorous pests on native trees, reducing the growth of these trees and dampening the ecosystem services they provide (Meineke *et al.*, 2016; Dixon, 1971; Coffelt *et al.*, 1993; Dale and Frank, 2014a). Increasingly, exotic plants are recognized as sustainable alternatives in pest-prone planting situations since they generally have fewer herbivores and less herbivory (Kendle and Rose, 2000; Sjöman *et al.*, 2016; Frank *et al.*, 2019).

B. Exotic trees may be resistant or highly susceptible to native insect herbivores

As exotic plants and insects continue to be introduced to new habitats worldwide (Aukema *et al.*, 2010; Seebens *et al.*, 2017), the potential for novel species interactions, i.e. interactions among species with little recent coevolutionary history, continues to increase (Pearse and Altermatt, 2013). New plants are often difficult for non-coevolved herbivorous insects to utilize. As such, plants may present novel physical or chemical defences and lack necessary host plant recognition cues (Ehrlich and Raven, 1964; Tahvanainen and Niemelä, 1987; Futuyma and Agrawal, 2009).

One of the major hypotheses regarding plant invasions is that many plants are introduced without herbivores, releasing them from top-down control (e.g. the Enemy Release Hypothesis, Keane and Crawley, 2002). The number of herbivorous insects utilizing an exotic plant is often related to how long ago it was introduced, its abundance, or the geographic range of the exotic plant where it has been introduced. This suggests that colonization of novel hosts is not immediate and is positively related to the spatial and temporal opportunities for colonization (Kennedy and Southwood, 1984; Brändle *et al.*, 2008). The aforementioned factors and the often narrow diet breadth of herbivorous insects (Forister *et al.*, 2015) may lead to herbivore communities on exotic plants that are species-poor, skewed toward generalists, and which lack coevolutionary history with the plants on which they feed.

Consistent with this prediction, Roques *et al.* (2006) found that herbivore species richness on conifers introduced to Europe was lower than on the same plants in their native ranges. For example, Douglas fir, *Pseudotsuga menziesii*, which has been cultivated in plantations in Europe for over 130 years, has acquired only 33.9% of the species richness of herbivores in its native North American range. Furthermore, the herbivore communities in Europe were dominated by polyphagous species and external feeders. Similarly, in a study of the accumulation of herbivores in central Europe, Brändle *et al.* (2008) found that exotic plants supported a lower overall species richness of butterflies and moths (Lepidoptera) and leafhoppers (Hemiptera: Auchenorrhyncha) compared to native woody plants (Brändle *et al.*, 2008). Further, the diet breadth of herbivores on the exotic species was much broader than on the native species. In common gardens of native and exotic woody trees and shrubs, native plants supported higher abundances and a greater species richness of butterfly and moth caterpillars (Burghardt *et al.*, 2010). Furthermore, fewer than 6% of the herbivores on exotic plants were specialists (i.e. feeding on fewer than three families), whereas up to 33% of the species found in native-only plantings were specialists.

In addition to driving differences in herbivore richness, a lack of colonization of novel plants by native herbivores may also result in lower levels of herbivory on exotic plants. For example, in wild and cultivated settings in Florida, native members of the genus *Eugenia* (plants in the myrtle family) experienced higher levels of herbivory than did exotic species (Liu *et al.*, 2007). Norway maple (*Acer platanoides*) can experience three times less herbivory in its exotic range in North America compared to its native range in Europe (Adams *et al.*, 2009). Similarly, higher levels of herbivory have been recorded on native plants compared to exotic plants in both ornamental and natural settings in and around Cincinnati, Ohio (Matter *et al.*, 2012). Several recent studies, however, have demonstrated that this pattern of lower herbivory on exotic plants is far from universal. Scale insects were more abundant on some but not all native maple and oak species compared to exotic congeners in the south-eastern USA (Frank *et al.*, 2019). Furthermore, a recent study in experimental landscapes made entirely of either native species or their exotic congeners found similar herbivory on native and exotic trees (Parsons, 2019), highlighting that more work needs to be done to understand the mechanistic basis of these patterns.

Reductions to the abundance and species richness of herbivores on exotic plants may have consequences for the larger ecological community and contribute to unstable herbivore dynamics in cities. When exotic plants increase in prevalence in cities and few herbivores are able to colonize these plants, the resulting insect communities may be simplified. Further, they may be dominated by generalists and lack the biomass necessary to sustain higher-order consumers such as predatory arthropods, parasitic arthropods and birds (Tallamy, 2004). In a common garden experiment, exotic plants were found to support fewer individuals and less diverse herbivore communities compared to native plants (Burghardt and Tallamy, 2013). Specific components of the herbivore community were sensitive to plant geographic origin, such that exotic plants supported fewer immature individuals, leaf chewers and internal feeders. Furthermore, such changes to the arthropod community led to lower beta diversity for immature herbivores on exotic trees, both among individual trees and among sites (Burghardt and Tallamy, 2015). Exotic experimental landscapes supported herbivores with lower host specificity; and among sites, herbivores were a redundant subset of species on natives. Thus, community simplification and biotic homogenization may occur with the planting of exotic trees in cities, especially when phylogenetically distinct exotic plants replace natives.

Not all research finds that exotic street trees have fewer herbivores or less herbivory than natives. Whether exotics have lower, equal or higher arthropod abundance or diversity likely depends on the tree species and the herbivore taxa selected for study. In several recent studies, herbivore abundance, diversity or herbivory were similar among native and exotic trees and other woody plants (Backe, 2019; Parsons, 2019). In addition, herbivore and natural enemy communities on native and exotic plants depend strongly on plant composition in the surrounding landscape. For example, in addition to finding that exotic maple street trees can have greater arthropod diversity than native congeners, Backe (2019) found that at high levels of impervious surface cover, the exotic trees maintained their diverse arthropod communities while communities on natives were diminished. Finally, in contrast to research focused on herbivores, many studies have found similar densities of natural enemies on native and exotic tree species (Proches *et al.*, 2008; Hartley *et al.*, 2010; Backe, 2019; Frank *et al.*, 2019; Parsons, 2019).

When herbivores can utilize novel host plants (i.e. those plants with which they share little coevolutionary history) and when they are not limited by natural enemies, dramatic increases in herbivore densities and damage to exotic plants can occur. For example, the North American pine needle scale, *Chionaspis pinifoliae*, produces more eggs, survives better and achieves higher abundances on the exotic Scots pine (*Pinus sylvestris*) than on the North American red pine, *P. resinosa* in field plots in Ohio (Glynn and Herms, 2004). Similarly, in a 20-year common garden experiment, three of four exotic birch species suffered 100% mortality due to the native bronze birch borer, *Agrilus anxius* (Nielsen *et al.*, 2011). In contrast, the three native birch species in the common garden persisted, with greater than 70% survival over the 20-year period. In general, a mismatch in the origins of plants and herbivorous insects indicates a lack of coevolutionary history and a resulting lack of specific resistance mechanisms. This defence-free space can lead to outbreaks of herbivorous insects and damage to the plants on which they feed (Gandhi and Herms, 2010; Showalter *et al.*, 2018).

The likelihood of herbivores experiencing defence-free space versus not being able to utilize a novel host may be related to the feeding ecology of the herbivore and the relatedness of novel plants to those in the local plant community. In a meta-analysis of the fitness of herbivores on novel compared to native hosts, Bertheau *et al.* (2010) found that, overall, the use of novel hosts depresses measures of insect fitness. Two aspects of the biology of the insects and the plants emerged as important explanatory factors in that analysis: the diet breadth of the herbivore and the phylogenetic relatedness of the plants. The most negative effects of novel plants on herbivore fitness were on herbivores feeding on only one host plant (i.e. monophagous species), whereas the effect on the fitness of generalist (i.e. polyphagous) herbivores was neutral. Second, the effect on fitness of using a novel but closely related plant was minimal compared to the effect of using a distantly related novel host; this effect was notable for species with narrow diet breadths (i.e. mono- and oligophagous species), although there was no effect of phylogenetic relatedness for polyphagous species. Thus, we would expect the herbivore community on exotic plants in cities to be comprised largely of generalist species, or of specialists if the plant is closely related to native hosts. In addition to evolutionary and ecological aspects influencing herbivore–plant interactions, a large proportion of urban plants are chosen and planted by people. Often, ornamental plant introductions and the choices made by consumers are premised, at least in part, on low damage by herbivores and pathogens.

C. Native trees may be severely damaged by exotic herbivores in cities

In much the same way that exotic trees may be colonized and consumed by native insects, so too may native trees be colonized and consumed by exotic insects. In either case, herbivores able to feed on non-coevolved plants may experience defence-free space and inflict substantial damage to trees (Showalter *et al.*, 2018). Examples in natural and ornamental systems abound and include historical and recent exotic herbivore outbreaks. Gypsy moth has spread dramatically from its introduction in New England across much of the north-east and mid-west, defoliating oak-dominated forests as it spreads (Liebhold *et al.*, 1992, 1995). The exotic hemlock woolly adelgid, *Adelges tsugae*, has devastated native Appalachian hemlocks (*Tsuga canadensis* and *T. caroliniana*) since its introduction from Japan prior to 1951 (Havill *et al.*, 2011), and the introduction of the balsam woolly adelgid led to massive declines in the native Fraser fir, *Abies fraseri*, within five years of first detection in North Carolina (Witter and Ragenovich, 1986; Hollingsworth and Hain, 1991).

The consequences of exotic herbivore introduction, establishment and spread can be dire for native plant communities in natural and urban settings, such as when exotic herbivores threaten the persistence of entire plant species and genera. As the emerald ash borer spreads across the USA, several species of ash trees (*Fraxinus* spp.) are threatened with extinction (Herms and McCullough, 2014). The economic losses of susceptible native trees to exotic pests in natural and urban areas can be astounding (Aukema *et al.*, 2011; Lovett *et al.*, 2016). In the case of the emerald ash borer, the projected cost of management in US municipalities, where an estimated 38 million ash trees occur, is greater than $10 billion (Kovacs *et al.*, 2010).

A similar phenomenon of devastation of native plants by introduced natural enemies is evident in plant diseases, which may additionally be vectored by exotic insects or facilitated by the feeding activity of herbivores (Lovett *et al.*, 2016). Such is the case for beech bark scale, *Cryptococcus fagisuga*, which facilitates the infection of American beech, *Fagus grandifolia*, by ascomycete fungi in the genus *Nectria* (Houston, 1994). In one of the most notable examples of an exotic natural enemy devastating naïve native trees, the chestnut blight, *Cryphonectria parasitica*, has all but eliminated American chestnut, *Castanea dentata*, from the eastern forests in which it had previously been a dominant canopy tree (Anagnostakis, 1987; Griffin, 2000). With respect to urban forests, the loss of tens of millions of American elms, *Ulmus americana*, in the USA and English elms, *U. procera*, in Europe to Dutch elm disease is the quintessential example of how homogenization of street tree communities can result in catastrophic tree loss (Raupp *et al.*, 2012).

Where controlled experiments have been undertaken, exotic herbivores are often reported to achieve high levels of fitness and inflict great damage on non-coevolved host plants. These effects have been tied to specific defences present in coevolved plants that are lacking or insufficient in non-coevolved plants. In the case of the emerald ash borer, the coevolved *Fraxinus mandshurica* exhibits greater constitutive and induced defences. Higher concentrations of bark lignins and a faster oxidation of phenolic compounds lead to lower rates of oviposition, adult feeding and larval damage compared to North American ash species (Rebek *et al.*, 2008; Herms and McCullough, 2014; Martinson *et al.*, 2014; Villari *et al.*, 2016). In a study of the introduced European viburnum leaf beetle, *Pyrrhalta viburni*, successful herbivory of North American *Viburnum* spp. can be attributed to the absence of the egg-crushing response found in their European counterparts (Desurmont *et al.*, 2011).

Many of the above examples include relatively specialized herbivores or phytopathogens that shift onto novel plants closely related to the hosts from their original range (Rebek *et al.*, 2008; Bertheau *et al.*, 2010; Desurmont *et al.*, 2011; Havill *et al.*, 2011), but generalist herbivores may also colonize native urban trees and shrubs. Outside of their native Mediterranean region, pine processionary moths colonize some, though not all, species of both native and exotic conifers (Rossi *et al.*, 2016). In woody ornamental nurseries in the eastern USA, the Asian brown marmorated stink bug, *Halyomorpha halys*, was less abundant on Asian plants compared to non-Asian plants (Martinson *et al.*, 2016). Differences based on host plant origin were more extreme for plants that were abundant in those landscapes. The extent to which generalist herbivores preferentially feed on abundant plants (Mason *et al.*, 2011) and those with which they share little evolutionary history (Parker and Hay, 2005) is not yet fully known. However, this may have implications for the health of trees in cities, where herbivores and plants from various geographical provenances typically co-occur and whose abundance distributions are often heavily skewed.

D. Exotic herbivores may damage exotic plants in cities

Because exotic plants are often introduced without their herbivores (Keane and Crawley, 2002) and because native herbivores may be unable to utilize such novel plants (Bertheau *et al.*, 2010; Forister *et al.*, 2015), exotic plants well-suited to urban environments may experience very little herbivory in cities (Backe, 2019; Frank *et al.*, 2019). For example, *Ginkgo biloba* is widely planted in urban landscapes worldwide and is relatively free of injury by insects (Wheeler, 1975). However, introductions of exotic herbivores along with their exotic host plant, or sometime afterward, may serve to undermine enemy release. Often, natural enemies of these exotic herbivores do not accompany them to the invaded realm, and top-down pressure from predators, parasitoids and pathogens is relaxed. Examples of co-introductions include the tree of heaven (*Ailanthus altissima*) and the Cynthia moth (*Samia cynthia*; Pyle, 1975), azalea and azalea lace bug (Shrewsbury and Raupp, 2000), euonymous and euonymous scale (Van Driesche *et al.*, 1998), and eucalyptus and several of its coevolved insect pests (Paine *et al.*, 2010). Similar co-introductions have occurred in Europe and Asia, such as the boxtree moth on boxtree (*Buxus* spp.) and the lime leaf miner on *Tilia* spp.

Crape myrtles, native to Asia, provide an interesting case study of the potential outcomes of introducing exotic trees and their exotic herbivores (Chappell *et al.*, 2012). Crape myrtles have been grown in the southern US for nearly 200 years as street and landscape trees. For most of this time, they have had a single co-introduced herbivore, the crape myrtle aphid (*Tinocallis kahawaluokalani*). Crape myrtle aphids are innocuous and well regulated by generalist natural enemies like lady beetles, lace wings and hoverfly larvae, except when plants are severely stressed. Thus, the crape myrtle has been a relatively pest-free, low-maintenance plant (Mizell and Schiffhauer, 1987; Chappell *et al.*, 2012; Parsons, 2019). However, a new pest, crape myrtle bark scale (*Acanthococcus lagerstroemiae*), was first detected in the USA in 2004 (Wang *et al.*, 2016). Since 2004, crape myrtle bark scale has spread rapidly throughout the south-eastern USA and, in contrast to crape myrtle aphid, has caused severe damage to urban crape myrtle trees and has led to the removal of many crape myrtles (Wang *et al.*, 2016). Additionally, crape myrtle bark scale has a wide host range and has spread to other plant species, so the total economic damage of this pest is yet to be seen (Wang *et al.*, 2016; Schultz and Szalanski, 2019).

Herbivores co-introduced with their host plant to a new range often inflict more damage to their host plant in the exotic range (Roques *et al.*, 2006). This suggests that a lack of competition, a reduction in natural enemies and/or altered abiotic conditions may facilitate increases to herbivory even among exotic herbivore–exotic plant interactions. Where the reduction in insect biomass, the simplification of arthropod food webs and changes to abiotic conditions in cities favour herbivores, but not natural enemies, exotic herbivores may exhibit outbreaks on exotic plants (Raupp *et al.*, 2012). More broadly, the relatively high overall diversity of plants in cities may itself provide important opportunities for exotic herbivores to colonize both native and exotic plants (Liebhold *et al.*, 2018). Where the urban forest is made of a relatively high diversity of both native and exotic plants, exotic herbivores may be expected to establish.

Summary

In summary, urban forests generally consist of many native and exotic tree species, which are encountered by herbivores from a variety of geographic origins. In addition to changes in plant density, forest structure and composition, and impervious surface cover, the increase in the proportion of exotic trees in cities can simplify arthropod communities and change patterns of herbivory relative to natural forests. Herbivory on urban trees becomes more severe compared to on forest trees for many taxa, either consistently like scale insects, or episodically like many Lepidoptera. However, total herbivory or the specific types of herbivory can also decrease on urban trees due to lower herbivore diversity. Many factors, including tree density and diversity, habitat fragmentation, temperature, tree quality and natural enemy abundance, contribute to these patterns. The geographic origin of trees and herbivores are important drivers of patterns of herbivory on urban trees. Combinations of native and exotic trees and herbivores lead to four primary outcomes for urban trees. Exotic trees may be relatively immune to native herbivores and thus low maintenance or, if a coevolved exotic herbivore is introduced, trees could be heavily attacked and damaged, and thereby become unsustainable for use in urban areas. Native trees are typically subject to a suite of native herbivores, though in some cases these herbivores are not severe enough to warrant management. By contrast, urban conditions such as high temperature and drought can cause both native and exotic herbivores to become severe, damaging pests. The most extreme threat to native trees is generally from exotic herbivores against which they are undefended. Increasing research suggests that arthropod communities and levels of herbivory on urban trees are determined by complex associations between plant provenance and myriad biotic and abiotic conditions surrounding the plants on which these herbivores live and feed.

Acknowledgments

We wish to thank Pedro Barbosa for inviting us to write this chapter and for providing editorial comments. We are grateful to Kristi Marie Backe, Michael Just and Paula Shrewsbury, whose thoughtful comments greatly improved this work, and to Chris Sargent for technical assistance. Components of referenced and ongoing work by the authors described in this chapter have been supported by McDaniel College, the USDA McIntire-Stennis Grant (MD-ENTM-8757), USDA NIFA (Award numbers 2016-70006-25827 and 2013–02476 to SDF), and the USA Geological Survey (Cooperative Agreement no. G15AP00153 to SDF). The contents of this work are solely the responsibility of the authors and do not necessarily represent the views of the South-east Climate Science Center or the USGS.

References

Abdollahi, K.K. and Ning, Z.H. (1996) Research update on environmental influences of trees on mitigating air pollution effects and improving urban air quality. *Sustaining People, Sustaining Forests, Proceedings of the 1995 Society of American Foresters Convention*, Society of American Foresters, Bethesda, Maryland, pp. 177–185.

Adams, J.M., Fang, W., Callaway, R.M., Cipollini, D., Newell, E. *et al.* (2009) A cross-continental test of the Enemy Release Hypothesis: leaf herbivory on *Acer platanoides* (L.) is three times lower in North America than in its native Europe. *Biological Invasions* 11(4), 1005–1016. DOI: 10.1007/s10530-008-9312-4.

Anagnostakis, S.L. (1987) Chestnut blight: the classical problem of an introduced pathogen. *Mycologia* 79(1), 23–37. DOI: 10.1080/00275514.1987.12025367.

Aronson, M.F.J., La Sorte, F.A., Nilon, C.H., Katti, M., Goddard, M.A. *et al.* (2014) A global analysis of the impacts of urbanization on bird and plant diversity reveals key anthropogenic drivers. *Proceedings of the Royal Society B: Biological Sciences* 281(1780), 20133330. DOI: 10.1098/rspb.2013.3330.

Aronson, M.F.J., Handel, S.N., La Puma, I.P. and Clemants, S.E. (2015) Urbanization promotes non-native woody species and diverse plant assemblages in the New York metropolitan region. *Urban Ecosystems* 18(1), 31–45. DOI: 10.1007/s11252-014-0382-z.

Asaro, C. and Chamberlin, L.A. (2015) Outbreak history (1953–2014) of spring defoliators impacting oak-dominated forests in Virginia, with emphasis on gypsy moth (*Lymantria dispar* L.) and fall cankerworm (*Alsophila pometaria* Harris). *American Entomologist* 61(3), 174–185. DOI: 10.1093/ae/tmv043.

Aukema, J.E., McCullough, D.G., Von Holle, B., Liebhold, A.M., Britton, K. *et al.* (2010) Historical accumulation of nonindigenous forest pests in the continental United States. *BioScience* 60(11), 886–897. DOI: 10.1525/bio.2010.60.11.5.

Aukema, J.E., Leung, B., Kovacs, K., Chivers, C., Britton, K.O. *et al.* (2011) Economic impacts of non-native forest insects in the continental United States. *PLOS ONE* 6(9), e24587. DOI: 10.1371/journal.pone.0024587.

Avolio, M.L., Pataki, D.E., Trammell, T.L.E. and Endter-Wada, J. (2018) Biodiverse cities: the nursery industry, homeowners, and neighborhood differences drive urban tree composition. *Ecological Monographs* 88(2), 259–276. DOI: 10.1002/ecm.1290.

Backe, K.M. (2019) The effects of urban environments on trees and their herbivores. PhD dissertation. North Carolina State University. URI. Available at: http://www.lib.ncsu.edu/resolver/1840.20/36423

Bale, J.S., Harrington, R. and Clough, M.S. (1988) Low temperature mortality of the peach-potato aphid *Myzus persicae*. *Ecological Entomology* 13(2), 121–129. DOI: 10.1111/j.1365-2311.1988.tb00340.x.

Bale, J.S., Masters, G.J., Hodkinson, I.D., Awmack, C., Bezemer, T.M. *et al.* (2002) Herbivory in global climate change research: direct effects of rising temperature on insect herbivores. *Global Change Biology* 8(1), 1–16. DOI: 10.1046/j.1365-2486.2002.00451.x.

Barbosa, P., Hines, J., Kaplan, I., Martinson, H., Szczepaniec, A. *et al.* (2009) Associational resistance and associational susceptibility: having right or wrong neighbors. *Annual Review of Ecology, Evolution, and Systematics* 40(1), 1–20. DOI: 10.1146/annurev.ecolsys.110308.120242.

Beninde, J., Veith, M. and Hochkirch, A. (2015) Biodiversity in cities needs space: a meta-analysis of factors determining intra-urban biodiversity variation. *Ecology Letters* 18(6), 581–592. DOI: 10.1111/ele.12427.

Berland, A. (2012) Long-term urbanization effects on tree canopy cover along an urban–rural gradient. *Urban Ecosystems* 15(3), 721–738. DOI: 10.1007/s11252-012-0224-9.

Bertheau, C., Brockerhoff, E.G., Roux-Morabito, G., Lieutier, F. and Jactel, H. (2010) Novel insect-tree associations resulting from accidental and intentional biological 'invasions': a meta-analysis of effects on insect fitness. *Ecology Letters* 13(4), 506–515. DOI: 10.1111/j.1461-0248.2010.01445.x.

Blair, R.B. and Launer, A.E. (1997) Butterfly diversity and human land use: species assemblages along an urban gradient. *Biological Conservation* 80(1), 113–125. DOI: 10.1016/S0006-3207(96)00056-0.

Brändle, M., Kühn, I., Klotz, S., Belle, C. and Brandl, R. (2008) Species richness of herbivores on exotic host plants increases with time since introduction of the host. *Diversity and Distributions* 14(6), 905–912. DOI: 10.1111/j.1472-4642.2008.00511.x.

Burghardt, K.T. and Tallamy, D.W. (2013) Plant origin asymmetrically impacts feeding guilds and life stages driving community structure of herbivorous arthropods. *Diversity and Distributions* 19(12), 1553–1565. DOI: 10.1111/ddi.12122.

Burghardt, K.T. and Tallamy, D.W. (2015) Not all non-natives are equally unequal: reductions in herbivore β-diversity depend on phylogenetic similarity to native plant community. *Ecology Letters* 18(10), 1087–1098. DOI: 10.1111/ele.12492.

Burghardt, K.T., Tallamy, D.W., Philips, C. and Shropshire, K.J. (2010) Non-native plants reduce abundance, richness, and host specialization in lepidopteran communities. *Ecosphere* 1(5), art11–22. DOI: 10.1890/ES10-00032.1.

Cardinale, B.J., Matulich, K.L., Hooper, D.U., Byrnes, J.E., Duffy, E. *et al.* (2011) The functional role of producer diversity in ecosystems. *American Journal of Botany* 98(3), 572–592. DOI: 10.3732/ajb.1000364.

Chalker-Scott, L. (2015) Nonnative, noninvasive woody species can enhance urban landscape biodiversity. *Arboriculture & Urban Forestry* 41, 173–186.

Chappell, M.R., Braman, S.K., Williams-Woodward, J. and Knox, G. (2012) Optimizing plant health and pest management of *Lagerstroemia* spp. in commercial production and landscape situations in the southeastern United States: a review. *Journal of Environmental Horticulture* 30(3), 161–172.

Christie, F.J. and Hochuli, D.F. (2005) Elevated levels of herbivory in urban landscapes: are declines in tree health more than an edge effect? *Ecology and Society* 10(1), 10. DOI: 10.5751/ES-00704-100110.

Coffelt, M.A., Schultz, P.B. and Wolf, D.D. (1993) Impact of late-season orangestriped oakworm (Lepidoptera: Saturniidae) defoliation on oak growth and vigor. *Environmental Entomology* 22(6), 1318–1324. DOI: 10.1093/ee/22.6.1318.

Concepción, E.D., Moretti, M., Altermatt, F., Nobis, M.P. and Obrist, M.K. (2015) Impacts of urbanisation on biodiversity: the role of species mobility, degree of specialisation and spatial scale. *Oikos* 124(12), 1571–1582. DOI: 10.1111/oik.02166.

Connor, E.F., Courtney, A.C. and Yoder, J.M. (2000) Individuals-area relationships: the relationship between animal population density and area. *Ecology* 81(3), 734–748.

Cowett, F.D. and Bassuk, N.L. (2014) Statewide assessment of street trees in New York state, USA. *Urban Forestry & Urban Greening* 13(2), 213–220. DOI: 10.1016/j.ufug.2014.02.001.

Cregg, B.M. and Dix, M.E. (2001) Tree moisture stress and insect damage in urban areas in relation to heat island effects. *Journal of Arboriculture* 27, 8–17.

Dale, A.G. and Frank, S.D. (2014a) The effects of urban warming on herbivore abundance and street tree condition. *PLoS ONE* 9(7), e102996. DOI: 10.1371/journal.pone.0102996.

Dale, A.G. and Frank, S.D. (2014b) Urban warming trumps natural enemy regulation of herbivorous pests. *Ecological Applications* 24(7), 1596–1607. DOI: 10.1890/13-1961.1.

Dale, A.G. and Frank, S.D. (2017) Warming and drought combine to increase pest insect fitness on urban trees. *Plos One* 12(3), e0173844. DOI: 10.1371/journal.pone.0173844.

Dale, A.G. and Frank, S.D. (2018) Urban plants and climate drive unique arthropod interactions with unpredictable consequences. *Current Opinion in Insect Science* 29, 27–33. DOI: 10.1016/j.cois.2018.06.001.

Desurmont, G.A., Donoghue, M.J., Clement, W.L. and Agrawal, A.A. (2011) Evolutionary history predicts plant defense against an invasive pest. *Proceedings of the National Academy of Sciences* 108(17), 7070–7074. DOI: 10.1073/pnas.1102891108.

Dixon, A.F.G. (1971) The role of aphids in wood formation. I. The effect of the sycamore aphid, *Dreopanosiphum platanoides* (Schr.) (Aphididae), on the growth of sycamore, *Acer pseudoplatanus* (L.). *The Journal of Applied Ecology* 8(1), 165–179. DOI: 10.2307/2402135.

Donovan, G.H., Butry, D.T., Michael, Y.L., Prestemon, J.P., Liebhold, A.M. *et al.* (2013) The relationship between trees and human health: evidence from the spread of the emerald ash borer. *American Journal of Preventative Medicine* 44, 139–145.

Dreistadt, S.H., Dahlsten, D.L. and Frankie, G.W. (1990) Urban forests and insect ecology: complex interactions among trees, insects, and people. *BioScience* 40(3), 192–198.

Ehrlich, P.R. and Raven, P.H. (1964) Butterflies and plants: a study in coevolution. *Evolution* 18(4), 586–608. DOI: 10.1111/j.1558-5646.1964.tb01674.x.

Forister, M.L., Novotny, V., Panorska, A.K., Baje, L., Basset, Y. *et al.* (2015) The global distribution of diet breadth in insect herbivores. *Proceedings of the National Academy of Sciences* 112(2), 442–447. DOI: 10.1073/pnas.1423042112.

Frank, S.D. (2014) Bad neighbors: urban habitats increase cankerworm damage to non-host understory plants. *Urban Ecosystems* 17(4), 1135–1145. DOI: 10.1007/s11252-014-0368-x.

Frank, S.D., Backe, K.M., McDaniel, C., Green, M., Widney, S. *et al.* (2019) Exotic urban trees conserve similar natural enemy communities to native congeners but have fewer pests. *PeerJ* 7(4), e6531. DOI: 10.7717/peerj.6531.

Futuyma, D.J. and Agrawal, A.A. (2009) Macroevolution and the biological diversity of plants and herbivores. *Proceedings of the National Academy of Sciences* 106(43), 18054–18061. DOI: 10.1073/pnas.0904106106.

Gandhi, K.J.K. and Herms, D.A. (2010) Direct and indirect effects of alien insect herbivores on ecological processes and interactions in forests of eastern North America. *Biological Invasions* 12(2), 389–405. DOI: 10.1007/s10530-009-9627-9.

Glynn, C. and Herms, D.A. (2004) Local adaptation in pine needle scale (*Chionaspis pinifoliae*): natal and novel host quality as tests for specialization within and among red and Scots pine. *Environmental Entomology* 33(3), 748–755. DOI: 10.1603/0046-225X-33.3.748.

Godefroid, S. and Koedam, N. (2003) Distribution pattern of the flora in a peri-urban forest: an effect of the city–forest ecotone. *Landscape and Urban Planning* 65(4), 169–185. DOI: 10.1016/S0169-2046(03)00013-6.

Griffin, G.J. (2000) Blight control and restoration of the American chestnut. *Journal of Forestry* 98, 22–27.

Gwedla, N. and Shackleton, C.M. (2017) Population size and development history determine street tree distribution and composition within and between eastern Cape towns, South Africa. *Urban Forestry & Urban Greening* 25, 11–18. DOI: 10.1016/j.ufug.2017.04.014.

Hambäck, P.A. and Englund, G. (2005) Patch area, population density and the scaling of migration rates: the resource concentration hypothesis revisited. *Ecology Letters* 8(10), 1057–1065. DOI: 10.1111/j.1461-0248.2005.00811.x.

Hanks, L.M. and Denno, R.F. (1993) Natural enemies and plant water relations influence the distribution of an armored scale insect. *Ecology* 74(4), 1081–1091. DOI: 10.2307/1940478.

Hart, E.R., Miller Jr, F.D. and Bastian, R.A. (1986) Tree location and winter temperature influence on Mimosa webworm populations in a northern urban environment. *Journal of Arboriculture* 12, 237–240.

Hartley, M.K., Rogers, W.E. and Siemann, E. (2010) Comparisons of arthropod assemblages on an invasive and native trees: abundance, diversity and damage. *Arthropod-Plant Interactions* 4, 237–245. DOI: 10.1007/s11829-010-9105-4.

Hauer, R.J., Miller, R.W., Werner, L.P. and van den Bosch, C.C.K. (2017) The history of trees in the city. In: Ferrini, F., Konijnendijk van den Bosch, C.C. and Fini, A. (eds) *Routledge Handbook of Urban Forestry*, 1st edn. Routledge, London, pp. 17–32.

Havill, N., Montgomery, M. and Keena, M. (2011) Hemlock woolly adelgid and its hemlock hosts: a global perspective. In: Onken, B. and Readon, R. (eds) *Implementation and Status of Biological Control of the Hemlock Woolly Adelgid*. FHTET-2011-04. U.S. Department of Agriculture, Forest Service Forest Health Technology Enterprise Team, Morgantown, WV, pp. 3–14.

Herms, D.A. and McCullough, D.G. (2014) Emerald ash borer invasion of North America: history, biology, ecology, impacts, and management. *Annual Review of Entomology* 59(1), 13–30. DOI: 10.1146/annurev-ento-011613-162051.

Hogsden, K.L. and Hutchinson, T.C. (2004) Butterfly assemblages along a human disturbance gradient in Ontario, Canada. *Canadian Journal of Zoology* 82(5), 739–748. DOI: 10.1139/z04-048.

Hollingsworth, R.G. and Hain, F.P. (1991) Balsam woolly adelgid (Homoptera: Adelgidae) and spruce-fir decline in the southern Appalachians: assessing pest relevance in a damaged ecosystem. *The Florida Entomologist* 74(2), 179–187. DOI: 10.2307/3495294.

Holmes, T.P., Aukema, J.E., Von Holle, B., Liebhold, A. and Sills, E. (2009) Economic impacts of invasive species in forests. *Annals of the New York Academy of Sciences* 1162(1), 18–38. DOI: 10.1111/j.1749-6632.2009.04446.x.

Houston, D.R. (1994) Major new tree disease epidemics: beech bark disease. *Annual Review of Phytopathology* 32(1), 75–87. DOI: 10.1146/annurev.py.32.090194.000451.

Huberty, A.F. and Denno, R.F. (2004) Plant water stress and its consequences for herbivorous insects: a new synthesis. *Ecology* 85(5), 1383–1398. DOI: 10.1890/03-0352.

Jefferies, M.J. and Lawton, J.H. (1984) Enemy free space and the structure of ecological communities. *Biological Journal of the Linnean Society* 23(4), 269–286. DOI: 10.1111/j.1095-8312.1984.tb00145.x.

Just, M.G., Frank, S.D. and Dale, A.G. (2018) Impervious surface thresholds for urban tree site selection. *Urban Forestry & Urban Greening* 34, 141–146. DOI: 10.1016/j.ufug.2018.06.008.

Just, M.G., Dale, A.G., Long, L.C. and Frank, S.D. (2019) Urbanization drives unique latitudinal patterns of insect herbivory and tree condition. *Oikos* 128(7), 984–993. DOI: 10.1111/oik.05874.

Keane, R. and Crawley, M.J. (2002) Exotic plant invasions and the enemy release hypothesis. *Trends in Ecology & Evolution* 17(4), 164–170. DOI: 10.1016/S0169-5347(02)02499-0.

Kendle, A.D. and Rose, J.E. (2000) The aliens have landed! What are the justifications for 'native only' policies in landscape plantings? *Landscape and Urban Planning* 47(1-2), 19–31. DOI: 10.1016/S0169-2046(99)00070-5.

Kennedy, C.E.J. and Southwood, T.R.E. (1984) The number of species of insects associated with British trees: a re-analysis. *The Journal of Animal Ecology* 53(2), 455–478. DOI: 10.2307/4528.

Konijnendijk, C.C., Ricard, R.M., Kenney, A. and Randrup, T.B. (2006) Defining urban forestry – a comparative perspective of North America and Europe. *Urban Forestry & Urban Greening* 4(3-4), 93–103. DOI: 10.1016/j.ufug.2005.11.003.

Koricheva, J., Larsson, S. and Haukioja, E. (1998) Insect performance on experimentally stressed woody plants: a meta-analysis. *Annual Review of Entomology* 43(1), 195–216. DOI: 10.1146/annurev.ento.43.1.195.

Kovacs, K.F., Haight, R.G., McCullough, D.G., Mercader, R.J., Siegert, N.W. *et al.* (2010) Cost of potential emerald ash borer damage in U.S. communities, 2009–2019. *Ecological Economics* 69(3), 569–578. DOI: 10.1016/j.ecolecon.2009.09.004.

Kowarik, I. (2008) On the role of alien species in urban flora and vegetation. In: Marzluff, J.M., Shulenberger, E., Endlicher, W., Alberti, M., Bradley, G., *et al.* (eds) *Urban Ecology*. Springer, Boston, Massachusetts, pp. 321–338.

Kowarik, I. (2011) Novel urban ecosystems, biodiversity, and conservation. *Environmental Pollution* 159(8-9), 1974–1983. DOI: 10.1016/j.envpol.2011.02.022.

Kozlov, M.V., Lanta, V., Zverev, V., Rainio, K., Kunavin, M.A. *et al.* (2017) Decreased losses of woody plant foliage to insects in large urban areas are explained by bird predation. *Global Change Biology* 23(10), 4354–4364. DOI: 10.1111/gcb.13692.

Kühn, I. and Klotz, S. (2006) Urbanization and homogenization – comparing the floras of urban and rural areas in Germany. *Biological Conservation* 127(3), 292–300. DOI: 10.1016/j.biocon.2005.06.033.

Liebhold, A.M., Halverson, J.A. and Elmes, G.A. (1992) Gypsy moth invasion in North America: a quantitative analysis. *Journal of Biogeography* 19(5), 513–520. DOI: 10.2307/2845770.

Liebhold, A.M., MacDonald, W.L., Bergdahl, D. and Mastro, V.C. (1995) Invasion by exotic forest pests: a threat to forest ecosystems. *Forest Science Monographs* 30, 1–49.

Liebhold, A.M., Yamanaka, T., Roques, A., Augustin, S., Chown, S.L. *et al.* (2018) Plant diversity drives global patterns of insect invasions. *Scientific Reports* 8(1), 12095. DOI: 10.1038/s41598-018-30605-4.

Liu, H., Stiling, P. and Pemberton, R.W. (2007) Does enemy release matter for invasive plants? Evidence from a comparison of insect herbivore damage among invasive, non-invasive and native congeners. *Biological Invasions* 9(7), 773–781. DOI: 10.1007/s10530-006-9074-9.

Long, L.C. (2019) Effects of urbanization and warming on trees, pests, and urban birds. PhD dissertation. North Carolina State University. Available at: http://www.lib.ncsu.edu/resolver/1840.20/36436

Long, L.C., D'Amico, V. and Frank, S.D. (2019) Urban forest fragments buffer trees from warming and pests. *Science of The Total Environment* 658, 1523–1530. DOI: 10.1016/j.scitotenv.2018.12.293.

Lovett, G.M., Weiss, M., Liebhold, A.M., Holmes, T.P., Leung, B. *et al.* (2016) Nonnative forest insects and pathogens in the United States: impacts and policy options. *Ecological Applications* 26(5), 1437–1455. DOI: 10.1890/15-1176.

Luck, G.W. and Smallbone, L.T. (2010) Species diversity in urban landscapes: patterns, drivers and implications. In: Gaston, K. (ed.) *Urban Ecology*. University Press, Cambridge, Cambridge, pp. 88–119.

Martinson, H., Sargent, C.S. and Raupp, M.J. (2014) Tree water stress and insect geographic origin influence patterns of herbivory in green (*Fraxinus pennsylvanica*) and Manchurian (*F. mandshurica*) ash. *Arboriculture & Urban Forestry* 40, 332–344.

Martinson, H.M., Bergmann, E.J., Venugopal, P.D., Riley, C.B., Shrewsbury, P.M. *et al.* (2016) Invasive stink bug favors naïve plants: testing the role of plant geographic origin in diverse, managed environments. *Scientific Reports* 6(1), 32646. DOI: 10.1038/srep32646.

Mason, P.A., Wilkes, S.R., Lill, J.T. and Singer, M.S. (2011) Abundance trumps quality: bi-trophic performance and parasitism risk fail to explain host use in the fall webworm. *Oikos* 120(10), 1509–1518. DOI: 10.1111/j.1600-0706.2011.19053.x.

Matter, S.F., Brzyski, J.R., Harrison, C.J., Hyams, S., Loo, C. *et al.* (2012) Invading from the garden? A comparison of leaf herbivory for exotic and native plants in natural and ornamental settings. *Insect Science* 19(6), 677–682. DOI: 10.1111/j.1744-7917.2012.01524.x.

Mattson, W.J., Herms, D.A., Witter, J.A. and Allen, D.C. (1991) Woody plant grazing systems: North American outbreak folivores and their host plants. In: Baranchikov, Y.N., Mattson, W.J., Hain, F.P. and Payne, T.L. (eds) *Forest Insect Guilds: Patterns of Interaction with Host Trees*. USDA-FS General Technical Report NE-153, Radnor, Pennsylvania, pp. 53–84.

McBride, J. and Jacobs, D. (1976) Urban forest development: a case study, Menlo Park, California. *Urban Ecology* 2(1), 1–14. DOI: 10.1016/0304-4009(76)90002-4.

McCluney, K.E., George, T. and Frank, S.D. (2018) Water availability influences arthropod water demand, hydration and community composition on urban trees. *Journal of Urban Ecology* 4(1), juy003. DOI: 10.1093/jue/juy003.

McClure, M.S. (1980) Foliar nitrogen: a basis for host suitability for elongate hemlock scale, *Fiorinia externa* (Homoptera: Diaspididae). *Ecology* 61(1), 72–79. DOI: 10.2307/1937157.

McDonnell, M.J. and Pickett, S.T.A. (1990) Ecosystem structure and function along urban-rural gradients: an unexploited opportunity for ecology. *Ecology* 71(4), 1232–1237. DOI: 10.2307/1938259.

McIntyre, N.E. (2000) Ecology of urban arthropods: a review and a call to action. *Annals of the Entomological Society of America* 93(4), 825–835. DOI: 10.1603/0013-8746(2000)093[0825:EOUAAR]2.0.CO;2.

McKinney, M.L. (2002) Urbanization, biodiversity, and conservation. *BioScience* 52(10), 883–890. DOI: 10.1641/0006-3568(2002)052[0883:UBAC]2.0.CO;2.

McKinney, M.L. (2004) Measuring floristic homogenization by non-native plants in North America. *Global Ecology and Biogeography* 13(1), 47–53. DOI: 10.1111/j.1466-882X.2004.00059.x.

McKinney, M.L. (2006) Urbanization as a major cause of biotic homogenization. *Biological Conservation* 127(3), 247–260. DOI: 10.1016/j.biocon.2005.09.005.

McKinney, M.L. and Lockwood, J.L. (1999) Biotic homogenization: a few winners replacing many losers in the next mass extinction. *Trends in Ecology & Evolution* 14(11), 450–453. DOI: 10.1016/S0169-5347(99)01679-1.

McPherson, E.G. and Nowak, D.J. (1993) Value of urban greenspace for air quality improvement: Lincoln Park, Chicago. *Arborist News* 2, 30–32.

Meineke, E.K. and Frank, S.D. (2018) Water availability drives urban tree growth responses to herbivory and warming. *Journal of Applied Ecology* 55(4), 1701–1713. DOI: 10.1111/1365-2664.13130.

Meineke, E.K., Dunn, R.R., Sexton, J.O. and Frank, S.D. (2013) Urban warming drives insect pest abundance on street trees. *PLOS ONE* 8(3), e59687. DOI: 10.1371/journal.pone.0059687.

Meineke, E., Youngsteadt, E., Dunn, R.R. and Frank, S.D. (2016) Urban warming reduces aboveground carbon storage. *Proceedings of the Royal Society B: Biological Sciences* 283(1840), 20161574. DOI: 10.1098/rspb.2016.1574.

Meineke, E.K., Holmquist, A.J., Wimp, G.M. and Frank, S.D. (2017) Changes in spider community composition are associated with urban temperature, not herbivore abundance. *Journal of Urban Ecology* 3(1), juw010. DOI: 10.1093/jue/juw010.

Melliger, R.L., Braschler, B., Rusterholz, H.-P. and Baur, B. (2018) Diverse effects of degree of urbanisation and forest size on species richness and functional diversity of plants, and ground surface-active ants and spiders. *Plos One* 13(6), e0199245. DOI: 10.1371/journal.pone.0199245.

Menéndez, R. (2007) How are insects responding to global warming? *Tijdschrift voor Entomologie* 150(2), 355–365.

Merckx, T., Kaiser, A. and Van Dyck, H. (2018) Increased body size along urbanization gradients at both community and intraspecific level in macro-moths. *Global Change Biology* 24(8), 3837–3848. DOI: 10.1111/gcb.14151.

Miller, F.D. and Hart, E.R. (1987) Overwintering survivorship of pupae of the mimosa web worm, *Homadaula anisocentra* (Lepidoptera: Plutellidae), in an urban landscape. *Ecological Entomology* 12(1), 41–50. DOI: 10.1111/j.1365-2311.1987.tb00983.x.

Mitton, J.B. and Ferrenberg, S.M. (2012) Mountain pine beetle develops an unprecedented summer generation in response to climate warming. *The American Naturalist* 179(5), E163–E171. DOI: 10.1086/665007.

Mizell, R.F. and Schiffhauer, D.E. (1987) Seasonal abundance of the crapemyrtle aphid, *Sarucallis kahawaluokalani*, in relation to the pecan aphids, *Monellia caryella and Monelliopsis pecanis* and their common predators. *Entomophaga* 32(5), 511–520. DOI: 10.1007/BF02373520.

Mlynarek, J.J., Moffat, C.E., Edwards, S., Einfeldt, A.L., Heustis, A. *et al.* (2017) Enemy escape: a general phenomenon in a fragmented literature? *FACETS* 2, 1015–1044. DOI: 10.1139/facets-2017-0041.

Moll, G. and Ebenreck, S. (1989) Shading our cities: a resource guide for urban and community forests. Island Press, Washington, DC

Moreira, X., Abdala-Roberts, L., Berny Mier y Teran, J.C., Covelo, F., de la Mata, R. *et al.* (2019) Impacts of urbanization on insect herbivory and plant defences in oak trees. *Oikos* 128(1), 113–123. DOI: 10.1111/oik.05497.

Moriyama, M. and Numata, H. (2008) Diapause and prolonged development in the embryo and their ecological significance in two cicadas, *Cryptotympana facialis* and *Graptopsaltria nigrofuscata*. *Journal of Insect Physiology* 54(12), 1487–1494. DOI: 10.1016/j.jinsphys.2008.08.008.

Moro, M.F., Westerkamp, C. and de Araújo, F.S. (2014) How much importance is given to native plants in cities' treescape? A case study in Fortaleza, Brazil. *Urban Forestry & Urban Greening* 13(2), 365–374. DOI: 10.1016/j.ufug.2014.01.005.

Netherer, S. and Schopf, A. (2010) Potential effects of climate change on insect herbivores in European forests—General aspects and the pine processionary moth as specific example. *Forest Ecology and Management* 259(4), 831–838. DOI: 10.1016/j.foreco.2009.07.034.

Nguyen, H.Q., Andersen, D.K., Kim, Y. and Jang, Y. (2018) Urban heat island effect on cicada densities in metropolitan Seoul. *PeerJ* 6, e4238. DOI: 10.7717/peerj.4238.

Nielsen, D.G., Muilenburg, V.L. and Herms, D.A. (2011) Interspecific variation in resistance of Asian, European, and North American birches (*Betula* spp.) to bronze birch borer (Coleoptera: Buprestidae). *Environmental Entomology* 40(3), 648–653. DOI: 10.1603/EN10227.

Nock, C.A., Paquette, A., Follett, M., Nowak, D.J. and Messier, C. (2013) Effects of urbanization on tree species functional diversity in eastern North America. *Ecosystems* 16(8), 1487–1497. DOI: 10.1007/s10021-013-9697-5.

Nowak, D.J., Rowntree, R.A., McPherson, E.G., Sisinni, S.M., Kerkmann, E.R. *et al.* (1996) Measuring and analyzing urban tree cover. *Landscape and Urban Planning* 36(1), 49–57. DOI: 10.1016/S0169-2046(96)00324-6.

Nowak, D.J., Civerolo, K.L., Trivikrama Rao, S.,Luley, C.J., Gopal, S. *et al* . (2000) A modeling study of the impact of urban trees on ozone. *Atmospheric Environment* 34(10), 1601–1613. DOI: 10.1016/S1352-2310(99)00394-5.

Nowak, D.J., Pasek, J.E., Sequeira, R.A., Crane, D.E. and Mastro, V.C. (2001) Potential effect of *Anoplophora glabripennis* (Coleoptera: Cerambycidae) on urban trees in the United States. *Journal of Economic Entomology* 94(1), 116–122. DOI: 10.1603/0022-0493-94.1.116.

Nowak, D.J., Crane, D.E. and Dwyer, J.F. (2002) Compensatory value of urban trees in the United States. *Journal of Arboriculture* 28, 194–199.

Oke, T.R. (1973) City size and the urban heat island. *Atmospheric Environment* 7(8), 769–779. DOI: 10.1016/0004-6981(73)90140-6.

Ortega-Álvarez, R., Rodríguez-Correa, H.A. and MacGregor-Fors, I. (2011) Trees and the city: diversity and composition along a neotropical gradient of urbanization. *International Journal of Ecology* 2011(3), 1–8. DOI: 10.1155/2011/704084.

Paine, T.D., Millar, J.G. and Daane, K.M. (2010) Accumulation of pest insects on Eucalyptus in California: random process or smoking gun. *Journal of Economic Entomology* 103(6), 1943–1949. DOI: 10.1603/EC10214.

Parker, J.D. and Hay, M.E. (2005) Biotic resistance to plant invasions? native herbivores prefer non-native plants. *Ecology Letters* 8(9), 959–967. DOI: 10.1111/j.1461-0248.2005.00799.x.

Parsons, S.E. (2019) Effects of urbanization on Pest-Natural enemy systems in ornamental landscapes. PhD dissertation. North Carolina State University. Available at: https://repository.lib.ncsu.edu/bitstream/handle/1840.20/36583/etd.pdf?sequence=1

Pearse, I.S. and Altermatt, F. (2013) Predicting novel trophic interactions in a non-native world. *Ecology Letters* 16(8), 1088–1094. DOI: 10.1111/ele.12143.

Pimentel, D., Zuniga, R. and Morrison, D. (2005) Update on the environmental and economic costs associated with alien-invasive species in the United States. *Ecological Economics* 52(3), 273–288. DOI: 10.1016/j.ecolecon.2004.10.002.

Proches, E., Wilson, J.R.U., Richardson, D.M. and Chown, S.L. (2008) Herbivores, but not other insects, are scarce on alien plants. *Austral Ecology* 33(5), 691–700. DOI: 10.1111/j.1442-9993.2008.01836.x.

Pyle, R.M. (1975) Silk moth of railroad yards. *Natural History* 84(5), 45–51.

Pyle, R.M. (1983) Urbanization and endangered insect populations. In: Frankie, G.W. and Koehler, C.S. (eds) *Urban Entomology: Interdisciplinary Perspectives*. Praeger, New York, pp. 367–394.

Pyšek, P. (1998) Alien and native species in central European urban floras: a quantitative comparison. *Journal of Biogeography* 25(1), 155–163. DOI: 10.1046/j.1365-2699.1998.251177.x.

Raupp, M.J. and Noland, R.M. (1984) Implementing plant management programs in residential and institutional settings. *Journal of Arboriculture* 10, 161–169.

Raupp, M.J., Cumming, A.B. and Raupp, E.C. (2006) Street tree diversity in eastern North America and its potential for tree loss to exotic borers. *Arboriculture and Urban Forestry* 32(6), 297–304.

Raupp, M.J., Shrewsbury, P.M. and Herms, D.A. (2010) Ecology of herbivorous arthropods in urban landscapes. *Annual Review of Entomology* 55, 19–38. DOI: 10.1146/annurev-ento-112408-085351.

Raupp, M.J., Shrewsbury, P.M. and Herms, D.A. (2012) Disasters by design: outbreaks along urban gradients. In: Barbosa, P., Letourneau, D.K. and Agrawal, A.A. (eds) *Insect Outbreaks Revisited*. Wiley-Blackwell, West Sussex, United Kingdom, pp. 313–340.

Raupp, M.J., Shrewsbury, P.M., Holmes, J.J. and Davidson, J.A. (2001) Plant species diversity and abundance affects the number of arthropod pests in residential landscapes. *Journal of Arboriculture* 27(4), 222–229.

Rebek, E.J., Herms, D.A. and Smitley, D.R. (2008) Interspecific variation in resistance to emerald ash borer (Coleoptera: Buprestidae) among North American and Asian ash (*Fraxinus* spp.). *Environmental Entomology* 37(1), 242–246. DOI: 10.1603/0046-225X(2008)37[242:IVIRTE]2.0.CO;2.

Rickman, J. and Connor, E. (2003) The effect of urbanization on the quality of remnant habitats for leaf-mining Lepidoptera on *Quercus agrifolia*. *Ecography* 26(6), 777–787. DOI: 10.1111/j.0906-7590.2003.03345.x.

Riley, C.B., Herms, D.A. and Gardiner, M.M. (2018) Exotic trees contribute to urban forest diversity and ecosystem services in inner-city Cleveland, Ohio. *Urban Forestry & Urban Greening* 29, 367–376. DOI: 10.1016/j.ufug.2017.01.004.

Root, R.B. (1973) Organization of a plant-arthropod association in simple and diverse habitats: the fauna of collards (*Brassicaoleraceae*). *Ecological Monographs* 43(1), 95–124. DOI: 10.2307/1942161.

Roques, A., Auger-Rozenberg, M.-A. and Boivin, S. (2006) A lack of native congeners may limit colonization of introduced conifers by indigenous insects in Europe. *Canadian Journal of Forest Research* 36(2), 299–313. DOI: 10.1139/x05-277.

Rossi, J.-P., Imbault, V., Lamant, T. and Rousselet, J. (2016) A citywide survey of the pine processionary moth *Thaumetopoea pityocampa* spatial distribution in Orléans (France). *Urban Forestry & Urban Greening* 20, 71–80. DOI: 10.1016/j.ufug.2016.07.015.

Schultz, P.B. and Szalanski, A.L. (2019) *Hypericum kalmianum* (St. Johnswort) confirmed as a new host of the crapemyrtle bark scale in Virginia, USA. *Journal of Agricultural and Urban Entomology* 35(1), 12–15. DOI: 10.3954/1523-5475-35.1.12.

Seebens, H., Blackburn, T.M., Dyer, E.E., Genovesi, P., Hulme, P.E. *et al.* (2017) No saturation in the accumulation of alien species worldwide. *Nature Communications* 8, 14435. DOI: 10.1038/ncomms14435.

Showalter, D.N., Raffa, K.F., Sniezko, R.A., Herms, D.A., Liebhold, A.M. *et al.* (2018) Strategic development of tree resistance against forest pathogen and insect invasions in defense-free space. *Frontiers in Ecology and Evolution* 6, Article 124. DOI: 10.3389/fevo.2018.00124.

Shrewsbury, P.M. and Raupp, M.J. (2000) Evaluation of components of vegetational texture for predicting azalea lace bug, *Stephanitis pyrioides* (Heteroptera: Tingidae), abundance in managed landscapes. *Environmental Entomology* 29(5), 919–926. DOI: 10.1603/0046-225X-29.5.919.

Sjöman, H. and Östberg, J. (2019) Vulnerability of ten major Nordic cities to potential tree losses caused by longhorned beetles. *Urban Ecosystems* 22(2), 385–395. DOI: 10.1007/s11252-019-0824-8.

Sjöman, H., Östberg, J. and Bühler, O. (2012) Diversity and distribution of the urban tree population in ten major Nordic cities. *Urban Forestry & Urban Greening* 11(1), 31–39. DOI: 10.1016/j.ufug.2011.09.004.

Sjöman, H., Morgenroth, J., Sjöman, J.D., Sæbø, A. and Kowarik, I. (2016) Diversification of the urban forest—Can we afford to exclude exotic tree species? *Urban Forestry & Urban Greening* 18, 237–241. DOI: 10.1016/j.ufug.2016.06.011.

Slade, E.M., Merckx, T., Riutta, T., Bebber, D.P., Redhead, D. *et al.* (2013) Life-history traits and landscape characteristics predict macro-moth responses to forest fragmentation. *Ecology* 94(7), 1519–1530. DOI: 10.1890/12-1366.1.

Tahvanainen, J. and Niemelä, P. (1987) Biogeographical and evolutionary aspects of insect herbivory. *Annales Zoologici Fennici* 24, 239–247.

Tait, C.J., Daniels, C.B. and Hill, R.S. (2005) Changes in species assemblages within the Adelaide metropolitan area, Australia, 1836–2002. *Ecological Applications* 15(1), 346–359. DOI: 10.1890/04-0920.

Tallamy, D.W. (2004) Do alien plants reduce insect biomass? *Conservation Biology* 18(6), 1689–1692. DOI: 10.1111/j.1523-1739.2004.00512.x.

Thomas, S.M., Simmons, G.S. and Daugherty, M.P. (2017) Spatiotemporal distribution of an invasive insect in an urban landscape: introduction, establishment and impact. *Landscape Ecology* 32(10), 2041–2057. DOI: 10.1007/s10980-017-0565-0.

Trân, J.K., Ylioja, T., Billings, R.F., Régnière, J. and Ayres, M.P. (2007) Impact of minimum winter temperatures on the population dynamics of *Dendroctonus frontalis*. *Ecological Applications* 17(3), 882–899. DOI: 10.1890/06-0512.

Uka, U.N. and Belford, E.J.D. (2016) Inventory of street tree population and diversity in the Kumasi Metropolis, Ghana. *Journal of Forest and Environmental Science* 32(4), 367–376. DOI: 10.7747/JFES.2016.32.4.367.

Van Driesche, R., Kingsley, P., Rose, M. and Bryan, M. (1998) Effect of *euonymus* scale (Homoptera: Diaspididae) on *Euonymus* spp. survival in southern New England, with estimates of economic costs of pest damage. *Environmental Entomology* 27(2), 217–220. DOI: 10.1093/ee/27.2.217.

Van Nuland, M.E. and Whitlow, W.L. (2014) Temporal effects on biodiversity and composition of arthropod communities along an urban–rural gradient. *Urban Ecosystems* 17(4), 1047–1060. DOI: 10.1007/s11252-014-0358-z.

Villari, C., Herms, D.A., Whitehill, J.G.A., Cipollini, D. and Bonello, P. (2016) Progress and gaps in understanding mechanisms of ash tree resistance to emerald ash borer, a model for wood-boring insects that kill angiosperms. *The New Phytologist* 209(1), 63-79. DOI: 10.1111/nph.13604.

Wang, Z., Chen, Y., Gu, M., Vafaie, E., Merchant, M. *et al.* (2016) Crapemyrtle bark scale: a new threat for Crapemyrtles, a popular landscape plant in the US. *Insects* 7(4), 78. DOI: 10.3390/insects7040078.

Wheeler, A.G. (1975) Insect associates of *Ginkgo biloba*. *Entomological News* 86, 37–44.

Williams, N.S.G., Schwartz, M.W., Vesk, P.A., McCarthy, M.A., Hahs, A.K. *et al.* (2009) A conceptual framework for predicting the effects of urban environments on floras. *Journal of Ecology* 97(1), 4–9. DOI: 10.1111/j.1365-2745.2008.01460.x.

Witter, J.A. and Ragenovich, I.R. (1986) Regeneration of Fraser fir at Mt. Mitchell, North Carolina, after depredations by the balsam woolly adelgid. *Forest Science* 32, 585–594.

Yamamura, K. and Kiritani, K. (1998) A simple method to estimate the potential increase in the number of generations under global warming in temperate zones. *Applied Entomology and Zoology* 33(2), 289–298. DOI: 10.1303/aez.33.289.

Youngsteadt, E., Dale, A.G., Terando, A.J., Dunn, R.R. and Frank, S.D. (2015) Do cities simulate climate change? A comparison of herbivore response to urban and global warming. *Global Change Biology* 21(1), 97–105. DOI: 10.1111/gcb.12692.

Zhao, J., Ouyang, Z., Zheng, H., Zhou, W., Wang, X. *et al.* (2010) Plant species composition in green spaces within the built-up areas of Beijing, China. *Plant Ecology* 209(2), 189–204. DOI: 10.1007/s11258-009-9675-3.

3 Wildlife in the City: Human Drivers and Human Consequences

Susannah B. Lerman[1]*, Desiree L. Narango[2], Riley Andrade[3], Paige S. Warren[4], Aaron M. Grade[5] and Katherine Straley[5]

[1]USDA Forest Service Northern Research Station, Amherst, Massachusetts, USA; [2]Advanced Science Research Center, City University of New York, New York, New York, USA; [3]School of Geographical Sciences and Urban Planning, Arizona State University, Tempe, Arizona, USA; [4]Department of Environmental Conservation, University of Massachusetts, Amherst, Massachusetts, USA; [5]Graduate Program in Organismic and Evolutionary Biology, University of Massachusetts, Amherst, Massachusetts, USA

Abstract

The urban development process results in the removal, alteration and fragmentation of natural vegetation and environmental features, which have negatively impacted many wildlife species. With the loss of large tracts of intact wildlands (e.g. forests, deserts and grasslands), and the demise of specific habitat features (e.g. early successional habitat or native plants), many specialist species are filtered out from urban ecosystems. As a result, some argue that urbanization has a homogenizing effect on wildlife communities. However, these general patterns belie a high degree of variability in urban biodiversity patterns. In this chapter, we focus on vertebrate and invertebrate species that contribute to urban fauna (hereafter 'wildlife'). We review how wildlife species have responded to altered conditions of the urban environment, with a focus on the environmental features and species traits that filter wildlife communities from the regional scale to the city scale. We also focus on how built structures, species interactions and socio-cultural factors further influence the local species pool. Within this context, we assess the ecosystem services and disservices provided by urban wildlife, how management decisions are shaped by attitudes and exposure to wildlife, and how these decisions then feed back to the local species pool. By understanding why some animals are better able to persist in human modified landscapes than others, land managers, city planners, private homeowners and other stakeholders can make better-informed decisions when managing properties in ways that also conserve and promote wildlife.

Introduction

More than half the world's population lives in cities and suburbs (Grimm *et al.*, 2008), and an estimated 80% of the world's population will live in urbanized areas by 2050 (United Nations Department of Economic and Social Affairs,

*Corresponding author: susannah.b.lerman@usda.gov

Population Division, 2019). A consequence of this migration to and expansion of cities includes the degradation of wildlife habitat, with implications for local and regional biodiversity (Grimm *et al.*, 2008; Seto *et al.*, 2012). The urban development process results in the removal, alteration and fragmentation of natural vegetation and environmental features, which have negatively impacted many wildlife species (McKinney, 2002; Grimm *et al.*, 2008). Natural habitats transform into human-dominated environments, which include residential developments, public parks, commercial/industrial districts, transportation corridors and impervious surfaces (e.g. buildings and roads), and become a prominent feature that supports a growing urban population. Urban wildlife studies commonly document a depauperate fauna in cities compared to their non-urban habitats, resulting in lower species diversity (Emlen, 1974; Beissinger and Osborne, 1982; Mills *et al.*, 1989; Blair, 1996; Marzluff, 2001; Donnelly and Marzluff, 2004; Chace and Walsh, 2006; Croci *et al.*, 2008). With the loss of large tracts of intact wildlands (e.g. forests, deserts and grasslands), and the demise of specific habitat features such as early successional habitat or native plants, many specialist species are filtered out from urban ecosystems (Blair, 1996; Aronson *et al.*, 2016; Evans *et al.*, 2018). As a result, some argue that urbanization has a homogenizing effect on wildlife communities (McKinney, 2006; Sol *et al.*, 2014; Knop, 2016; Morelli *et al.*, 2016; Murthy *et al.*, 2016; Salomão *et al.*, 2019), wherein invasive species (e.g. pigeons, Norway rats and cockroaches) dominate. However, these general patterns belie a high degree of variability in urban biodiversity patterns.

Negative effects of urbanization on biodiversity are not ubiquitous across taxa. Indeed, some urban areas can support high levels of biodiversity, and for some taxa, like birds, native species largely dominate urban communities (Aronson *et al.*, 2014). Some invertebrates and mammals also respond positively to urban areas. For example, in two separate studies, one from Phoenix, Arizona, and the other a meta-analysis that included various cities in Europe, Japan and Canada, investigators found that ground arthropod diversity did not differ between natural areas and various urban land areas, owing to the turnover of communities from habitat-specialist to non-specialist species (McIntyre *et al.*, 2001; Magura *et al.*, 2010). In addition, some taxonomic groups, like pollinators, may also thrive in urban landscapes due to increased resources. For example, a high diversity of bees can occur in urban areas (see Chapter 6, Protecting Pollinators in the Urban Environment), likely due to abundant flowers (Baldock *et al.*, 2015; Hall *et al.*, 2017; but see McIntyre and Hostetler, 2001), which are available in untreated lawns (Lerman and Milam, 2016) and in cultivated gardens (Levé *et al.*, 2019). Herbivorous and predatory insects also may be abundant in certain urban habitats when resources are available, such as appropriate host plants and vegetative complexity (Raupp *et al.*, 2010). However, despite high diversity, the species that colonize urban areas tend to consist of distinctly different suites of species from those that remain in natural habitats (e.g. Bang and Faeth, 2011; LaSorte *et al.*, 2018; Collado *et al.*, 2019). For mammals, generalist species such as raccoons (*Procyon lotor*) may thrive at high densities and occupancies in some types of urban green space, but not in others (Crooks and Soulé, 1999; Parker and Nilon, 2008; Gallo *et al.*, 2017). Thus, both the degree of urbanization and the quality of habitat within urban systems can have complementary forces driving the particular animal species that is able to occupy and thrive in urban systems.

In this chapter, we focus on vertebrate and invertebrate species that contribute to urban fauna (hereafter referred to as 'wildlife'). We review how wildlife species have responded to altered conditions of the urban environment, with a focus on the environmental features and species traits that filter wildlife communities from the regional scale to the city scale (Fig. 3.1). We also focus on how built structures, species interactions and socio-cultural factors further influence the local species pool. Within this context, we assess the ecosystem services and disservices provided by urban wildlife, how management decisions are shaped by attitudes and exposure to wildlife, and how these decisions then feed back to the local species pool (Fig. 3.1). Our taxonomic focus includes birds, bees, lepidopterans, ground arthropods and mammals, as the majority of research focuses on these groups (Pickett *et al.*, 2011; Beninde *et al.*, 2015). We provide a global perspective when data and examples are available, though much of the available

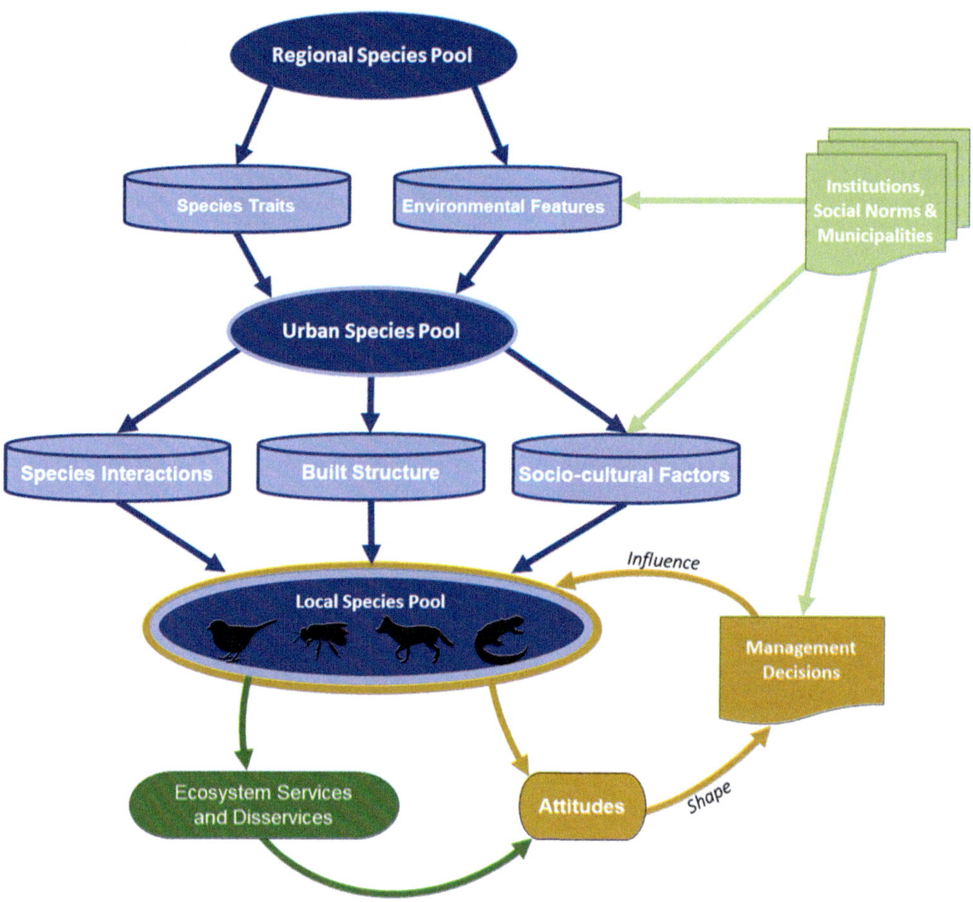

Fig. 3.1. A series of interacting ecological and social factors determine the local species pool in urban systems. The potential wildlife species that occupy a particular city is first determined from the regional species pool, which is primarily dictated by climate and latitude. Species traits (e.g. diet and body size) and specific environmental features (e.g. urban form and land use) filter the regional pool to the urban species pool. Species interactions (e.g. predator–prey dynamics), the built structure (e.g. roads and buildings), and socio-cultural factors (e.g. income and attitudes) further filter which species persist into the local species pool. Overarching each of these hierarchical layers are governance structures (i.e. institutions, social norms and municipalities) that interact at various levels and degrees with the factors shaping the urban and local species pool. It is at the local species pool level that people have regular contact with wildlife species and experience ecosystem services and disservices. These interactions further affect attitudes towards local wildlife, which subsequently shape management decisions to either encourage or discourage certain species, which ultimately feed back into influencing the local species pool. (Animal images by Creazilla.com are licensed under CC BY 4.0 and were recoloured.)

literature comes from North America, western Europe, Australia and New Zealand (Magle *et al.*, 2012; McDonnell and Hahs, 2013; Goddard *et al.*, 2017). We review the literature on habitat use of urban green spaces within the urban and suburban matrix itself rather than the response of animals to urban development compared to intact natural areas. A unique contribution of this chapter is that we integrate a description of the human drivers influencing urban wildlife communities with discussion of potential feedbacks to humans, particularly those at local scales, which can further alter management actions. By understanding why some animals are better able to persist in human modified landscapes than others, land managers, city planners, private homeowners and other stakeholders can make better-informed decisions when managing properties in ways that also conserve and promote wildlife.

Urbanization Shapes Species Pools

After the initial disturbance from urban development, novel habitats emerge in urban ecosystems (Kowarik, 2011; Swan *et al.*, 2011; Grimm *et al.*, 2017). The resulting altered landscape constitutes a complex mosaic of invasive (i.e. species that have a negative effect on the ecosystem), non-native (i.e. species that did not originate in a given habitat, but may have a neutral or even positive effect on the ecosystem) and native plants, large swaths of lawns, isolated trees and remnant habitat patches of varying sizes, all of which are interspersed with built structures, impervious surfaces and artificial sensory inputs (e.g. noise, light) that define urbanization for most observers. The green spaces in this mosaic frequently support the persistence of a number of wildlife species, populations and communities (Pickett *et al.*, 2011; Lepczyk *et al.*, 2017a). Since urban green space can encompass everything from residential yards, gardens, recreational parks, cemeteries, vacant lots and remnant patches of native vegetation, the quality and composition of habitat are tremendously variable among green space types (Lepczyk *et al.*, 2017a). In addition, within any type of green space there is considerable variation in management decision criteria (e.g. native plantings, having outdoor cats, and differences in mowing frequency; Loss *et al.*, 2013; Narango *et al.*, 2017; Lerman *et al.*, 2018) and management goals (i.e. regarding aesthetics, ease of maintenance and safety; Larson *et al.*, 2016) that can have far-reaching implications on habitat quality for wildlife. This results in heterogeneous patches of urban green space that support a range of both generalist and specialist native species, as well as non-native and invasive species (Shochat *et al.*, 2010; Aronson *et al.*, 2014; Lepczyk *et al.*, 2017b).

The communities of animals that inhabit urban landscapes are shaped by a series of interacting processes, often called filters, operating at scales from the regional to the local (Aronson *et al.*, 2016). Regional filters encompass eco-regional factors (e.g. climate and latitude) that dictate which species constitute the regional species pool and therefore have the potential to occur in a given city or metropolitan area (Aronson *et al.*, 2016). Urban form, urban land use, and species traits further filter species from the regional to the urban species pool. The built environment (e.g. roads and buildings), species interactions, and socio-cultural features (e.g. income and attitudes) act as further filters from the urban to the local species pool. Typically, it is these local species, encountered on a day-to-day basis by humans that deliver vital ecosystem services (Fig. 3.1). All of the filtering processes reflect human decisions and actions. However, the actors shaping broad-scale environmental filters are often institutions (e.g. municipal governments, regional planning boards, developers) rather than individual urban residents (Fig. 3.1; Warren *et al.*, 2010; Pickett *et al.*, 2011). Previous reviews have covered much of the literature on broader-scale regional and environmental filters (i.e. urban form and urban land use), as well as species traits filters (e.g. Aronson *et al.*, 2016; Evans *et al.*, 2018). Thus, we start by briefly summarizing the factors that lead to filtering at these broader scales, before delving into greater depth on species interactions and socio-cultural filters, which typically operate at smaller spatial scales. These finer-scale filters are therefore shaped strongly by individual human actions at the parcel level, e.g. in pocket parks, commercial parks and residential yards and gardens (Warren *et al.*, 2010; Swan *et al.*, 2011; Aronson *et al.*, 2017).

Environmental Features that Filter from the Regional to Urban Species Pool

At a coarse scale, wildlife communities frequently differentiate themselves among land-use types, largely due to differences in physical structure that shape habitat suitability (Ortega-Álvarez and MacGregor-Fors, 2009; Gallo *et al.*, 2017; Andrade *et al.*, 2018). For example, in Chicago, Illinois, coyotes, *Canis latrans*, have a higher detection rate in open golf courses, whereby white-tailed deer, *Odocoileus virginianus*, are more likely to persist in wooded cemeteries and natural areas embedded within the urban matrix (Gallo *et al.*, 2017). In the UK, allotment gardens (i.e. plots of land available for individual, non-commercial gardening or growing food plants) and private gardens support the highest pollinator (e.g. bees, hoverflies and non-syrphid *Diptera*) abundances when compared with cemeteries, nature reserves, parks, verges (green space associated with urban roads, typically 2–5 m wide, with trees sometimes being present) and other green spaces. Urban land uses that include habitat features and contribute to ecological function can also bolster connectivity in the urban landscape. For example, land uses with large proportions of vegetation, such as residential gardens, wooded streets and utility rights-of-way can link disparate habitat fragments into larger networks (Rudd *et al.*, 2002) and support movement, meta-community dynamics and increased genetic connectivity (Unfried *et al.*, 2013; Padilla and Rodewald, 2015; Gallo *et al.*, 2017).

Although different cities may have similar land-use categories, the particular urban form and structure of these land uses may differ. For instance, city age, amount of remnant vegetation and amount of impervious surfaces can vary dramatically both within and among cities, and wildlife communities respond to these differences (Ramalho and Hobbs, 2012). In Chicago, newer neighbourhoods supported higher bird richness (Loss *et al.*, 2009), contrasting with other studies that found higher bird richness in older neighbourhoods with more mature vegetation and extensive canopy cover (e.g. Palomino and Carrascal, 2005). However, context matters, and land-use legacies interact with urban development history. For instance, remnant vegetation in residential landscapes can differ based on former land use (e.g. cleared agricultural systems versus forests or woodlands; Hahs *et al.*, 2009), which can influence wildlife communities (DeGraaf and Wentworth, 1986; Loss *et al.*, 2009). Development patterns in cities often follow an urban-to-rural gradient, with more urban areas associated with higher percentages of impervious surface (McDonnell and Pickett, 1990; Blair, 1996; Pickett *et al.*, 2011). Wildlife communities respond to this gradient with a general pattern of lower species diversity; in particular, a lower representation of certain native species, but higher population densities of animals in the urban core versus more rural and wildland settings (Fortel *et al.*, 2014; Marzluff, 2017; Sol *et al.*, 2017). In some cases, maximum species richness occurs at intermediate levels of human development, i.e. suburban landscapes (Blair, 1996; Marzluff and Rodewald, 2008; Parsons *et al.*, 2018). High species richness in the suburbs may be due to the simultaneous local colonization of synanthropic species (i.e. species extremely tolerant of people) and the persistence of some habitat specialists as well as the high heterogeneity of habitat types in these systems (Marzluff and Rodewald, 2008).

Species Traits that Filter from the Regional to Urban Species Pool

Species assemblages that colonize urban habitats are also filtered according to their traits, which ultimately determine their success as well as their relationships with people (Aronson *et al.*, 2016). For some taxa, functional diversity tends to homogenize with increasing urbanization (Devictor *et al.*, 2008; Deguines *et al.*, 2016). For example, observations of plant–pollinator interactions across France showed that urbanization promoted a functional homogenization toward more generalist foraging insects (Deguines *et al.*, 2016), but increasing garden space generally positively benefitted pollinator diversity (Normandin *et al.*, 2017; Levé *et al.*, 2019). However, the identification of a common set of species traits associated with urbanization has proved elusive with conflicting findings in different studies (Croci *et al.*, 2008; Evans *et al.*, 2011; Reif *et al.*, 2011; Francis and Chadwick, 2012; Leveau, 2013).

One general trend, however, is a broad distinction between responses of generalists versus specialists. Urban areas have higher diversity and abundance of generalist and synanthropic species, whereas habitat specialists tend to decline (McIntyre *et al.*, 2001; Martinson and Raupp, 2013, p. 2013; Deguines *et al.*, 2016; Evans *et al.*, 2018; Callaghan *et al.*, 2019). Non-native and invasive species often become common in urban areas because of their synanthropic nature (e.g. house sparrows, *Passer domesticus*, nesting in buildings) or their introduction via global trade routes (Aronson *et al.*, 2016). For example, an invasive insect pest, the emerald ash borer, *Agrilus planipennis*, was introduced via the horticultural industry and has decimated urban *Fraxinus* trees (Cregg and Dix, 2001). Over-abundant invasive species, such as the Argentine ant (*Linepithema humile*) thrive in urban areas and reduce species richness and turnover by exclusion of native ant species (Holway and Suarez, 2006). Although urban areas support diverse communities of bees (Lerman and Milam, 2016; Hall *et al.*, 2017; Baldock *et al.*, 2019), insects of other functional groups like predators and parasites decline with increasing urbanization (Guenat *et al.*, 2019). Bat responses to urbanization are highly species-specific; some species strongly associate with human habitation, whereas others are absent from urban areas (Jung and Kalko, 2011), possibly due to species partitioning into different diet and foraging guilds. Thus the responses of different taxonomic groups to urbanization are complex and often taxon-specific due in part to differences in ecological function and linked to mechanisms that either enhance or limit population growth.

The availability of specific food items and nesting substrates vary within and among urban systems, and contribute to the filtering of species by their traits. For example, urban bird communities lose species with specialist dietary niches, like insectivores (Chace and Walsh, 2006; Rodewald and Bakermans, 2006; Evans *et al.*, 2018). The increase in generalist species may be due to complementary responses of animals that have the flexibility to utilize limited resources, as well as animals that respond positively to human resource supplements such as bird-feeding or refuse (Galbraith *et al.*, 2014). Specialist species such as herbivorous insects rely on native host plants, which are often uncommon in urban and suburban areas (Burghardt and Tallamy, 2013). Specialists are also more sensitive to introduced competitors or predators (Shochat *et al.*, 2010). Likewise, because of abundant nesting substrates, cavity-nesting species of birds (Chace and Walsh, 2006; Evans *et al.*, 2011) and bees (Matteson *et al.*, 2008) are abundant in urban areas, whereas ground-nesting birds (Evans *et al.*, 2018) and ground-nesting bees (Matteson *et al.*, 2008) lack nesting opportunities due to the loss of snags and deadwood, and bare, permeable ground. However, when nesting resources are present, some urban areas (e.g. suburban gardens) can support high densities of ground-nesting bees (Fetridge *et al.*, 2008; Lerman and Milam, 2016). Further, wildlife species with life-history characteristics that promote reproduction, e.g. large clutch sizes (Callaghan *et al.*, 2019) and/or dispersal, e.g. large wing morphology (Piano *et al.*, 2017), also thrive in urban areas. Although larger body size is often negatively associated with urbanization, the interaction between urban heat islands and habitat fragmentation might negate clear patterns. For example, in an investigation of ten taxonomic groups of aquatic and terrestrial invertebrates that persist in urban environments, ground spiders and ground beetles were found to decrease in size, following Atkinson's temperature-size rule (Atkinson, 1994), whereas species with a positive size-dispersal link increased in size, including orthopterans, macromoths and rotifers (Merckx *et al.*, 2018). Additional examples of large-bodied organisms persisting in urban areas can be found in mammals (Murray and St. Clair, 2017), beetles (Martinson and Raupp, 2013, p. 2013) and birds (Callaghan *et al.*, 2019). In addition to declines in taxonomic diversity, some urban wildlife communities have reduced evolutionary and phylogenetic diversity (Morelli *et al.*, 2016; Ibáñez-Álamo *et al.*, 2017; Sol *et al.*, 2017). These losses in key components of ecological processes could result in reduced ecosystem function.

Built Structural Features Shape Local Species Composition

A defining feature of urbanization is the built structure (e.g. buildings, roads and other transportation infrastructure). Collectively, these features yield a landscape with large swaths of impervious surfaces and fragmented natural areas (McDonnell and Pickett, 1990). Roads cross the landscape, rendering the remaining habitat unsuitable for species that require large tracts of unfragmented habitat (Forman and Alexander, 1998; Crooks, 2002) and create a barrier to movement or dispersal for many species (Orłowski, 2008; Beebee, 2013; Grilo *et al.*,

2014; Murray and St. Clair, 2015; Keilsohn *et al.*, 2018). Roads also alter local scale processes through mortality or behavioural changes for a variety of taxa, ranging from highly mobile animals, such as birds, insects and coyotes, to those with more restrictive movement, such as amphibians (Shepard *et al.*, 2008; Beebee, 2013; Loss *et al.*, 2014a; Murray and St. Clair, 2015; Keilsohn *et al.*, 2018). In terms of scale, an estimated 89–340 million birds (Loss *et al.*, 2014a) and billions of insects (Baxter-Gilbert *et al.*, 2015) collide with vehicles each year, a significant source of mortality in wildlife populations. Roads also have sublethal effects by altering animal behaviour and movement. In the northern city of Edmonton, Canada, urban coyotes changed their behaviour by avoiding roads or shifting the timing of their peak activity to times when road traffic had higher survival rates (Murray and St. Clair, 2015).

Vehicles travelling on roads also contribute to a significant source of noise. The peak amplitudes of highway or roadside noise have been shown to overlap in frequency with bird, frog and insect vocal signals, causing some individuals to shift their communication frequency in order to be heard by rivals or potential mates (Warren *et al.*, 2006; Barber *et al.*, 2010; Narango and Rodewald, 2016). Road noise also reduces detectability of important signals, such as anti-predator alarm calls (Grade and Sieving, 2016) or nestling begging (Leonard *et al.*, 2015). In addition to effects on communication, road noise also has detrimental impacts to individuals, populations and communities. For example, to experimentally isolate the effects of road noise from the physical effects of noise, McClure *et al.* (2013) created a 'phantom highway' by placing speakers across an intact forest and played the sounds of a busy highway. Migrating birds that experienced the noise treatment had lower body conditions and stop-over efficiencies compared to birds who experienced the 'road-less' control (Ware *et al.*, 2015). In addition, the abundance of birds significantly decreased when the speakers were on, and some species demonstrated high sensitivity to noise by avoiding the area completely during playback (McClure *et al.*, 2013). Noise effects from roads and highways extend one kilometre, or beyond the footprint of the road itself, amplifying their impacts even in areas with intact natural habitat (Grade and Sieving, 2016). Anthropogenic noise originates from a variety of sources, not just roadways. Construction, airplane engines and drilling, all predominantly located in urban environments, also interrupt settlement, animal communication, predator detection and breeding success (Shannon *et al.*, 2016; Swaddle *et al.*, 2016).

Buildings (e.g. residential and business structures) serve as another hazardous feature for wildlife to navigate, with direct and indirect consequences. It is estimated that 365–988 million birds collide with buildings each year in the USA, and roughly 99% of these collisions occur on low-rise buildings and residences (Loss *et al.*, 2014b). The reflections of vegetation on window surfaces disorientate birds, and they fly into the window as if they were simply moving from tree to tree. Although window strikes present a large source of mortality, particularly for migratory species (Loss *et al.*, 2015), buildings also provide nesting cavities or roosts for birds and bats (e.g. Tomasevic and Marzluff, 2017).

Because roads and buildings often replace vegetation in urban and suburban environments, the built environment influences the local and regional climate (Hondula *et al.*, 2017). As a result, factors such as the urban heat island (UHI) have the potential to affect wildlife and other urban organisms. The effect of the built environment on climate is partly due to the reduced evapotranspiration and shading by trees, in combination with the use of building materials that retain heat (Oke, 1995). Subsequently, urban heat can drive phenological patterns, behaviour, population cycles and species distributions of wildlife (Aronson *et al.*, 2014; Dale and Frank, 2018). For instance, when the effects of the UHI were isolated from habitat features and species interactions, the warmer conditions were responsible for driving pest insect outbreaks in Raleigh, North Carolina (Meineke *et al.*, 2013). Additionally, changes in local microclimates and the effect of the UHI can potentially influence activity time of thermo-sensitive wildlife such as herpetofauna (Ackley *et al.*, 2015a).

One of the most iconic images depicting the extent of urbanization is the satellite imagery of the earth at night. Artificial light has become ubiquitous with the urbanization of the planet (see Chapter 10). Light pollution can influence the phenology of plant (Škvareninová *et al.*, 2017) and animal life history events, and change foraging (Stone *et al.*, 2015), reproduction (Silva *et al.*, 2017), migration (Van Doren *et al.*, 2017; McLaren *et al.*, 2018) and sleep behaviours (Raap *et al.*, 2015; Ouyang *et al.*, 2017) across taxa (Gaston *et al.*, 2017). Artificial light pollution can also influence navigation, as seen in the case of the 'Tribute in Light' in New York City, which researchers estimate disorientate over a million birds during migration (Van Doren *et al.*, 2017). Artificial light at night can also influence predator–prey dynamics, in that some insectivorous bat species take advantage of artificial light and congregate around lit areas for foraging (Minnaar *et al.*, 2015; Stone *et al.*, 2015). In contrast, frugivorous bats avoid lit areas while foraging (Lewanzik and Voigt, 2014). This in turn has negative consequences for ecosystem services, particularly in tropical cities, since frugivorous bat avoidance interrupts the dispersal of fruiting plant seeds.

Solutions have been proposed to address many of these challenges imposed on wildlife by roads, buildings and fragmentation (Yanes *et al.*, 1995; Glista *et al.*, 2009). For example, wildlife crossings, bridges and culverts can minimize roadway collisions and wildlife mortality (Dodd *et al.*, 2004; Mata *et al.*, 2008). Policies that incentivize compact development (Farr *et al.*, 2018), reduction of sensory pollution (e.g. Audubon's Lights Out campaign; https://www.audubon.org/conservation/project/lights-out) or broad adoption of bird-safe glass in combination with altering vegetation and bird-feeder placement (Kummer *et al.*, 2016) provide other potential mechanisms to increase the capacity of urban areas to support wildlife species. But it is likely that the physical structure of cities will remain a barrier to colonization by many species.

Species Interactions Shape Local Species Composition

As described, thus far urban systems differ in their structural components and abiotic factors from non-urbanized systems. This affects biotic factors such as the quality and quantity of resources, as well as biotic interactions such as predator–prey dynamics, competition and host–parasite relationships. Predator–prey dynamics and urban food webs influence the success or failure of urban wildlife. Successful urban mesopredators (i.e. species that occupy the middle trophic level and serve as both predator and prey), such as raccoons (*Procyon lotor*), foxes (*Vulpes* spp.) and striped skunks (*Mephitis mephitis*), increase in abundance and occupancy relative to apex predators (i.e. predators at the top of a food chain, with no natural predators) (Crooks and Soulé, 1999; Prange *et al.*, 2003; Ellington and Gehrt, 2019; Santini *et al.*, 2019). Their success is partly due to the extirpation of apex predators through direct culling by humans or through the loss of habitat (Gompper, 2002; Gehrt, 2004; Estes *et al.*, 2011; Ellington and Gehrt, 2019). This extirpation 'releases' mesopredators from predation pressure (Crooks and Soulé, 1999) and has consequences that ripple throughout entire urban food webs (Newsome *et al.*, 2015; Santini *et al.*, 2019).

Sometimes, the interaction between 'bottom-up' (resources) and 'top-down' (predation) trophic dynamics in urban habitats results in counterintuitive patterns (Shochat *et al.*, 2006). One such example is the 'predation paradox', in which urban habitats have higher densities of potential generalist predators, yet lower per capita predation rates (Shochat, 2004; Stracey, 2011; Fischer *et al.*, 2012). Specifically, the input of alternative food resources for avian and mammalian mesopredators may dampen nest predation pressure (Shochat *et al.*, 2004; Rodewald *et al.*, 2011; Newsome *et al.*, 2015; Malpass *et al.*, 2017). Still, elevated populations of potential predators present a serious threat to songbirds and small mammals, and may alter wildlife behaviour, foraging and long-term reproductive success through non-lethal or fear effects (Creel and Christianson, 2008; Martin, 2011; Lerman *et al.*, 2012; LaManna and Martin, 2016). In natural systems, mesopredator populations that are not regulated by higher trophic levels tend to suppress populations of their prey. In some cases, this leads to local extinction of species, and therefore reduced biodiversity (Ritchie and Johnson, 2009; Estes *et al.*, 2011). Despite this tendency in natural systems, urban ecologists have found a different dynamic in urban systems. Few studies have attempted to understand the interactions between trophic levels in urban environments, particularly those related to direct human-directed management.

These interactions are complex and dynamic. For example, private gardens landscaped with non-native ornamental plants can have negative impacts on insect populations which, in turn, have cascading consequences for birds that rely on insect prey (Narango *et al.*, 2018). Alternatively, artificial light, which promotes accelerated plant growth, can both promote herbivorous insect abundance via bottom-up effects and reduce abundance via top-down predation by visual predators (Bennie *et al.*, 2016).

Although the overall effect of increased mesopredator populations in urban habitats is still unclear, the introduction of domesticated mammals, especially domestic cats (*Felis catus*), has had a significant impact on urban wildlife. It has been estimated that cats kill 1.4–3.7 billion birds and 6.9–20.7 billion mammals annually in the USA alone (Loss *et al.*, 2013). Cats also induce fear and alter behaviour, causing sublethal effects. For example, the fear instilled by the mere presence of a cat reduces bird fecundity by one offspring per year, resulting in up to a 95% reduction in bird abundances (Bonnington *et al.*, 2013). Outdoor cats, whether pets or feral, often receive supplemental food, potentially decoupling cats from predator–prey relationships, and allowing them to kill prey *ad libitum* without corresponding feedback to outdoor cat populations (Sims *et al.*, 2008). Yet, unlike native mesopredators, cats are beloved household companions, which makes the control of feral and outdoor cats in urban areas a particularly complex conservation challenge.

Humans as Producers and Providers of Resources

Environmental and biotic factors partially explain why some species successfully colonize urban habitats. However, one key factor that sets the urban environment apart from the surrounding wildlands is the introduction of anthropogenic (i.e. human-provided) resource subsidies (Shochat *et al.*, 2006; Tryjanowski *et al.*, 2015; Ciach and Fröhlich, 2017). People and their land management activities in urban green spaces have manipulated urban resources, primarily for their benefit. Thus, humans have had a heavy hand in determining habitat characteristics, and consequently have controlled patterns of urban wildlife biodiversity (Shochat *et al.*, 2006; Aronson *et al.*, 2017; Reed and Bonter, 2018). Though many land management decisions have been made without regard to wildlife, some of these management decisions have been made intentionally to enhance perceived habitat value for wildlife. Examples include the addition or retention of specific vegetation features (e.g. trees, shrubs, grasses and flowering plants), supplemental feeding, avoiding the use of pesticides, adding structural features such as ponds, and erecting nest-boxes (Belaire *et al.*, 2014). Despite considerable guidance to householders, we lack consensus regarding the effectiveness of wildlife-friendly gardening strategies (but see Gaston *et al.*, 2005) since wildlife populations and communities have responded to these human subsidies in both positive and negative ways, depending on the specific action (Prange *et al.*, 2003; Newsome *et al.*, 2015; Santini *et al.*, 2019). These actions, which have a social and cultural context (Fig. 3.1), help shape local species pools. The composition of local species pools has the most direct influence on human–wildlife interactions. Further, it is at the local level where the delivery of ecosystem services is the most direct.

Local Landscaping Decisions

Perhaps the largest structural change in urban systems is the shift in vegetation composition. Not only is vegetation less abundant, but it is also often comprised of invasive, non-native and ornamental plant species (Burghardt *et al.*, 2009; Goddard *et al.*, 2017). Non-native plants may differ in the timing of leaf-out (McEwan *et al.*, 2009; Shustack *et al.*, 2009; Wolkovich and Cleland, 2011), in their fruit or flower production (Corlett, 2005), and in the abundance and diversity of arthropods that they host, compared to native plants (Burghardt and Tallamy, 2013; Litt *et al.*, 2014; Narango *et al.*, 2017). Consequently, the dominance of non-native plant species can have an impact on wildlife communities and resource availability for higher trophic levels that use these plants for foraging and reproduction (Narango *et al.*, 2018). For example, Phoenix, Arizona, and Chicago, Illinois, neighbourhoods have strong and positive relationships between abundances of native plants and native birds (Lerman and Warren, 2011; Belaire *et al.*, 2014). Further, native plants support more foraging activity by both insectivorous (Narango *et al.*, 2017) and nectivorous bird species (French *et al.*, 2005).

Native plants may also increase pollinator diversity by supporting species with host-specific specialized diets (Pardee and Philpott, 2014), though several naturalized non-native species appear to be highly attractive to many pollinators (Lerman and Milam, 2016; Baldock *et al.*, 2019). However, even when a preferred foraging plant is present, novel non-native species can disrupt natural processes because of differences in resources or phenology. For example, migratory monarch butterflies (*Danaus plexippus*) exposed to blooming non-native milkweed (*Asclepias*) species, found in southern-US gardens, may forgo southward movements in favour of breeding, with the unintentional consequence of increasing pathogen prevalence in the population (Satterfield *et al.*, 2015). Many ornamental plants introduced via the horticultural trade produce fruits used extensively by generalist omnivorous and frugivorous birds, despite low nutritional values. Further, dispersal of non-native seeds from birds can facilitate invasion throughout both urban and natural systems that can have cascading impacts on plant communities, succession, and trophic interactions with consumers (Gosper *et al.*, 2005). Fruit and vegetable gardens also directly provide important food sources to many animals, from pollinators to large mammals (Contesse *et al.*, 2004; Daniels and Kirkpatrick, 2006; Baldock *et al.*, 2019), although the wildlife resources that result from these gardens may or may not be intentionally planted to attract wildlife (Goddard *et al.*, 2013; Mumaw and Bekessy, 2017).

Landscaping requires laborious upkeep, which, often, is driven by aesthetics (Cook *et al.*, 2012; Larson *et al.*, 2016). Humans provide inputs that are important and often naturally limited such as water, carbon, nitrogen and phosphorus (Kaye *et al.*, 2005; Trammell *et al.*, 2016; Palta *et al.*, 2017; Souto *et al.*, 2019), while some inputs such as the application of herbicides and pesticides can significantly degrade habitat (Aronson *et al.*, 2017; Sievers *et al.*, 2019). Humans also eliminate certain resources such as standing deadwood, spontaneous lawn flowers and leaf litter (Blewett and Marzluff, 2005; Templer *et al.*, 2015; Aronson *et al.*, 2017; Lerman *et al.*, 2018), which can further degrade the potential habitat. For example, frequent lawn mowing results in an aesthetically pleasing garden (Robbins, 2007), but removes floral resources, which in turn has a negative influence on bee abundance in suburban gardens due to their role as potential nectar sources (Lerman *et al.*, 2018). Regardless of the particular landscaping decision, the resulting landscapes have profound influence on wildlife communities (Goddard *et al.*, 2010).

Feeding Wildlife

Another distinguishing characteristic of urban habitats is the artificially inflated abundance of food resources (Prange *et al.*, 2003; Shochat *et al.*, 2006; Newsome *et al.*, 2015). Humans provide these resources directly or indirectly, and intentionally as well as unintentionally. Actively feeding wildlife, in particular birds, has become the most popular and widespread intentional resource-provisioning activity, which influences wildlife populations at multiple spatial scales (Reynolds *et al.*, 2017; Cox and Gaston, 2018; Baverstock *et al.*, 2019). Typically, resource (i.e. food) provision includes sugar water, commercial seed, meat, suet or bread (Rollinson *et al.*, 2003; Lepczyk *et al.*, 2012). In the UK, nearly 50% of householders feed birds, equating to approximately one bird feeder for every nine UK birds (Davies *et al.*, 2009). A similar percentage of households feed birds in the USA (approximately 50 million people; U.S. Department of the Interior, U.S. Fish and Wildlife Service, and U.S. Department of Commerce, US. Census Bureau, 2014) as well as in Australia (Rollinson *et al.*, 2003; Jones *et al.*, 2008) and New Zealand (Galbraith *et al.*, 2014). Households in New Zealand put out a staggering estimated 5.1 million loaves of bread per year, which has caused a shift in the bird community, towards species tolerant of this food source (Galbraith *et al.*, 2014).

The benefits of feeding birds and other wildlife are more often articulated in terms of their benefits to humans, through increased well-being, which typically reflects a state characterized by health, happiness and prosperity (Fuller *et al.*, 2007; Goddard *et al.*, 2013), the opportunities to interact with nature (Lepczyk *et al.*, 2012) and/or ecosystem services (Orros and Fellowes, 2012). Many people feed birds and other wildlife to enjoy wildlife viewing as well as for a personal sense of stewardship for their local fauna (Jones *et al.*, 2008; Cox and Gaston, 2018). Birds also benefit from the increased resources, as evidenced by higher species richness and abundance in areas with resource provisioning (Fuller *et al.*, 2008). Supplementary feeding in the city of Reading, UK, provided food for up to 320 red kites (*Milvus milvus*), contributing to their high abundance (Orros and Fellowes, 2015). Subsidizing food resources can increase winter survival or augment resources when food is scarce (Jones *et al.*, 2008), and can increase breeding success (Schoech and Bowman, 2001). Providing supplemental food for wildlife can improve survival and reproductive output (Brittingham and Temple, 1988; Robb *et al.*, 2008; Cox and Gaston, 2018; Santini *et al.*, 2019) as well as facilitate northward expansion of species distributions (Robb *et al.*, 2008; Greig *et al.*, 2017). However, negative effects might counteract some of the benefits. Frequent feeding or providing an unreliable food source might lead to dependence on human resources, and can also have detrimental effects on survival and productivity (Robb *et al.*, 2008). For example, food items with lower nutritional value might decrease individual fitness (Rollinson *et al.*, 2003), or high-fat and high-protein food items might cause earlier egg-laying. In the latter case, this might lead to a phenological mismatch whereby, for example, insect food required for nestlings is unavailable (Renner and Zohner, 2018). Feeding stations might elevate competition among the species attracted to feeders, leading to novel interspecific interactions. For example, when squirrels were present at feeders in Sheffield, UK, bird visitation and the amount of food consumed by birds decreased by more than 90% (Bonnington *et al.*, 2014). Human subsidies also attract higher densities of non-native and invasive wildlife species (Daniels and Kirkpatrick, 2006; Galbraith *et al.*, 2014) that could out-compete native species (Galbraith *et al.*, 2015). Increased aggregations of birds around feeding stations can facilitate the spread of disease and parasites, increase exposure to predation (Adelman *et al.*, 2015; Becker *et al.*, 2015; Galbraith *et al.*, 2017; Civitello *et al.*, 2018) and disease transmission (Robb *et al.*, 2008). If the negative implications of intentionally or unintentionally feeding wildlife outweigh positive aspects, then feeding wildlife might create an ecological trap (Schlaepfer *et al.*, 2002; Robb *et al.*, 2008; Plummer *et al.*, 2013). This is particularly true when supplementary food leads to false cues that the green spaces consist of high-quality habitat, but lack sufficient resources for breeding, or have increased stressors that lead to decreased fitness (Plummer *et al.*, 2013). It is still unclear whether wildlife feeding has a net benefit or a cost to urban wildlife populations, but what is clear is that it has a cultural and personal benefit to people around the world (Cox and Gaston, 2018).

Human Variation

Social and demographic differences in urban planning, landscaping and human values also correlate with wildlife communities, resulting in unequal exposure to urban biodiversity by people. For example, in some cities, affluent neighbourhoods have higher diversity of birds, mammals, reptiles and arthropods (Leong *et al.*, 2018). This may be due to closer proximities to green space, a positive correlation between income and tree canopy (Grove *et al.*, 2014) and/or the capacity to provide more wildlife-friendly features such as plants that provide resources in garden landscaping, or that provide bird seed (Lepczyk *et al.*, 2012; Goddard *et al.*, 2013; Belaire *et al.*, 2015; Baldock *et al.*, 2019). The biodiversity disparity is especially apparent in arid cities, where fewer species and fewer habitat specialists are found in low-income and minority neighbourhoods for both birds (Lerman and Warren, 2011; Warren *et al.*, 2019) and lizards (Ackley *et al.*, 2015b). In temperate systems, high-income neighbourhoods may also harbour high numbers of mammals (Magle *et al.*, 2016) and frog species (Smallbone *et al.*, 2011), in part related to these neighbourhoods being less densely developed with lower human population densities, and located on the outskirts of the urban core. In four UK cities, higher pollinator abundance was associated with higher-income neighbourhoods, a pattern primarily driven by higher quality of floral resources (Baldock *et al.*, 2019). However, in Chicago, Illinois, lower-income neighbourhoods had higher native and migratory bird diversity due to the higher presence of non-developed patches (i.e. >40% in a non-urban land-use class) of habitat (Loss *et al.*, 2009). Thus, generalizable relationships between income and biodiversity may not exist across cities but, rather, are driven by the specific context of an individual city's development pattern.

Humans are the primary drivers of environmental change in cities, and at times wildlife populations and communities benefit from these management decisions. However, management decisions are often facilitated or constrained by attitudinal, geographical and structural factors (Cook *et al.*, 2012), and must be considered in order to fully understand urban wildlife patterns and processes. In the remainder of the chapter, we describe how personal experiences shape attitudes, how attitudes shape management actions (with a focus on garden management), their feedbacks on wildlife species and people, and the ecosystem services, as well as disservices, associated with directly and indirectly interacting with urban wildlife.

Attitudinal Factors Associated with Urban Wildlife

Attitudinal factors are defined as the positive or negative evaluations of an object and are underpinned by values, beliefs and affect/emotion (Thurstone, 1928; Stern, 2000). The attitudinal literature on urban wildlife encompasses a variety of taxa, including birds, herpetofauna, arthropods and mammals. Attitudes are commonly used as both dependent variables (driven by personal characteristics and experiences with nature/wildlife) as well as explanatory variables (used to help explain behaviour, such as garden management decisions and support of conservation initiatives), and thus can be considered as a construct with dynamic feedback loops. For example, attitudes about the local bird community in Chicago neighbourhoods are correlated with a key determinant of ecological resource provisioning in residential back gardens (Belaire *et al.*, 2016). The resulting ecological resources, such as gardens with fruit- and berry-producing trees, were, in turn, important for determining actual bird species richness in the same residential neighbourhoods (Belaire *et al.*, 2014). This iterative feedback loop, in which people both affect and are affected by local wildlife, underscores that people and wildlife are connected through attitudinal factors (Fig. 3.1).

Negative attitudes are driven by perceptions of wildlife as threatening. For example, people often kill snakes indiscriminately due to their perception that they are 'dangerous', despite the fact that most laypeople are unable to distinguish venomous from non-venomous snakes (Alves *et al.*, 2012, 2014; Pandey *et al.*, 2016). Gender and personal vulnerability are important factors in the case of attitudes toward snakes. For instance, in Brazil, women were more likely to hold negative attitudes toward snakes, largely due to gender differences in risk perceptions (Alves *et al.*, 2014). A person's belief that West Nile virus would harm a family member can be directly related to the management practices used to decrease mosquito abundance in their residential garden (Tuiten *et al.*, 2009). Risk perceptions also influence the likelihood that a person will formally complain about alligators in Florida, despite positive attitudes toward alligators more generally (Hayman *et al.*, 2014). Concerns about coyote encounters being dangerous negatively influences support for the presence of coyotes in the urban areas of Washington, DC (Draheim *et al.*, 2013). Conversely, reptiles that are not perceived as dangerous, such as turtles, are more likeable (Tisdell, 2010).

The likeability (or popularity) of species or taxa is one of the most common measures of positive attitudes for urban wildlife (Schuetz and Johnston, 2019), with much of the literature focusing on garden and urban birds. These studies have found support for the connection between likeability and positive attitudes towards a species, such as songbirds being more likeable compared with other bird species (Cox and Gaston, 2015; Belaire *et al.*, 2015). In general, iconic species (e.g. sports mascots), readily encountered species and species with distinguishing physical characteristics represent traits related to positive attitudes (Ainsworth *et al.*, 2018; Schuetz and Johnston, 2019). Biodiversity and species richness are also associated with positive attitudes

(Lerman and Warren, 2011; Cox and Gaston, 2015; Warren *et al.*, 2019). For example, people would rather see ten individual birds of different species than ten similar looking individuals of the same species (Cox and Gaston, 2015), and higher satisfaction with the desert bird community in Phoenix, Arizona, neighbourhoods aligned with actual desert bird species richness (Lerman and Warren, 2011). However, the mechanisms driving the attitudinal–wildlife biodiversity relationship remain unclear, as people often do a poor job of assessing actual biodiversity levels (Dallimer *et al.*, 2012).

Tolerance Threshold

Contrasts in positive versus negative attitudes towards wildlife, and how these attitudes translate to human–wildlife interactions, emerge as one of the most interesting themes from the wildlife attitudinal literature. For example, residents in Bakersfield, California, had positive attitudes toward endangered San Joaquin kit foxes (*Vulpes macrotis mutica*) and supported their protection in urbanized areas (Bjurlin and Cypher, 2005). In contrast, the residents in a suburb of Munich, Germany, did not want red foxes in their residential gardens and neighbourhoods, largely due to concerns about the health risks, despite the fact that they agreed foxes were 'useful', 'beautiful' and 'had the right to live' (König, 2008). People living in the Munich suburbs held positive attitudes toward foxes in general, until it came to foxes being present in their own neighbourhoods. In contrast, the majority of people in Bakersfield, California, did not associate kit foxes in their neighbourhood with any potential risk. As a result, people in Bakersfield supported the presence of foxes whereas Munich residents did not, though people in both cities held generally positive attitudes towards foxes. Similarly, for arthropods, people are tolerant and even have positive attitudes about arthropods present in outdoor areas of their property, but are simultaneously intolerant of arthropods indoors (Hahn and Ascerno, 1991).

Tolerance of urban wildlife can easily turn to intolerance when personal welfare and property come into play (Clergeau *et al.*, 2001). Although attitudes toward urban birds are generally positive, attitudes toward specific species, such as house sparrows (*Passer domesticus*) and rock pigeons (*Columba livia*) are more negative, since many people perceive these birds as a 'nuisance' (Cox and Gaston, 2015). Species with abundant local populations are more likely to cross the 'threshold of tolerance', and are often considered as 'pests' (Clergeau *et al.*, 2001). In general, pest control at the household level is largely targeted at abundant species such as rats and arthropods, which can quickly cause noticeable property damage when overly abundant (Morzillo and Mertig, 2011; Schoelitsz *et al.*, 2019).

Individual and Household Structure

Given that attitudes have a strong tie to people's identity, it is important to consider how individual and household structure, or socio-cultural factors, influence attitudes (Heberlein, 2012). These variables associated with attitudes include income, age, education, cultural identity and gender. For example, older and wealthier individuals tend to hold more positive attitudes about the birds present in residential neighbourhoods and gardens (Clergeau *et al.*, 2001; Clucas *et al.*, 2011; Belaire *et al.*, 2015; Cox and Gaston, 2015). For attitudes towards species that carry a greater perceived risk, a person's social vulnerability measured by factors such as socio-economic status and being a cultural or racial minority become important attitudinal drivers. For instance, a person's identity and position within society influences attitudes toward tigers (*Panthera tigris*) in Nepal more than past experiences with tigers (Carter *et al.*, 2014). Variables related to individual and household structure also influence management decisions along with attitudinal judgements, which ultimately affect wildlife biodiversity.

Overall, attitudes have an important but complex relationship with behaviour. A better understanding of attitudes will improve our understanding of how people affect and are affected by urban wildlife. The benefits or negative impacts on well-being associated with interacting with urban wildlife occur in both indirect and direct ways and may be driven by how these services play out in an urban landscape. Ultimately, understanding the interaction between attitudinal and structural factors can inform our understanding of the interactions between people and wildlife in cities, and how these interactions help drive urban wildlife patterns and process.

Ecosystem Services and Disservices

Broadly speaking, biodiversity is imperative for functioning ecosystems, and the wildlife found in urban areas provides essential ecosystem services. Here we define ecosystem services as those services that provide benefits for people, including supporting, provisioning, regulating, and cultural and social services (Millennium Ecosystems Assessment, 2005; Larson *et al.*, 2019). We define disservices as the health and other hazards associated with interacting with 'nuisance' or 'pest' animals (e.g. mosquitoes; Lyytimäki *et al.*, 2008). In general, biodiversity correlates positively with ecosystem services, provided that as plants and animals increase in richness, so does the efficiency and efficacy of the ecological service (Schwarz *et al.*, 2017). However, urban areas, which are composed of novel ecological features, and animal and plant communities, may have different relationships between biodiversity and ecosystem services compared to natural areas (Ziter, 2016). Increased services may be driven by higher functional diversity within animal communities, or the presence of particular species that contribute disproportionately to services (Schwarz *et al.*, 2017). In either case, intentional and unintentional management activities can either promote or discourage a wide breadth of taxa that perform ecosystem services, as well as promote disservices in urban ecosystems. Conversely, the reception of wildlife-derived services may be a mechanism of encouragement to promote urban habitat restoration. Thus, managing urban green spaces in a manner that aims to maximize ecosystem services while minimizing disservices has gained considerable traction (Goddard *et al.*, 2013; Larson *et al.*, 2016). Here we describe the variety of ecological and cultural services that are derived from urban wildlife, and describe how feedback loops between public attitudes may contribute to the management of urban green space (Fig. 3.1).

Ecological Services (Regulating and Supporting)

The presence of wildlife can provide ecosystem services that both support and regulate ecosystems and provide provisions and goods. Pollinators (e.g. bees) and seed dispersers (e.g. birds) provide supporting services, which ultimately increase economic yields in urban agriculture and promote vegetation growth more generally (Sekercioğlu *et al.*, 2004; Luck *et al.*, 2009; Cardinale *et al.*, 2012). The role of pollinators and seed dispersers further supports other ecosystem functions, due to the role vegetation plays in mitigating the urban heat island, providing flood control and purifying water (Bolund and Hunhammar, 1999). In addition, agricultural opportunities in urban areas have increased in recent years as more derelict lands have been restored into functional food-producing spaces (Lin *et al.*, 2015; Ferreira *et al.*, 2018, but see Badami and Ramankutty, 2015).

With increased floral resources in certain urban green spaces (e.g. gardens and allotments; Baldock *et al.*, 2019) comes increased pollen deposition (Werrell *et al.*, 2009), which further enhances a positive feedback loop that increases the floral resources and abundances in urban green spaces. Urban areas also contain small fragments of retained natural lands, which serve as support networks for surrounding intact natural areas. These small fragments can act as sources of pollinators to the surrounding matrix of natural lands (Hall *et al.*, 2017). Thus, ensuring management of urban green space and cultivated floral resources can have cascading effects that benefit both people and wildlife. And finally, birds and mammals provide long-distance dispersal of seeds, aiding in restoration and forest growth (Corlett, 2005; Whelan *et al.*, 2015), though sometimes contributing to non-native plant invasions, which further degrades habitat quality (Gosper *et al.*, 2005; Minor and Gardner, 2011). Species that disperse long distances (e.g. migratory birds) or have expansive home ranges (e.g. large mammals) can be important transporters of nutrients such as nitrogen across space and in between habitat boundaries (Whelan *et al.*, 2008). Though not directly derived from animals themselves, the provision of habitat for wildlife by way of planting trees or conserving natural green space can also provide additive service by sequestering carbon, preventing erosion, reducing urban heat and improving air quality (Livesley *et al.*, 2016).

Wildlife species that scavenge carrion (e.g. crows, *Corvus* spp.) and consume vertebrate and invertebrate pests provide regulating services (Sekercioğlu *et al.*, 2004; Luck *et al.*, 2009; Cardinale *et al.*, 2012). These regulating services, in turn, provide pest control of insect and rodent outbreaks, and dispose of waste (Luck *et al.*, 2009). For example, insectivorous birds consume billions of insects, which improves the health of the urban forest, plant growth and survival, and increases crop production in urban gardens (Wenny *et al.*, 2011; Whelan *et al.*, 2015). Pest consumption by birds, along with other biocontrol agents, may also contribute to reducing populations of disease vectors, such as ticks and mosquitoes (Samish and Rehacek, 1999). Vertebrates are not the only taxa that provide regulating services. Although 'urban entomology' may invoke thoughts of pests and disease vectors, urban areas also harbour immense diversity of beneficial taxa and species of conservation importance. Urban arthropods perform important ecosystem services such as pest consumption and parasitization (Shrewsbury and Raupp, 2006) and decomposition and consumption of refuse (e.g. Youngsteadt *et al.*, 2015). Insects also serve as the crucial food resources for diverse taxonomic groups that provide other ecological and social values, such as insectivorous songbirds (Narango *et al.*, 2017). Predatory and parasitic arthropods reduce the impact of herbivorous consumers, like caterpillars, by reducing herbivory, stymieing disease vectors (i.e. presenting such difficulties as to discourage or defeat the effectiveness of the vector) and improving plant health (McIntyre, 2000).

Social and Cultural Services

Urban wildlife provides social and cultural ecosystem services that improve the quality of life for urban dwellers. Arguably, out of all ecosystem services, these social and cultural services have the broadest and most tangible impact on people's lives. For example, bird diversity is linked to the well-being of city residents through affective (emotional) mechanisms such as experiencing the restorative properties of hearing bird song (Ratcliffe *et al.*, 2016; Hedblom *et al.*, 2017). Further, songbirds provide value by their beauty, their song and as objects of educational study, connecting people with nearby nature (Belaire *et al.*, 2015). Likewise, park visitors reported 'feeling better' in environments that support greater species richness for trees, birds and butterflies (Fuller *et al.*, 2007). Bird feeding and birdwatching best exemplify cultural activities, which are both multi-million-dollar industries. These activities are structured so that people can have intimate and personal experiences with birds (Reynolds *et al.*, 2017). At the other end of the experiential spectrum with nature, feelings of disconnect from nature have been linked to negative environmental attitudes (Soga *et al.*, 2016) and the reduction of personal subjective well-being (Nisbet *et al.*, 2011). Ultimately, positive interactions with wildlife can mediate pro-environmental behaviours and people's support for public environmental initiatives, which hold important conservation implications (Soga and Gaston, 2016). Therefore, one of the most critical ecosystem services that urban wildlife provides is the ability to galvanize urban residents' support for conservation of the natural world (Hughes *et al.*, 2018).

Disservices

In addition to services, urban wildlife can also provide disservices, since not all wildlife is a desirable component of the landscape from the perspective of the public (Belaire *et al.*, 2015). Thus, by inviting wildlife to share our living and working spaces in cities and suburbs, there may be unintentional consequences. Despite tremendous ecological and economic importance, insects are often either overlooked or despised by the public (Kellert, 1993). For example, urban areas harbour increased abundance of insects that perform disservices, such as disease vectors (e.g. West Nile virus; LaDeau *et al.*, 2007), tree pests (e.g. insects such as the gloomy scale, *Melanaspis tenebricosa*; Frank *et al.*, 2013), or species that are generally despised due to aesthetic or cultural revulsion (e.g. cockroaches; Kellert, 1993). Invasive insects alter forest health and are particularly detrimental to urban forests, which are often planted in monocultures, or experience frequent drought-induced stress. For example, in 2007 it was estimated that the emerald ash borer (EAB; *Agrilus planipennis* Fairmaire) killed more than 53 million native ash trees (*Fraxinus* sp.) in Ohio, Michigan and Indiana, with the majority of mortality occurring in urban environments (Kovacs *et al.*, 2010). Given the importance of urban trees for regulating climate, stormwater mitigation, well-being and habitat (Bolund and Hunhammar, 1999), the impact from EAB and other invasive insects is far-reaching.

Urban generalist mammals and birds can also invoke negative reactions from the public when interactions with private land become more apparent or detrimental to human well-being. For example, predatory carnivores (e.g. coyotes, and mountain lions, *Puma concolor*) are seen as threats to children and pets (Soulsbury and White, 2015) and vectors of disease (e.g. rabies; Riley *et al.*, 1998). Generalist omnivores (e.g. black bears, *Ursus americanus*, raccoons, gulls (*Larridae* spp.)) often scavenge in human refuse and garbage, and are therefore considered a nuisance (Newsome and Van Eeden, 2017). Large flocks of birds (e.g. starlings (*Sturnus vulgaris*), crows (*Corvus brachyrhynchos*) and geese (various species in the family *Anatidae*)) also can be seen as problems because of the noise and mess they create *en masse* when congregating in forage or roosting locations (Whelan *et al.*, 2015). Surprisingly, although white-tailed deer (*Odocoileus virginianus*) populations were once threatened due to over-hunting, deer populations have rebounded and enjoy some of the highest densities around urban metropolitan areas where hunting opportunities may be limited and their natural predators have long since been extirpated (Indiana Division of Fish & Wildlife, 2019). Consequently, deer populations perform disservices by over-browsing managed gardens and natural plant communities, degrading habitat quality for songbirds and other wildlife, disrupting forest succession, causing car accidents and serving as a long-distance dispersal agent for Lyme disease (Conover, 1997; McShea and Rappole, 2000; Kilpatrick *et al.*, 2007; Indiana Division of Fish & Wildlife, 2019). Yet deer are also a well-recognized representation of 'nature,' creating a conundrum for managers who need to appease public stakeholders, which may have conflicting attitudes towards deer (Rutberg, 1997). It is clear that urban wildlife provides both ecosystem services and disservices, sometimes simultaneously. As urbanization continues to encroach on natural areas, urban managers and residents will have to cope with the trade-offs associated with the management of urban green spaces for wildlife.

Directions for Future Research in Urban Wildlife Ecology

In this chapter, some of the ecological and social filtering factors that affect wildlife patterns in urban environments have been highlighted (Fig. 3.1). Although we attempted to include examples from a range of taxa, a few groups have received the majority of research interest. In addition to advancing our understanding of species with broad public appeal, like songbirds (Marzluff, 2017), bees and butterflies (Ramírez-Restrepo and MacGregor-Fors, 2017), we suggest that future studies continue the recent trend of investigating less 'charismatic' taxa such as ground arthropods, flies, moths, reptiles, amphibians and small mammals (e.g. Merckx *et al.*, 2018; McCary *et al.*, 2018). Similarly, expanding research in neo-, Afro- and Indo-tropical regions can expand our level of inference of urban wildlife patterns (McDonnell and Hahs, 2013), particularly with regard to social and cultural contexts. These understudied tropical regions are also critical for future investigations since they contain both the most rapidly growing metropolitan regions in the world and high overlap with biodiversity hotspots (Seto *et al.*, 2012). Recent reviews of urban biodiversity in urban green spaces have highlighted avenues ripe for future research (Beninde *et al.*, 2015). For example, a deeper understanding of the thresholds of green space size, and whether green spaces serve as population sources or sinks, can help inform managers who require detailed information for a variety of different species (Aronson *et al.*, 2017; Lepczyk *et al.*, 2017a).

There has been an increased interest in how management can enhance the conservation value of urban green spaces (e.g. Goddard *et al.*, 2013; Lerman *et al.*, 2014; Aronson *et al.*, 2017). Thus, additional detailed studies that identify particular recommendations and actions that managers, urban planners and the public can take to restore ecological integrity and improve biodiversity in urban green spaces can further advance biodiversity goals. Experimental manipulative studies are necessary to isolate potential mechanistic drivers, but are curiously absent in this field (Felson *et al.*, 2013; Marzluff, 2017). Recent research that has assessed specific mechanisms includes identifying simple solutions for promoting pollinators via less frequent lawn mowing (Lerman *et al.*, 2018) and how to manage urban green spaces to reduce ticks and Lyme disease by planting native plants (Adalsteinsson *et al.*, 2016). In addition, identifying specific thresholds, such as the amount of native plants necessary to support stable insectivorous bird populations (Narango *et al.*, 2018), can help guide restoration efforts and provide empirical support for planning goals. These and future studies will help homeowners, developers, urban green space managers and designers make informed decisions that may simultaneously promote biodiversity and improve human health and ecosystem services (Aronson *et al.*, 2017).

We also encourage further exploration of the links between wildlife and ecosystem services that are particular to novel urban systems. Increasing our understanding of the interactions and feedback loops between management decisions, their impact on a variety of different species, and how interactions and subsequent attitudes, in turn, further shape decisions, can assist with future management recommendations that aim to enhance urban habitats for wildlife and people. With this information, we could begin to uncover the answers to ultimate questions in the field, such as how does the taxonomic and functional homogenization of some taxa impact ecosystem services? Do people recognize the value of biodiverse animal communities, and how can we highlight opportunities for enhancing communication strategies that affect management decisions? Finally, what are the primary drivers of personal conservation stewardship, and what scale is necessary to support sustainable and biodiverse urban wildlife communities?

Conclusion

Urban areas represent a fascinating and novel mixture of human-provided resource inputs, altered ecological communities and human–wildlife interactions. We have highlighted some key differences between vertebrate and invertebrate responses to the urban landscape, the unique barriers associated with filtering urban communities, and the dynamism within this coupled social and ecological system. Urban settings also present unique opportunities for conservation, education and stewardship. In particular, recommendations for managing urban green spaces can empower local communities to promote features that enhance habitat, improve biodiversity and ensure beneficial services for people.

References

Ackley, J.W., Angilletta, M.J., DeNardo, D., Sullivan, B. and Wu, J. (2015a) Urban heat island mitigation strategies and lizard thermal ecology: landscaping can quadruple potential activity time in an arid city. *Urban Ecosystems* 18(4), 1447–1459. DOI: 10.1007/s11252-015-0460-x.

Ackley, J.W., Wu, J., Angilletta, M.J., Myint, S.W. and Sullivan, B. (2015b) Rich lizards: how affluence and land cover influence the diversity and abundance of desert reptiles persisting in an urban landscape. *Biological Conservation* 182, 87–92. DOI: 10.1016/j.biocon.2014.11.009.

Adalsteinsson, S.A., D'Amico, V., Shriver, W.G., Brisson, D. and Buler, J.J. (2016) Scale-dependent effects of nonnative plant invasion on host-seeking tick abundance. *Ecosphere* 7(3), e01317. DOI: 10.1002/ecs2.1317.

Adelman, J.S., Moyers, S.C., Farine, D.R. and Hawley, D.M. (2015) Feeder use predicts both acquisition and transmission of a contagious pathogen in a North American songbird. *Proceedings of the Royal Society B: Biological Sciences* 282(1815), 20151429. DOI: 10.1098/rspb.2015.1429.

Ainsworth, G.B., Fitzsimons, J.A., Weston, M.A. and Garnett, S.T. (2018) The culture of bird conservation: Australian stakeholder values regarding iconic, flagship and rare birds. *Biodiversity and Conservation* 27(2), 345–363. DOI: 10.1007/s10531-017-1438-1.

Alves, R.R.N., Vieira, K.S., Santana, G.G., Vieira, W.L.S., Almeida, W.O. *et al.* (2012) A review on human attitudes towards reptiles in Brazil. *Environmental Monitoring and Assessment* 184(11), 6877–6901. DOI: 10.1007/s10661-011-2465-0.

Alves, R.R.N., Silva, V.N., Trovão, D.M.B.M., Oliveira, J.V., Mourão, J.S. *et al.* (2014) Students' attitudes toward and knowledge about snakes in the semiarid region of Northeastern Brazil. *Journal of Ethnobiology and Ethnomedicine* 10(1), 30–37. DOI: 10.1186/1746-4269-10-30.

Andrade, R., Bateman, H.L., Franklin, J. and Allen, D. (2018) Waterbird community composition, abundance, and diversity along an urban gradient. *Landscape and Urban Planning* 170, 103–111. DOI: 10.1016/j.landurbplan.2017.11.003.

Aronson, M.F.J., La Sorte, F.A., Nilon, C.H., Katti, M., Goddard, M.A. *et al.* (2014) A global analysis of the impacts of urbanization on bird and plant diversity reveals key anthropogenic drivers. *Proceedings of the Royal Society B: Biological Sciences* 281(1780), 20133330. DOI: 10.1098/rspb.2013.3330.

Aronson, M.F.J., Nilon, C.H., Lepczyk, C.A., Parker, T.S., Warren, P.S. *et al.* (2016) Hierarchical filters determine community assembly of urban species pools. *Ecology* 97(11), 2952–2963. DOI: 10.1002/ecy.1535.

Aronson, M.F.J., Lepczyk, C.A., Evans, K.L., Goddard, M.A., Lerman, S.B. *et al.* (2017) Biodiversity in the city: key challenges for urban green space management. *Frontiers in Ecology and the Environment* 15(4), 189–196. DOI: 10.1002/fee.1480.

Atkinson, D. (1994) Temperature and organism size : a biological law for ectotherms? *Advances in Ecological Research* 25, 1–58.

Badami, M.G. and Ramankutty, N. (2015) Urban agriculture and food security: a critique based on an assessment of urban land constraints. *Global Food Security* 4, 8–15. DOI: 10.1016/j.gfs.2014.10.003.

Baldock, K.C., Goddard, M.A., Hicks, D.M., Kunin, W.E., Mitschunas, N. *et al.* (2015) Where is the UK's pollinator biodiversity? The importance of urban areas for flower-visiting insects. *Proceedings of the Royal Society B Biological Sciences* 282, 20142849.

Baldock, K.C.R., Goddard, M.A., Hicks, D.M., Kunin, W.E., Mitschunas, N. *et al.* (2019) A systems approach reveals urban pollinator hotspots and conservation opportunities. *Nature Ecology & Evolution* 3, 363–373. DOI: 10.1038/s41559-018-0769-y.

Bang, C. and Faeth, S.H. (2011) Variation in arthropod communities in response to urbanization: seven years of arthropod monitoring in a desert city. *Landscape and Urban Planning* 103(3-4), 383–399. DOI: 10.1016/j.landurbplan.2011.08.013.

Barber, J.R., Crooks, K.R. and Fristrup, K.M. (2010) The costs of chronic noise exposure for terrestrial organisms. *Trends in Ecology & Evolution* 25(3), 180–189. DOI: 10.1016/j.tree.2009.08.002.

Baverstock, S., Weston, M.A. and Miller, K.K. (2019) A global paucity of wild bird feeding policy. *Science of the Total Environment* 653, 105–111. DOI: 10.1016/j.scitotenv.2018.10.338.

Baxter-Gilbert, J.H., Riley, J.L., Neufeld, C.J.H., Litzgus, J.D., Lesbarrères, D. *et al.* (2015) Road mortality potentially responsible for billions of pollinating insect deaths annually. *Journal of Insect Conservation* 19(5), 1029–1035. DOI: 10.1007/s10841-015-9808-z.

Becker, D.J., Streicker, D.G. and Altizer, S. (2015) Linking anthropogenic resources to wildlife-pathogen dynamics: a review and meta-analysis. *Ecology Letters* 18(5), 483–495. DOI: 10.1111/ele.12428.

Beebee, T.J.C. (2013) Effects of road mortality and mitigation measures on amphibian populations. *Conservation Biology* 27(4), 657–668. DOI: 10.1111/cobi.12063.

Beissinger, S.R. and Osborne, D.R. (1982) Effects of urbanization on avian community organization. *The Condor* 84(1), 75–83. DOI: 10.2307/1367825.

Belaire, J.A., Whelan, C.J. and Minor, E.S. (2014) Having our yards and sharing them too: the collective effects of yards on native bird species in an urban landscape. *Ecological Applications* 24(8), 2132–2143. DOI: 10.1890/13-2259.1.

Belaire, J.A., Westphal, L.M., Whelan, C.J. and Minor, E.S. (2015) Urban residents' perceptions of birds in the neighborhood: biodiversity, cultural ecosystem services, and disservices. *The Condor* 117(2), 192–202. DOI: 10.1650/CONDOR-14-128.1.

Belaire, J.A., Westphal, L.M. and Minor, E.S. (2016) Different social drivers, including perceptions of urban wildlife, explain the ecological resources in residential landscapes. *Landscape Ecology* 31(2), 401–413. DOI: 10.1007/s10980-015-0256-7.

Beninde, J., Veith, M. and Hochkirch, A. (2015) Biodiversity in cities needs space: a meta-analysis of factors determining intra-urban biodiversity variation. *Ecology Letters* 18(6), 581–592. DOI: 10.1111/ele.12427.

Bennie, J., Davies, T.W., Cruse, D. and Gaston, K.J. (2016) Ecological effects of artificial light at night on wild plants. *Journal of Ecology* 104(3), 611–620. DOI: 10.1111/1365-2745.12551.

Bjurlin, C.D. and Cypher, B.L. (2005) Encounter frequency with the urbanized San Joaquin kit fox correlates with public beliefs and attitudes toward the species. *Endanger. Species Update* 22, 107–115.

Blair, R.B. (1996) Land use and avian species diversity along an urban gradient. *Ecological Applications* 6(2), 506–519. DOI: 10.2307/2269387.

Blewett, C.M. and Marzluff, J.M. (2005) Effects of urban sprawl on snags and the abundance and productivity of cavity-nesting birds. *The Condor* 107(3), 678–693. DOI: 10.1093/condor/107.3.678.

Bolund, P. and Hunhammar, S. (1999) Ecosystem services in urban areas. *Ecological Economics* 29(2), 293–301. DOI: 10.1016/S0921-8009(99)00013-0.

Bonnington, C., Gaston, K.J. and Evans, K.L. (2013) Fearing the feline: domestic cats reduce avian fecundity through trait-mediated indirect effects that increase nest predation by other species. *Journal of Applied Ecology* 50(1), 15–24. DOI: 10.1111/1365-2664.12025.

Bonnington, C., Gaston, K.J. and Evans, K.L. (2014) Squirrels in suburbia: influence of urbanisation on the occurrence and distribution of a common exotic mammal. *Urban Ecosystems* 17(2), 533–546. DOI: 10.1007/s11252-013-0331-2.

Brittingham, M.C. and Temple, S.A. (1988) Impacts of supplemental feeding on survival rates of black-capped chickadees. *Ecology* 69(3), 581–589. DOI: 10.2307/1941007.

Burghardt, K.T. and Tallamy, D.W. (2013) Plant origin asymmetrically impacts feeding guilds and life stages driving community structure of herbivorous arthropods. *Diversity and Distributions* 19(12), 1553–1565. DOI: 10.1111/ddi.12122.

Burghardt, K.T., Tallamy, D.W. Gregory Shriver, W, Shriver, G. (2009) Impact of native plants on bird and butterfly biodiversity in suburban landscapes. *Conservation Biology* 23(1), 219–224. DOI: 10.1111/j.1523-1739.2008.01076.x.

Callaghan, C.T., Major, R.E., Wilshire, J.H., Martin, J.M., Kingsford, R.T. *et al.* (2019) Generalists are the most urban-tolerant of birds: a phylogenetically controlled analysis of ecological and life history traits using a novel continuous measure of bird responses to urbanization. *Oikos* 128(6), 845–858. DOI: 10.1111/oik.06158.

Cardinale, B.J., Duffy, J.E., Gonzalez, A., Hooper, D.U., Perrings, C. *et al.* (2012) Biodiversity loss and its impact on humanity. *Nature* 486(7401), 59–67. DOI: 10.1038/nature11148.

Carter, N.H., Riley, S.J., Shortridge, A., Shrestha, B.K. and Liu, J. (2014) Spatial assessment of attitudes toward tigers in Nepal. *AMBIO* 43(2), 125–137. DOI: 10.1007/s13280-013-0421-7.

Chace, J.F. and Walsh, J.J. (2006) Urban effects on native avifauna: a review. *Landscape and Urban Planning* 74(1), 46–69. DOI: 10.1016/j.landurbplan.2004.08.007.

Ciach, M. and Fröhlich, A. (2017) Habitat type, food resources, noise and light pollution explain the species composition, abundance and stability of a winter bird assemblage in an urban environment. *Urban Ecosystems* 20(3), 547–559. DOI: 10.1007/s11252-016-0613-6.

Civitello, D.J., Allman, B.E., Morozumi, C. and Rohr, J.R. (2018) Assessing the direct and indirect effects of food provisioning and nutrient enrichment on wildlife infectious disease dynamics. *Philosophical Transactions of the Royal Society B: Biological Sciences* 373(1745), 20170101. DOI: 10.1098/rstb.2017.0101.

Clergeau, P., Mennechez, G., Sauvage, A. and Lemoine, A. (2001) Human perception and appreciation of birds: a motivation for wildlife conservation in urban environments of France. In: Marzluff, J.M., Bowman, R. and Donnelly, R. (eds) *Avian Ecology and Conservation in an Urbanizing World*. Springer, Boston, Massachusetts, pp. 69–88.

Clucas, B., Marzluff, J.M., Kübler, S. and Meffert, P. (2011) New directions in urban avian ecology: reciprocal connections between birds and humans in cities. In: Endlicher, W. (ed.) *Perspectives in Urban Ecology*. Springer, Berlin, Heidelberg, pp. 167–195.

Collado, Miguel Á.., Sol, D. and Bartomeus, I. (2019) Bees use anthropogenic habitats despite strong natural habitat preferences. *Diversity and Distributions* 25(6), 924–935. DOI: 10.1111/ddi.12899.

Conover, M.R. (1997) Monetary and intangible valuation of deer in the United States. *Wildlife Society B* 25, 298–305.

Contesse, P., Hegglin, D., Gloor, S., Bontadina, F. and Deplazes, P. (2004) The diet of urban foxes (*Vulpes vulpes*) and the availability of anthropogenic food in the city of Zurich, Switzerland. *Mammalian Biology* 69(2), 81–95. DOI: 10.1078/1616-5047-00123.

Cook, E.M., Hall, S.J. and Larson, K.L. (2012) Residential landscapes as social-ecological systems: a synthesis of multi-scalar interactions between people and their home environment. *Urban Ecosystems* 15(1), 19–52. DOI: 10.1007/s11252-011-0197-0.

Corlett, R.T. (2005) Interactions between birds, fruit bats and exotic plants in urban Hong Kong, South China. *Urban Ecosystems* 8(3-4), 275–283. DOI: 10.1007/s11252-005-3260-x.

Cox, D.T.C. and Gaston, K.J. (2015) Likeability of garden birds: importance of species knowledge & richness in connecting people to nature. *PLOS ONE* 10(11), e0141505–0141514. DOI: 10.1371/journal.pone.0141505.

Cox, D.T.C. and Gaston, K.J. (2018) Human–nature interactions and the consequences and drivers of provisioning wildlife. *Philosophical Transactions of the Royal Society B: Biological Sciences* 373(1745), 20170092. DOI: 10.1098/rstb.2017.0092.

Creel, S. and Christianson, D. (2008) Relationships between direct predation and risk effects. *Trends in Ecology & Evolution* 23(4), 194–201. DOI: 10.1016/j.tree.2007.12.004.

Cregg, B.M. and Dix, M.E. (2001) Tree moisture stress and insect damage in urban areas in relation to heat island effects. *Journal of Arboriculture* 27, 8–17.

Croci, S., Butet, A. and Clergeau, P. (2008) Does urbanization filter birds on the basis of their biological traits? *The Condor* 110(2), 223–240. DOI: 10.1525/cond.2008.8409.

Crooks, K.R. (2002) Relative sensitivities of mammalian carnivores to habitat fragmentation. *Conservation Biology* 16(2), 488–502. DOI: 10.1046/j.1523-1739.2002.00386.x.

Crooks, K.R. and Soulé, M.E. (1999) Mesopredator release and avifaunal extinctions in a fragmented system. *Nature* 400(6744), 563–566. DOI: 10.1038/23028.

Dale, A.G. and Frank, S.D. (2018) Urban plants and climate drive unique arthropod interactions with unpredictable consequences. *Current Opinion in Insect Science* 29, 27–33.

Dallimer, M., Irvine, K.N., Skinner, A.M.J., Davies, Z.G., Rouquette, J.R. *et al.* (2012) Biodiversity and the feel-good factor: understanding associations between self-reported human well-being and species richness. *BioScience* 62(1), 47–55. DOI: 10.1525/bio.2012.62.1.9.

Daniels, G.D. and Kirkpatrick, J.B. (2006) Does variation in garden characteristics influence the conservation of birds in suburbia? *Biological Conservation* 133(3), 326–335. DOI: 10.1016/j.biocon.2006.06.011.

Davies, Z.G., Fuller, R.A., Loram, A., Irvine, K.N., Sims, V. *et al.* (2009) A national scale inventory of resource provision for biodiversity within domestic gardens. *Biological Conservation* 142(4), 761–771. DOI: 10.1016/j.biocon.2008.12.016.

DeGraaf, R.M. and Wentworth, J.M. (1986) Avian guild structure and habitat associations in suburban bird communities. *Urban Ecology* 9(3-4), 399–412. DOI: 10.1016/0304-4009(86)90012-4.

Deguines, N., Julliard, R., de Flores, M. and Fontaine, C. (2016) Functional homogenization of flower visitor communities with urbanization. *Ecology and Evolution* 6(7), 1967–1976. DOI: 10.1002/ece3.2009.

Devictor, V., Julliard, R., Clavel, J., Jiguet, F., Lee, A. *et al.* (2008) Functional biotic homogenization of bird communities in disturbed landscapes. *Global Ecology and Biogeography* 17(2), 252–261. DOI: 10.1111/j.1466-8238.2007.00364.x.

Dodd,K.C., Barichivich, W.J. and Smith, L.L. (2004) Effectiveness of a barrier wall and culverts in reducing wildlife mortality on a heavily traveled highway in Florida. *Biological Conservation* 118(5), 619–631. DOI: 10.1016/j.biocon.2003.10.011.

Donnelly, R. and Marzluff, J.M. (2004) Importance of reserve size and landscape context to urban bird conservation. *Conservation Biology* 18(3), 733–745. DOI: 10.1111/j.1523-1739.2004.00032.x.

Draheim, M.M., Patterson, K.W., Rockwood, L.L., Guagnano, G.A. and Parsons, E.C.M. (2013) Attitudes of college undergraduates towards coyotes (*Canis latrans*) in an urban landscape: management and public outreach implications. *Animals* 3(1), 1–18. DOI: 10.3390/ani3010001.

Ellington, E.H. and Gehrt, S.D. (2019) Behavioral responses by an apex predator to urbanization. *Behavioral Ecology* 30(3), 821–829. DOI: 10.1093/beheco/arz019.

Emlen, J.T. (1974) An urban bird community in Tucson, Arizona: derivation, structure, regulation. *The Condor* 76(2), 184–197. DOI: 10.2307/1366729.

Estes, J.A., Terborgh, J., Brashares, J.S., Power, M.E., Berger, J. *et al.* (2011) Trophic downgrading of planet earth. *Science* 333(6040), 301–306. DOI: 10.1126/science.1205106.

Evans, K.L., Chamberlain, D.E., Hatchwell, B.J., Gregory, R.D. and Gaston, K.J. (2011) What makes an urban bird? *Global Change Biology* 17, 32–44.

Evans, B.S., Reitsma, R., Hurlbert, A.H. and Marra, P.P. (2018) Environmental filtering of avian communities along a rural-to-urban gradient in Greater Washington, D.C., USA. *Ecosphere* 9(11), e02402. DOI: 10.1002/ecs2.2402.

Farr, C.M., Reed, S.E. and Pejchar, L. (2018) How often are conservation developments managed for biodiversity protection? A case study in Colorado, USA. *Landscape and Urban Planning* 169, 105–114. DOI: 10.1016/j.landurbplan.2017.08.010.

Felson, A.J., Oldfield, E.E. and Bradford, M.A. (2013) Involving ecologists in shaping large-scale green infrastructure projects. *BioScience* 63, 882–890.

Ferreira, A.J.D., Guilherme, R.I.M.M., Ferreira, C.S.S. and de Oliveira, M.F.M.L. (2018) Urban agriculture, a tool towards more resilient urban communities? *Current Opinion in Environmental Science & Health* 5, 93–97. DOI: 10.1016/j.coesh.2018.06.004.

Fetridge, E.D., Ascher, J.S. and Langellotto, G.A. (2008) The bee fauna of residential gardens in a suburb of New York City (Hymenoptera: Apoidea). *Annals of the Entomological Society of America* 101(6), 1067–1077. DOI: 10.1603/0013-8746-101.6.1067.

Fischer, J.D., Cleeton, S.H., Lyons, T.P. and Miller, J.R. (2012) Urbanization and the predation paradox: the role of trophic dynamics in structuring vertebrate communities. *BioScience* 62(9), 809–818. DOI: 10.1525/bio.2012.62.9.6.

Forman, R.T.T. and Alexander, L.E. (1998) Roads and their major ecological effects. *Annual Review of Ecology and Systematics* 29(1), 207–231. DOI: 10.1146/annurev.ecolsys.29.1.207.

Fortel, L., Henry, M., Guilbaud, L., Guirao, A.L., Kuhlmann, M. *et al.* (2014) Decreasing abundance, increasing diversity and changing structure of the wild bee community (Hymenoptera: anthophila) along an urbanization gradient. *PLOS ONE* 9(8), e104679. DOI: 10.1371/journal.pone.0104679.

Francis, R.A. and Chadwick, M.A. (2012) What makes a species synurbic? *Applied Geography* 32(2), 514–521. DOI: 10.1016/j.apgeog.2011.06.013.

Frank, S.D., Klingeman, W.E., White, S.A. and Fulcher, A. (2013) Biology, injury, and management of maple tree pests in nurseries and urban landscapes. *Journal of Integrated Pest Management* 4(1), 1–14. DOI: 10.1603/IPM12007.

French, K., Major, R. and Hely, K. (2005) Use of native and exotic garden plants by suburban nectarivorous birds. *Biological Conservation* 121(4), 545–559. DOI: 10.1016/j.biocon.2004.06.004.

Fuller, R.A., Irvine, K.N., Devine-Wright, P., Warren, P.H. and Gaston, K.J. (2007) Psychological benefits of greenspace increase with biodiversity. *Biology Letters* 3(4), 390–394. DOI: 10.1098/rsbl.2007.0149.

Fuller, R.A., Warren, P.H., Armsworth, P.R., Barbosa, O. and Gaston, K.J. (2008) Garden bird feeding predicts the structure of urban avian assemblages. *Diversity and Distributions* 14(1), 131–137. DOI: 10.1111/j.1472-4642.2007.00439.x.

Galbraith, J.A., Beggs, J.R., Jones, D.N., McNaughton, E.J., Krull, C.R. *et al.* (2014) Risks and drivers of wild bird feeding in urban areas of New Zealand. *Biological Conservation* 180, 64–74. DOI: 10.1016/j.biocon.2014.09.038.

Galbraith, J.A., Beggs, J.R., Jones, D.N. and Stanley, M.C. (2015) Supplementary feeding restructures urban bird communities. *Proceedings of the National Academy of Sciences* 112(20), E2648–E2657. DOI: 10.1073/pnas.1501489112.

Galbraith, J.A., Stanley, M.C., Jones, D.N. and Beggs, J.R. (2017) Experimental feeding regime influences urban bird disease dynamics. *Journal of Avian Biology* 48(5), 700–713. DOI: 10.1111/jav.01076.

Gallo, T., Fidino, M., Lehrer, E.W. and Magle, S.B. (2017) Mammal diversity and metacommunity dynamics in urban green spaces: implications for urban wildlife conservation. *Ecological Applications* 27(8), 2330–2341. DOI: 10.1002/eap.1611.

Gaston, K.J., Warren, P.H., Thompson, K. and Smith, R.M. (2005) Urban domestic gardens (IV): the extent of the resource and its associated features. *Biodiversity and Conservation* 14, 3327–3349. DOI: 10.1007/s10531-004-9513-9.

Gaston, K.J., Davies, T.W., Nedelec, S.L. and Holt, L.A. (2017) Impacts of artificial light at night on biological timings. *Annual Review of Ecology, Evolution, and Systematics* 48(1), 49–68. DOI: 10.1146/annurev-ecolsys-110316-022745.

Gehrt, S.D. (2004) Ecology and management of striped skunks, raccoons, and coyotes in urban landscapes. Chapter 4. In: Fascione, N., Delach, A. and Smith, M. (eds) *People and Predators: From Conflict to Coexistence*. Defenders of Wildlife, Island Press, Washington, D.C.

Glista, D.J., DeVault, T.L. and DeWoody, J.A. (2009) A review of mitigation measures for reducing wildlife mortality on roadways. *Landscape and Urban Planning* 91(1), 1–7. DOI: 10.1016/j.landurbplan.2008.11.001.

Goddard, M.A., Dougill, A.J. and Benton, T.G. (2010) Scaling up from gardens: biodiversity conservation in urban environments. *Trends in Ecology & Evolution* 25(2), 90–98. DOI: 10.1016/j.tree.2009.07.016.

Goddard, M.A., Dougill, A.J. and Benton, T.G. (2013) Why garden for wildlife? Social and ecological drivers, motivations and barriers for biodiversity management in residential landscapes. *Ecological Economics* 86, 258–273. DOI: 10.1016/j.ecolecon.2012.07.016.

Goddard, M.A., Ikin, K. and Lerman, S.B. (2017) Ecological and social factors determining the diversity of birds in residential yards and gardens. In: Murgui, E. and Hedblom, M. (eds) *Ecology and Conservation of Birds in Urban Environments*. Springer International Publishing, pp. 371–397

Gompper, M.E. (2002) Top carnivores in the suburbs? ecological and conservation issues raised by colonization of north eastern North America by coyotes. *BioScience* 52(2), 185–190. DOI: 10.1641/0006-3568(2002)052[0185:TCITSE]2.0.CO;2.

Gosper, C.R., Stansbury, C.D. and Vivian-Smith, G. (2005) Seed dispersal of fleshy-fruited invasive plants by birds: contributing factors and management options. *Diversity <html_ent glyph="@amp;" ascii="&"/> Distributions* 11(6), 549–558. DOI: 10.1111/j.1366-9516.2005.00195.x.

Grade, A.M. and Sieving, K.E. (2016) When the birds go unheard: highway noise disrupts information transfer between bird species. *Biology Letters* 12(4), 20160113. DOI: 10.1098/rsbl.2016.0113.

Greig, E.I., Wood, E.M. and Bonter, D.N. (2017) Winter range expansion of a hummingbird is associated with urbanization and supplementary feeding. *Proceedings of the Royal Society B: Biological Sciences* 284(1852), 20170256. DOI: 10.1098/rspb.2017.0256.

Grilo, C., Reto, D., Filipe, J., Ascensão, F. and Revilla, E. (2014) Understanding the mechanisms behind road effects: linking occurrence with road mortality in owls. *Animal Conservation* 17(6), 555–564. DOI: 10.1111/acv.12120.

Grimm, N.B., Faeth, S.H., Golubiewski, N.E., Redman, C.L., Wu, J. *et al.* (2008) Global change and the ecology of cities. *Science* 319(5864), 756–760. DOI: 10.1126/science.1150195.

Grimm, N.B., Pickett, S.T.A., Hale, R.L. and Cadenasso, M.L. (2017) Does the ecological concept of disturbance have utility in urban social–ecological–technological systems? *Ecosystem Health and Sustainability* 3(1), e01255. DOI: 10.1002/ehs2.1255.

Grove, J.M., Locke, D.H. and O'Neil-Dunne, J.P.M. (2014) An ecology of prestige in New York City: examining the relationships among population density, socio-economic status, group identity, and residential canopy cover. *Environmental Management* 54(3), 402–419. DOI: 10.1007/s00267-014-0310-2.

Guenat, S., Kunin, W.E., Dougill, A.J. and Dallimer, M. (2019) Effects of urbanisation and management practices on pollinators in tropical Africa. *Journal of Applied Ecology* 56(1), 214–224. DOI: 10.1111/1365-2664.13270.

Hahn, J.D. and Ascerno, M.E. (1991) Public attitudes toward urban arthropods in Minnesota. *American Entomologist* 37(3), 179–185. DOI: 10.1093/ae/37.3.179.

Hahs, A.K., McDonnell, M.J., McCarthy, M.A., Vesk, P.A., Corlett, R.T. *et al.* (2009) A global synthesis of plant extinction rates in urban areas. *Ecology Letters* 12(11), 1165–1173. DOI: 10.1111/j.1461-0248.2009.01372.x.

Hall, D.M., Camilo, G.R., Tonietto, R.K., Ollerton, J., Ahrné, K. *et al.* (2017) The city as a refuge for insect pollinators. *Conservation Biology* 31(1), 24–29. DOI: 10.1111/cobi.12840.

Hayman, R.B., Harvey, R.G., Mazzotti, F.J., Israel, G.D. and Woodward, A.R. (2014) Who complains about alligators? Cognitive and situational factors influence behavior toward wildlife. *Human Dimensions of Wildlife* 19(6), 481–497. DOI: 10.1080/10871209.2014.918218.

Heberlein, T.A. (2012) Navigating environmental attitudes. *Conservation Biology* 26(4), 583–585. DOI: 10.1111/j.1523-1739.2012.01892.x.

Hedblom, M., Knez, I. and Gunnarsson, B. (2017) Bird diversity improves the well-being of city residents.. In: Murgui, E. and Hedblom, M. (eds) *Ecology and Conservation of Birds in Urban Environments*. Springer International Publishing, pp. 287–306.

Holway, D.A. and Suarez, A.V. (2006) Homogenization of ant communities in Mediterranean California: the effects of urbanization and invasion. *Biological Conservation* 127(3), 319–326. DOI: 10.1016/j.biocon.2005.05.016.

Hondula, D.M., Balling, R.C., Andrade, R., Scott Krayenhoff, E., Middel, A, Krayenhoff, E.S.M., Urban, A. *et al.* (2017) Biometeorology for cities. *International Journal of Biometeorology* 61(S1), 59–69. DOI: 10.1007/s00484-017-1412-3.

Hughes, J., Richardson, M. and Lumber, R. (2018) Evaluating connection to nature and the relationship with conservation behaviour in children. *Journal for Nature Conservation* 45, 11–19. DOI: 10.1016/j.jnc.2018.07.004.

Ibáñez-Álamo, J.D., Rubio, E., Benedetti, Y. and Morelli, F. (2017) Global loss of avian evolutionary uniqueness in urban areas. *Global Change Biology* 23(8), 2990–2998. DOI: 10.1111/gcb.13567.
Indiana Division of Fish & Wildlife (2019) *Urban Deer: Technical Guide*.
Jones, D.N. and Reynolds, S.J. (2008) Feeding birds in our towns and cities: a global research opportunity. *Journal of avian biology* 39(3), 265–271. DOI: 10.1111/j.0908-8857.2008.04271.x.
Jung, K. and Kalko, E.K.V. (2011) Adaptability and vulnerability of high flying neotropical aerial insectivorous bats to urbanization. *Diversity and Distributions* 17(2), 262–274. DOI: 10.1111/j.1472-4642.2010.00738.x.
Kaye, J.P., McCulley, R.L. and Burke, I.C. (2005) Carbon fluxes, nitrogen cycling, and soil microbial communities in adjacent urban, native and agricultural ecosystems. *Global Change Biology* 11(4), 575–587. DOI: 10.1111/j.1365-2486.2005.00921.x.
Keilsohn, W., Narango, D.L. and Tallamy, D.W. (2018) Roadside habitat impacts insect traffic mortality. *Journal of Insect Conservation* 22(2), 183–188. DOI: 10.1007/s10841-018-0051-2.
Kellert, S.R. (1993) Values and perceptions of invertebrates. *Conservation Biology* 7(4), 845–855. DOI: 10.1046/j.1523-1739.1993.740845.x.
Kilpatrick, H.J., Labonte, A.M. and Barclay, J.S. (2007) Acceptance of deer management strategies by suburban homeowners and bowhunters. *Journal of Wildlife Management* 71(6), 2095–2101. DOI: 10.2193/2007-058.
Knop, E. (2016) Biotic homogenization of three insect groups due to urbanization. *Global Change Biology* 22(1), 228–236. DOI: 10.1111/gcb.13091.
König, A. (2008) Fears, attitudes and opinions of suburban residents with regards to their urban foxes. *European Journal of Wildlife Research* 54(1), 101–109. DOI: 10.1007/s10344-007-0117-z.
Kovacs, K.F., Haight, R.G., McCullough, D.G., Mercader, R.J., Siegert, N.W. *et al.* (2010) Cost of potential emerald ash borer damage in U.S. communities, 2009–2019. *Ecological Economics* 69(3), 569–578. DOI: 10.1016/j.ecolecon.2009.09.004.
Kowarik, I. (2011) Novel urban ecosystems, biodiversity, and conservation. *Environmental Pollution* 159(8-9), 1974–1983. DOI: 10.1016/j.envpol.2011.02.022.
Kummer, J.A., Bayne, E.M. and Machtans, C.S. (2016) Use of citizen science to identify factors affecting bird–window collision risk at houses. *The Condor* 118(3), 624–639. DOI: 10.1650/CONDOR-16-26.1.
LaDeau, S.L., Kilpatrick, A.M. and Marra, P.P. (2007) West Nile virus emergence and large-scale declines of North American bird populations. *Nature* 447(7145), 710–713. DOI: 10.1038/nature05829.
LaManna, J.A. and Martin, T.E. (2016) Costs of fear: behavioural and life-history responses to risk and their demographic consequences vary across species. *Ecology Letters* 19(4), 403–413. DOI: 10.1111/ele.12573.
La Sorte, F.A., Lepczyk, C.A., Aronson, M.F.J., Goddard, M.A., Hedblom, M. *et al.* (2018) The phylogenetic and functional diversity of regional breeding bird assemblages is reduced and constricted through urbanization. *Diversity and Distributions* 24(7), 928–938. DOI: 10.1111/ddi.12738.
Larson, K.L., Nelson, K.C., Samples, S.R., Hall, S.J., Bettez, N. *et al.* (2016) Ecosystem services in managing residential landscapes: priorities, value dimensions, and cross-regional patterns. *Urban Ecosystems* 19(1), 95–113. DOI: 10.1007/s11252-015-0477-1.
Larson, K.L., Corley, E.A., Andrade, R., Hall, S.J., York, A.M. *et al.* (2019) Subjective evaluations of ecosystem services and disservices: an approach to creating and analyzing robust survey scales. *Ecology and Society* 24(2), 7–12. DOI: 10.5751/ES-10888-240207.
Leonard, M.L., Horn, A.G., Oswald, K.N. and McIntyre, E. (2015) Effect of ambient noise on parent–offspring interactions in tree swallows. *Animal Behaviour* 109, 1–7. DOI: 10.1016/j.anbehav.2015.07.036.
Leong, M., Dunn, R.R. and Trautwein, M.D. (2018) Biodiversity and socioeconomics in the city: a review of the luxury effect. *Biology Letters* 14(5), 20180082. DOI: 10.1098/rsbl.2018.0082.
Lepczyk, C.A., Warren, P.S., Machabee, L., Kinzig, A.P. and Mertig, A.G. (2012) Who feeds the birds? A comparison across regions. *Urban Bird Ecology and Conservation* 45, 267–284.
Lepczyk, C.A., Aronson, M.F.J., Evans, K.L., Goddard, M.A., Lerman, S.B. *et al.* (2017a) Biodiversity in the city: fundamental questions for understanding the ecology of urban green spaces for biodiversity conservation. *BioScience* 67(9), 799–807. DOI: 10.1093/biosci/bix079.
Lepczyk, C.A., La Sorte, F.A., Aronson, M.F.J., Goddard, M.A., MacGregor-Fors, I. *et al.* (2017b) Global patterns and drivers of urban bird diversity. In: Murgui, E. and Hedblom, M. (eds) *Ecology and Conservation of Birds in Urban Environments*. Springer International Publishing, pp. 13–33

Lerman, S.B. and Milam, J. (2016) Bee fauna and floral abundance within lawn-dominated suburban yards in Springfield, MA. *Annals of the Entomological Society of America* 109(5), 713–723. DOI: 10.1093/aesa/saw043.

Lerman, S.B. and Warren, P.S. (2011) The conservation value of residential yards: linking birds and people. *Ecological Applications* 21(4), 1327–1339. DOI: 10.1890/10-0423.1.

Lerman, S.B., Warren, P.S., Gan, H. and Shochat, E. (2012) Linking foraging decisions to residential yard bird composition. *PLOS ONE* 7(8), e43497. DOI: 10.1371/journal.pone.0043497.

Lerman, S.B., Nislow, K.H., Nowak, D.J., DeStefano, S., King, D.I. *et al.* (2014) Using urban forest assessment tools to model bird habitat potential. *Landscape and Urban Planning* 122, 29–40. DOI: 10.1016/j.landurbplan.2013.10.006.

Lerman, S.B., Contosta, A.R., Milam, J. and Bang, C. (2018) To mow or to mow less: Lawn mowing frequency affects bee abundance and diversity in suburban yards. *Biological Conservation* 221, 160–174. DOI: 10.1016/j.biocon.2018.01.025.

Levé, M., Baudry, E. and Bessa-Gomes, C. (2019) Domestic gardens as favorable pollinator habitats in impervious landscapes. *Science of the Total Environment* 647, 420–430. DOI: 10.1016/j.scitotenv.2018.07.310.

Leveau, L.M. (2013) Bird traits in urban–rural gradients: how many functional groups are there? *Journal of Ornithology* 154(3), 655–662. DOI: 10.1007/s10336-012-0928-x.

Lewanzik, D. and Voigt, C.C. (2014) Artificial light puts ecosystem services of frugivorous bats at risk. *Journal of Applied Ecology* 51(2), 388–394. DOI: 10.1111/1365-2664.12206.

Lin, B.B., Philpott, S.M. and Jha, S. (2015) The future of urban agriculture and biodiversity-ecosystem services: challenges and next steps. *Basic and Applied Ecology* 16(3), 189–201. DOI: 10.1016/j.baae.2015.01.005.

Litt, A.R., Cord, E.E., Fulbright, T.E. and Schuster, G.L. (2014) Effects of invasive plants on arthropods. *Conservation Biology* 28(6), 1532–1549. DOI: 10.1111/cobi.12350.

Livesley, S.J., McPherson, G.M. and Calfapietra, C. (2016) The urban forest and ecosystem services: impacts on urban water, heat, and pollution cycles at the tree, street, and city scale. *Journal of Environmental Quality* 45(1), 119–124.

Loss, S.R., Ruiz, M.O. and Brawn, J.D. (2009) Relationships between avian diversity, neighborhood age, income, and environmental characteristics of an urban landscape. *Biological Conservation* 142(11), 2578–2585. DOI: 10.1016/j.biocon.2009.06.004.

Loss, S.R., Will, T. and Marra, P.P. (2013) The impact of free-ranging domestic cats on wildlife of the United States. *Nature Communications* 4(1), 1396. DOI: 10.1038/ncomms2380.

Loss, S.R., Will, T., Loss, S.S. and Marra, P.P. (2014a) Bird–building collisions in the United States: estimates of annual mortality and species vulnerability. *The Condor* 116(1), 8–23. DOI: 10.1650/CONDOR-13-090.1.

Loss, S.R., Will, T. and Marra, P.P. (2014b) Estimation of bird-vehicle collision mortality on U.S. roads. *The Journal of Wildlife Management* 78(5), 763–771. DOI: 10.1002/jwmg.721.

Loss, S.R., Will, T. and Marra, P.P. (2015) Direct mortality of birds from anthropogenic causes. *Annual Review of Ecology, Evolution, and Systematics* 46(1), 99–120. DOI: 10.1146/annurev-ecolsys-112414-054133.

Luck, G.W., Harrington, R., Harrison, P.A., Kremen, C., Berry, P.M. *et al.* (2009) Quantifying the contribution of organisms to the provision of ecosystem services. *BioScience* 59(3), 223–235. DOI: 10.1525/bio.2009.59.3.7.

Lyytimäki, J., Petersen, L.K., Normander, B. and Bezák, P. (2008) Nature as a nuisance? ecosystem services and disservices to urban lifestyle. *Environmental Sciences* 5(3), 161–172. DOI: 10.1080/15693430802055524.

Magle, S.B., Hunt, V.M., Vernon, M. and Crooks, K.R. (2012) Urban wildlife research: past, present, and future. *Biological Conservation* 155, 23–32. DOI: 10.1016/j.biocon.2012.06.018.

Magle, S.B., Lehrer, E.W. and Fidino, M. (2016) Urban mesopredator distribution: examining the relative effects of landscape and socioeconomic factors. *Animal Conservation* 19(2), 163–175. DOI: 10.1111/acv.12231.

Magura, T., Lövei, G.L. and Tóthmérész, B. (2010) Does urbanization decrease diversity in ground beetle (Carabidae) assemblages? *Global Ecology and Biogeography* 19(1), 16–26. DOI: 10.1111/j.1466-8238.2009.00499.x.

Malpass, J.S., Rodewald, A.D. and Matthews, S.N. (2017) Species-dependent effects of bird feeders on nest predators and nest survival of urban American robins and northern cardinals. *The Condor* 119(1), 1–16. DOI: 10.1650/CONDOR-16-72.1.

Martin, T.E. (2011) The cost of fear. *Science* 334(6061), 1353–1354. DOI: 10.1126/science.1216109.
Martinson, H.M. and Raupp, M.J. (2013) A meta-analysis of the effects of urbanization on ground beetle communities. *Ecosphere* 4(5), art60–24. DOI: 10.1890/ES12-00262.1.
Marzluff, J.M. (2017) A decadal review of urban ornithology and a prospectus for the future. *Ibis* 159(1), 1–13. DOI: 10.1111/ibi.12430.
Marzluff, J.M. and Rodewald, A.D. (2008) Conserving biodiversity in urbanizing areas: nontraditional views from a bird's perspective. *Cities and the Environment* 1(2), 1–27. DOI: 10.15365/cate.1262008.
Mata, C., Hervás, I., Herranz, J., Suárez, F. and Malo, J.E. (2008) Are motorway wildlife passages worth building? Vertebrate use of road-crossing structures on a Spanish motorway. *Journal of Environmental Management* 88(3), 407–415. DOI: 10.1016/j.jenvman.2007.03.014.
Marzluff, J.M. (2001) Worldwide urbanization and its effects on birds.. In: Marzluff, J.M., Bowman, R. and Donnelly, R. (eds) *Avian Ecology and Conservation in an Urbanizing World*. Springer US, Boston, Massachusetts, pp. 19–47.
Matteson, K.C., Ascher, J.S. and Langellotto, G.A. (2008) Bee richness and abundance in New York City urban gardens. *Annals of the Entomological Society of America* 101(1), 140–150. DOI: 10.1603/0013-8746(2008)101[140:BRAAIN]2.0.CO;2.
McCary, M.A., Minor, E. and Wise, D.H. (2018) Covariation between local and landscape factors influences the structure of ground-active arthropod communities in fragmented metropolitan woodlands. *Landscape Ecology* 33(2), 225–239. DOI: 10.1007/s10980-017-0593-9.
McClure, C.J.W., Ware, H.E., Carlisle, J., Kaltenecker, G. and Barber, J.R. (2013) An experimental investigation into the effects of traffic noise on distributions of birds: avoiding the phantom road. *Proceedings of the Royal Society B: Biological Sciences* 280(1773), 20132290. DOI: 10.1098/rspb.2013.2290.
McDonnell, M.J. and Hahs, A.K. (2013) The future of urban biodiversity research: Moving beyond the 'low-hanging fruit'. *Urban Ecosystems* 16(3), 397–409. DOI: 10.1007/s11252-013-0315-2.
McDonnell, M.J. and Pickett, S.T.A. (1990) Ecosystem structure and function along urban-rural gradients: an unexploited opportunity for ecology. *Ecology* 71(4), 1232–1237. DOI: 10.2307/1938259.
McEwan, R.W., Birchfield, M.K., Schoergendorfer, A. and Arthur, M.A. (2009) Leaf phenology and freeze tolerance of the invasive shrub Amur honeysuckle and potential native competitors. *The Journal of the Torrey Botanical Society* 136(2), 212–220. DOI: 10.3159/08-RA-109.1.
McIntyre, N.E. (2000) Ecology of urban arthropods: a review and a call to action. *Annals of the Entomological Society of America* 93(4), 825–835. DOI: 10.1603/0013-8746(2000)093[0825:EOUAAR]2.0.CO;2.
McIntyre, N.E. and Hostetler, M.E. (2001) Effects of urban land use on pollinator (Hymenoptera: Apoidea) communities in a desert Metropolis. *Basic Applied Ecology* 2, 209–218.
McIntyre, N.E., Rango, J., Fagan, W.F. and Faeth, S.H. (2001) Ground arthropod community structure in a heterogeneous urban environment. *Landscape and Urban Planning* 52(4), 257–274. DOI: 10.1016/S0169-2046(00)00122-5.
McKinney, M.L. (2002) Urbanization, biodiversity, and conservation. *BioScience* 52(10), 883–890. DOI: 10.1641/0006-3568(2002)052[0883:UBAC]2.0.CO;2.
McKinney, M.L. (2006) Urbanization as a major cause of biotic homogenization. *Biological Conservation* 127(3), 247–260. DOI: 10.1016/j.biocon.2005.09.005.
McLaren, J.D., Buler, J.J., Schreckengost, T., Smolinsky, J.A., Boone, M. *et al*. (2018) Artificial light at night confounds broad-scale habitat use by migrating birds. *Ecology Letters* 21(3), 356–364. DOI: 10.1111/ele.12902.
McShea, W.J. and Rappole, J.H. (2000) Managing the abundance and diversity of breeding bird populations through manipulation of deer populations. *Conservation Biology* 14(4), 1161–1170. DOI: 10.1046/j.1523-1739.2000.99210.x.
Meineke, E.K., Dunn, R.R., Sexton, J.O. and Frank, S.D. (2013) Urban warming drives insect pest abundance on street trees. *PLOS ONE* 8(3), e59687. DOI: 10.1371/journal.pone.0059687.
Merckx, T., Souffreau, C., Kaiser, A., Baardsen, L.F., Backeljau, T. *et al*. (2018) Body-size shifts in aquatic and terrestrial urban communities. *Nature* 558(7708), 113–116. DOI: 10.1038/s41586-018-0140-0.
Millennium Ecosystems Assessment (2005) *Ecosystems and Human Well-being*. World Resources Institute, Washington, DC
Mills, G.S., Dunning, J.B. and Bates, J.M. (1989) Effects of urbanization on breeding bird community structure in southwestern desert habitats. *The Condor* 91(2), 416–428. DOI: 10.2307/1368320.
Minnaar, C., Boyles, J.G., Minnaar, I.A., Sole, C.L. and McKechnie, A.E. (2015) Stacking the odds: light pollution may shift the balance in an ancient predator-prey arms race. *Journal of Applied Ecology* 52(2), 522–531. DOI: 10.1111/1365-2664.12381.

Minor, E.S. and Gardner, R.H. (2011) Landscape connectivity and seed dispersal characteristics inform the best management strategy for exotic plants. *Ecological Applications* 21(3), 739–749. DOI: 10.1890/10-0321.1.

Morelli, F., Benedetti, Y., Ibáñez-Álamo, J.D., Jokimäki, J., Mänd, R. *et al.* (2016) Evidence of evolutionary homogenization of bird communities in urban environments across Europe. *Global Ecology and Biogeography* 25(11), 1284–1293. DOI: 10.1111/geb.12486.

Morzillo, A.T. and Mertig, A.G. (2011) Urban resident attitudes toward rodents, rodent control products, and environmental effects. *Urban Ecosystems* 14(2), 243–260. DOI: 10.1007/s11252-010-0152-5.

Mumaw, L. and Bekessy, S. (2017) Wildlife gardening for collaborative public–private biodiversity conservation.. *Australasian Journal of Environmental Management* 24, 242–260.

Murray, M.H. and St. Clair, C.C. (2015) Individual flexibility in nocturnal activity reduces risk of road mortality for an urban carnivore. *Behavioral Ecology* 26(6), 1520–1527. DOI: 10.1093/beheco/arv102.

Murray, M.H. and St. Clair, C.C. (2017) Predictable features attract urban coyotes to residential yards. *The Journal of Wildlife Management* 81(4), 593–600. DOI: 10.1002/jwmg.21223.

Murthy, A.C., Fristoe, T.S. and Burger, J.R. (2016) Homogenizing effects of cities on North American winter bird diversity. *Ecosphere* 7(1), e01216. DOI: 10.1002/ecs2.1216.

Narango, D.L. and Rodewald, A.D. (2016) Urban-associated drivers of song variation along a rural–urban gradient. *Behavioral Ecology* 27(2), 608–616. DOI: 10.1093/beheco/arv197.

Narango, D.L., Tallamy, D.W. and Marra, P.P. (2017) Native plants improve breeding and foraging habitat for an insectivorous bird. *Biological Conservation* 213, 42–50. DOI: 10.1016/j.biocon.2017.06.029.

Narango, D.L., Tallamy, D.W. and Marra, P.P. (2018) Nonnative plants reduce population growth of an insectivorous bird. *Proceedings of the National Academy of Sciences* 115(45), 11549–11554. DOI: 10.1073/pnas.1809259115.

Newsome, T. and van Eeden, L. (2017) The effects of food waste on wildlife and humans. *Sustainability* 9(7), 1269–1277. DOI: 10.3390/su9071269.

Newsome, T.M., Dellinger, J.A., Pavey, C.R., Ripple, W.J., Shores, C.R. *et al.* (2015) The ecological effects of providing resource subsidies to predators. *Global Ecology and Biogeography* 24(1), 1–11. DOI: 10.1111/geb.12236.

Nisbet, E.K., Zelenski, J.M. and Murphy, S.A. (2011) Happiness is in our nature: exploring nature relatedness as a contributor to subjective well-being. *Journal of Happiness Studies* 12(2), 303–322. DOI: 10.1007/s10902-010-9197-7.

Normandin, Étienne., Vereecken, N.J., Buddle, C.M. and Fournier, V. (2017) Taxonomic and functional trait diversity of wild bees in different urban settings. *PeerJ* 5(1748), e3051. DOI: 10.7717/peerj.3051.

Oke, T.R. (1995) The heat island of the urban boundary layer: characteristics, causes and effects. In: Cermak, J.E., Davenport, A.G., Plate, E.J. and Viegas, D. X. (eds) *Wind Climate in Cities*. Springer, Netherlands, pp. 81–107.

Orros, M.E. and Fellowes, M.D.E. (2012) Supplementary feeding of wild birds indirectly affects the local abundance of arthropod prey. *Basic and Applied Ecology* 13(3), 286–293. DOI: 10.1016/j.baae.2012.03.001.

Orros, M.E. and Fellowes, M.D.E. (2015) Widespread supplementary feeding in domestic gardens explains the return of reintroduced red kites *Milvus milvus* to an urban area. *Ibis* 157(2), 230–238. DOI: 10.1111/ibi.12237.

Orłowski, G. (2008) Roadside hedgerows and trees as factors increasing road mortality of birds: implications for management of roadside vegetation in rural landscapes. *Landscape and Urban Planning* 86(2), 153–161. DOI: 10.1016/j.landurbplan.2008.02.003.

Ortega-Álvarez, R. and MacGregor-Fors, I. (2009) Living in the big city: effects of urban land-use on bird community structure, diversity, and composition. *Landscape and Urban Planning* 90(3-4), 189–195. DOI: 10.1016/j.landurbplan.2008.11.003.

Ouyang, J.Q., de Jong, M., van Grunsven, R.H.A., Matson, K.D., Haussmann, M.F. *et al.* (2017) Restless roosts: light pollution affects behavior, sleep, and physiology in a free-living songbird. *Global Change Biology* 23(11), 4987–4994. DOI: 10.1111/gcb.13756.

Padilla, B.J. and Rodewald, A.D. (2015) Avian metapopulation dynamics in a fragmented urbanizing landscape. *Urban Ecosystems* 18(1), 239–250. DOI: 10.1007/s11252-014-0390-z.

Palomino, D. and Carrascal, L.M. (2005) Birds on novel island environments. A case study with the urban avifauna of Tenerife (Canary Islands). *Ecological Research* 20(5), 611–617. DOI: 10.1007/s11284-005-0083-4.

Palta, M.M., Grimm, N.B. and Groffman, P.M. (2017) "Accidental" urban wetlands: ecosystem functions in unexpected places. *Frontiers in Ecology and the Environment* 15(5), 248–256. DOI: 10.1002/fee.1494.

Pandey, D.P., Subedi Pandey, G., Devkota, K. and Goode, M. (2016) Public perceptions of snakes and snakebite management: implications for conservation and human health in southern Nepal. *Journal of Ethnobiology and Ethnomedicine* 12(1), 1–24. DOI: 10.1186/s13002-016-0092-0.

Pardee, G.L. and Philpott, S.M. (2014) Native plants are the bee's knees: local and landscape predictors of bee richness and abundance in backyard gardens. *Urban Ecosystems* 17(3), 641–659. DOI: 10.1007/s11252-014-0349-0.

Parker, T.S. and Nilon, C.H. (2008) Gray squirrel density, habitat suitability, and behavior in urban parks. *Urban Ecosystems* 11(3), 243–255. DOI: 10.1007/s11252-008-0060-0.

Parsons, A.W., Forrester, T., Baker-Whatton, M.C., McShea, W.J., Rota, C.T. *et al.* (2018) Mammal communities are larger and more diverse in moderately developed areas. *eLife* 2018(7), e38012.

Piano, E., De Wolf, K., Bona, F., Bonte, D., Bowler, D.E. *et al.* (2017) Urbanization drives community shifts towards thermophilic and dispersive species at local and landscape scales. *Global Change Biology* 23(7), 2554–2564. DOI: 10.1111/gcb.13606.

Pickett, S.T.A., Cadenasso, M.L., Grove, J.M., Boone, C.G., Groffman, P.M. *et al.* (2011) Urban ecological systems: scientific foundations and a decade of progress. *Journal of Environmental Management* 92(3), 331–362. DOI: 10.1016/j.jenvman.2010.08.022.

Plummer, K.E., Bearhop, S., Leech, D.I., Chamberlain, D.E. and Blount, J.D. (2013) Winter food provisioning reduces future breeding performance in a wild bird. *Scientific Reports* 3(1), 1–6. DOI: 10.1038/srep02002.

Prange, S., Gehrt, S.D. and Wiggers, E.P. (2003) Demographic factors contributing to high raccoon densities in urban landscapes. *The Journal of Wildlife Management* 67(2), 324–333. DOI: 10.2307/3802774.

Raap, T., Pinxten, R. and Eens, M. (2015) Light pollution disrupts sleep in free-living animals. *Scientific Reports* 5, 13557. DOI: 10.1038/srep13557.

Ramalho, C.E. and Hobbs, R.J. (2012) Time for a change: dynamic urban ecology. *Trends in Ecology & Evolution* 27(3), 179–188. DOI: 10.1016/j.tree.2011.10.008.

Ramírez-Restrepo, L. and MacGregor-Fors, I. (2017) Butterflies in the city: a review of urban diurnal Lepidoptera. *Urban Ecosystems* 20(1), 171–182. DOI: 10.1007/s11252-016-0579-4.

Ratcliffe, E., Gatersleben, B. and Sowden, P.T. (2016) Associations with bird sounds: how do they relate to perceived restorative potential? *Journal of Environmental Psychology* 47, 136–144. DOI: 10.1016/j.jenvp.2016.05.009.

Raupp, M.J., Shrewsbury, P.M. and Herms, D.A. (2010) Ecology of herbivorous arthropods in urban landscapes. *Annual Review of Entomology* 55, 19–38. DOI: 10.1146/annurev-ento-112408-085351.

Reed, J.H. and Bonter, D.N. (2018) Supplementing non-target taxa: bird feeding alters the local distribution of mammals. *Ecological Applications: a publication of the Ecological Society of America* 28(3), 761–770. DOI: 10.1002/eap.1683.

Reif, J., Böhning-Gaese, K., Flade, M., Schwarz, J. and Schwager, M. (2011) Population trends of birds across the iron curtain: brain matters. *Biological Conservation* 144(10), 2524–2533. DOI: 10.1016/j.biocon.2011.07.009.

Renner, S.S. and Zohner, C.M. (2018) Climate change and phenological mismatch in trophic interactions among plants, insects, and vertebrates. *Annual Review of Ecology, Evolution, and Systematics* 49(1), 165–182. DOI: 10.1146/annurev-ecolsys-110617-062535.

Reynolds, S.J., Galbraith, J.A., Smith, J.A. and Jones, D.N. (2017) Garden bird feeding: insights and prospects from a north-south comparison of this global urban phenomenon. *Frontiers in Ecology and Evolution* 5, 24–38. DOI: 10.3389/fevo.2017.00024.

Riley, S.P.D., Hadidian, J. and Manski, D.A. (1998) Population density, survival, and rabies in raccoons in an urban National Park. *Canadian Journal of Zoology* 76(6), 1153–1164. DOI: 10.1139/z98-042.

Ritchie, E.G. and Johnson, C.N. (2009) Predator interactions, mesopredator release and biodiversity conservation. *Ecology Letters* 12(9), 982–998. DOI: 10.1111/j.1461-0248.2009.01347.x.

Robb, G.N., McDonald, R.A., Chamberlain, D.E. and Bearhop, S. (2008) Food for thought: supplementary feeding as a driver of ecological change in avian populations. *Frontiers in Ecology and the Environment* 6(9), 476–484. DOI: 10.1890/060152.

Robbins, P. (2007) *Lawn People: How grasses, weeds, and chemicals make us who we are*. Temple University Press.

Rodewald, A.D. and Bakermans, M.H. (2006) What is the appropriate paradigm for riparian forest conservation? *Biological Conservation* 128(2), 193–200. DOI: 10.1016/j.biocon.2005.09.041.

Rodewald, A.D., Kearns, L.J. and Shustack, D.P. (2011) Anthropogenic resource subsidies decouple predator-prey relationships. *Ecological Applications: a publication of the Ecological Society of America* 21(3), 936–943. DOI: 10.1890/10-0863.1.

Rollinson, D.J., O'Leary, R. and Jones, D.N. (2003) The practice of wildlife feeding in suburban Brisbane. *Corella* 27, 52–59.

Rudd, H., Vala, J. and Schaefer, V. (2002) Importance of backyard habitat in a comprehensive biodiversity conservation strategy: a connectivity analysis of urban green spaces. *Restoration Ecology* 10(2), 368–375. DOI: 10.1046/j.1526-100X.2002.02041.x.

Rutberg, A.T. (1997) Lessons from the urban deer battlefront: a plea for tolerance. *Wildlife Society Bulletin* 25, 520–523.

Salomão, R.P., Alvarado, F., Baena-Díaz, F., Favila, M.E., Iannuzzi, L. *et al.* (2019) Urbanization effects on dung beetle assemblages in a tropical city. *Ecological Indicators* 103, 665–675. DOI: 10.1016/j.ecolind.2019.04.045.

Samish, M. and Rehacek, J. (1999) Pathogens and predators of ticks and their potential in biological control. *Annual Review of Entomology* 44, 159–182. DOI: 10.1146/annurev.ento.44.1.159.

Santini, L., González-Suárez, M., Russo, D., Gonzalez-Voyer, A., von Hardenberg, A. *et al.* (2019) One strategy does not fit all: determinants of urban adaptation in mammals. *Ecology Letters* 22(2), 365–376. DOI: 10.1111/ele.13199.

Satterfield, D.A., Maerz, J.C. and Altizer, S. (2015) Loss of migratory behaviour increases infection risk for a butterfly host. *Proceedings of the Royal Biological Society B. Biological Sciences* 282(1801), 20141734. DOI: 10.1098/rspb.2014.1734.

Schlaepfer, M.A., Runge, M.C. and Sherman, P.W. (2002) Ecological and evolutionary traps. *Trends in Ecology & Evolution* 17(10), 474–480. DOI: 10.1016/S0169-5347(02)02580-6.

Schoech, S.J. and Bowman, R. (2001) Variation in the timing of breeding between suburban and wildland Florida Scrub-Jays: do physiologic measures reflect different environments?Marzluff, J.M., Bowman, R. and Donnelly, R. (eds) *Avian Ecology and Conservation in an Urbanizing World*. Springer, US, Boston, Massachusetts, pp. 289–306.

Schoelitsz, B., Meerburg, B.G. and Takken, W. (2019) Influence of the public's perception, attitudes, and knowledge on the implementation of integrated pest management for household insect pests. *Entomologia Experimentalis et Applicata* 167(1), 14–26. DOI: 10.1111/eea.12739.

Schuetz, J.G. and Johnston, A. (2019) Characterizing the cultural niches of North American birds. *Proceedings of the National Academy of Sciences of the United States of America* 116(22), 10868–10873. DOI: 10.1073/pnas.1820670116.

Schwarz, N., Moretti, M., Bugalho, M.N., Davies, Z.G., Haase, D. *et al.* (2017) Understanding biodiversity-ecosystem service relationships in urban areas: a comprehensive literature review. *Ecosystem Services* 27, 161–171. DOI: 10.1016/j.ecoser.2017.08.014.

Sekercioğlu, C.H., Daily, G.C. and Ehrlich, P.R. (2004) Ecosystem consequences of bird declines. *Proceedings of the National Academy of Sciences of the United States of America* 101(52), 18042–18047. DOI: 10.1073/pnas.0408049101.

Seto, K.C., Güneralp, B. and Hutyra, L.R. (2012) Global forecasts of urban expansion to 2030 and direct impacts on biodiversity and carbon pools. *Proceedings of the National Academy of Sciences of the United States of America* 109(40), 16083–16088. DOI: 10.1073/pnas.1211658109.

Shannon, G., McKenna, M.F., Angeloni, L.M., Crooks, K.R., Fristrup, K.M. *et al.* (2016) A synthesis of two decades of research documenting the effects of noise on wildlife. *Biological Reviews of the Cambridge Philosophical Society* 91(4), 982–1005. DOI: 10.1111/brv.12207.

Shepard, D.B., Kuhns, A.R., Dreslik, M.J. and Phillips, C.A. (2008) Roads as barriers to animal movement in fragmented landscapes. *Animal Conservation* 11(4), 288–296. DOI: 10.1111/j.1469-1795.2008.00183.x.

Shochat, E. (2004) Credit or debit? Resource input changes population dynamics of city-slicker birds. *Oikos* 106(3), 622–626. DOI: 10.1111/j.0030-1299.2004.13159.x.

Shochat, E., Lerman, S.B., Katti, M. and Lewis, D.B. (2004) Linking optimal foraging behavior to bird community structure in an urban-desert landscape: field experiments with artificial food patches. *The American Naturalist* 164(2), 232–243. DOI: 10.1086/422222.

Shochat, E., Warren, P.S., Faeth, S.H., McIntyre, N.E. and Hope, D. (2006) From patterns to emerging processes in mechanistic urban ecology. *Trends in Ecology & Evolution* 21(4), 186–191. DOI: 10.1016/j.tree.2005.11.019.

Shochat, E., Lerman, S.B., Anderies, J.M., Warren, P.S., Faeth, S.H. *et al.* (2010) Invasion, competition, and biodiversity loss in urban ecosystems. *BioScience* 60(3), 199–208. DOI: 10.1525/bio.2010.60.3.6.

Shrewsbury, P.M. and Raupp, M.J. (2006) Do top-down or bottom-up forces determine *Stephanitis pyrioides* abundance in urban landscapes? *Ecological Applications: a publication of the Ecological Society of America* 16(1), 262–272. DOI: 10.1890/04-1347.

Shustack, D.P., Rodewald, A.D. and Waite, T.A. (2009) Springtime in the city: exotic shrubs promote earlier greenup in urban forests. *Biological Invasions* 11(6), 1357–1371. DOI: 10.1007/s10530-008-9343-x.

Sievers, M., Hale, R., Swearer, S.E. and Parris, K.M. (2019) Frog occupancy of polluted wetlands in urban landscapes. *Conservation Biology: The Journal of the Society for Conservation Biology* 33(2), 389–402. DOI: 10.1111/cobi.13210.

Silva, E., Marco, A., da Graça, J., Pérez, H., Abella, E. *et al.* (2017) Light pollution affects nesting behavior of loggerhead turtles and predation risk of nests and hatchlings. *Journal of Photochemistry and Photobiology*173, 240–249. DOI: 10.1016/j.jphotobiol.2017.06.006.

Sims, V., Evans, K.L., Newson, S.E., Tratalos, J.A. and Gaston, K.J. (2008) Avian assemblage structure and domestic cat densities in urban environments. *Diversity and Distributions* 14(2), 387–399. DOI: 10.1111/j.1472-4642.2007.00444.x.

Škvareninová, J., Tuhárska, M., Škvarenina, J., Babálová, D., Slobodníková, L. *et al.* (2017) Effects of light pollution on tree phenology in the urban environment. *Moravian Geographical Reports* 25(4), 282–290. DOI: 10.1515/mgr-2017-0024.

Smallbone, L.T., Luck, G.W. and Wassens, S. (2011) Anuran species in urban landscapes: relationships with biophysical, built environment and socio-economic factors. *Landscape and Urban Planning* 101(1), 43–51. DOI: 10.1016/j.landurbplan.2011.01.002.

Soga, M. and Gaston, K.J. (2016) Extinction of experience: the loss of human-nature interactions. *Frontiers in Ecology and the Environment* 14(2), 94–101. DOI: 10.1002/fee.1225.

Soga, M., Gaston, K.J., Koyanagi, T.F., Kurisu, K. and Hanaki, K. (2016) Urban residents' perceptions of neighbourhood nature: does the extinction of experience matter? *Biological Conservation* 203, 143–150. DOI: 10.1016/j.biocon.2016.09.020.

Sol, D., González-Lagos, C., Moreira, D., Maspons, J. and Lapiedra, O. (2014) Urbanisation tolerance and the loss of avian diversity. *Ecology Letters* 17(8), 942–950. DOI: 10.1111/ele.12297.

Sol, D., Bartomeus, I., González-Lagos, C. and Pavoine, S. (2017) Urbanisation and the loss of phylogenetic diversity in birds. *Ecology Letters* 20(6), 721–729. DOI: 10.1111/ele.12769.

Soulsbury, C.D. and White, P.C.L. (2015) Human–wildlife interactions in urban areas: a review of conflicts, benefits and opportunities. *Wildlife Research* 42(7), 541–553. DOI: 10.1071/WR14229.

Souto, L.A., Listopad, C.M.C.S. and Bohlen, P.J. (2019) Forging linkages between social drivers and ecological processes in the residential landscape. *Landscape and Urban Planning* 185, 96–106. DOI: 10.1016/j.landurbplan.2019.01.002.

Stern, P.C. (2000) New environmental theories: toward a coherent theory of environmentally significant behavior. *Journal of Social Issues* 56(3), 407–424. DOI: 10.1111/0022-4537.00175.

Stone, E.L., Harris, S. and Jones, G. (2015) Impacts of artificial lighting on bats: a review of challenges and solutions. *Mammalian Biology* 80(3), 213–219. DOI: 10.1016/j.mambio.2015.02.004.

Stracey, C.M. (2011) Resolving the urban nest predator paradox: the role of alternative foods for nest predators. *Biological Conservation* 144(5), 1545–1552. DOI: 10.1016/j.biocon.2011.01.022.

Swaddle, J.P., Moseley, D.L., Hinders, M.K. and Smith, E.P. (2016) A sonic net excludes birds from an airfield: implications for reducing bird strike and crop losses. *Ecological Applications: a publication of the Ecological Society of America* 26(2), 339–345. DOI: 10.1890/15-0829.

Swan, C.M., Pickett, S.T., Szlavecz, K., Warren, P. and Willey, K.T. (2011) Biodiversity and community composition in urban ecosystems: coupled human, spatial, and metacommunity processes. In: *Handbook of Urban Ecology by Jari Niemela*. Oxford University Press, New York, pp. 179–186.

Templer, P.H., Toll, J.W., Hutyra, L.R. and Raciti, S.M. (2015) Nitrogen and carbon export from urban areas through removal and export of litterfall. *Environmental Pollution* 197, 256–261. DOI: 10.1016/j.envpol.2014.11.016.

Thurstone, L.L. (1928) Attitudes can be measured. *American Journal of Sociology* 33(4), 529–554. DOI: 10.1086/214483.

Tisdell, C.A. (2010) The influence of public attitudes on policies for conserving reptiles. Working Paper No. 166.. In: Baker, K.J. (ed.) *Reptiles: Biology, Behaviour and Conservation. Nova Science Publishers, Inc*, pp. 115–126.

Tomasevic, J.A. and Marzluff, J.M. (2017) Cavity nesting birds along an urban-wildland gradient: is human facilitation structuring the bird community? *Urban Ecosystems* 20(2), 435–448. DOI: 10.1007/s11252-016-0605-6.

Trammell, T.L.E., Pataki, D.E., Cavender-Bares, J., Groffman, P.M., Hall, S.J. *et al.* (2016) Plant nitrogen concentration and isotopic composition in residential lawns across seven us cities. *Oecologia* 181(1), 271–285. DOI: 10.1007/s00442-016-3566-9.

Tryjanowski, P., Skórka, P., Sparks, T.H., Biaduń, W., Brauze, T. *et al.* (2015) Urban and rural habitats differ in number and type of bird feeders and in bird species consuming supplementary food. *Environmental Science and Pollution Research* 22(19), 15097–15103. DOI: 10.1007/s11356-015-4723-0.

Tuiten, W., Koenraadt, C.J.M., McComas, K. and Harrington, L.C. (2009) The effect of West Nile virus perceptions and knowledge on protective behavior and mosquito breeding in residential yards in upstate New York. *EcoHealth* 6(1), 42–51. DOI: 10.1007/s10393-009-0219-z.

U.S. Department of the Interior, U.S. Fish and Wildlife Service, and U.S. Department of Commerce, US. Census Bureau (2014) *2011 National survey of fishing, hunting, and wildlife-associated recreation*. US Government Printing Office, Washington, DC.

Unfried, T.M., Hauser, L. and Marzluff, J.M. (2013) Effects of urbanization on song sparrow (*Melospiza melodia*) population connectivity. *Conservation Genetics* 14(1), 41–53. DOI: 10.1007/s10592-012-0422-2.

United Nations Department of Economic and Social Affairs, Population Division (2019) *World Urbanization Prospects: The 2018 Revision (ST/ESA/SER.A/420)*. United Nations, New York.

Van Doren, B.M., Horton, K.G., Dokter, A.M., Klinck, H., Elbin, S.B. *et al.* (2017) High-intensity urban light installation dramatically alters nocturnal bird migration. *Proceedings of the National Academy of Sciences of the United States of America* 114(42), 11175–11180. DOI: 10.1073/pnas.1708574114.

Ware, H.E., McClure, C.J.W., Carlisle, J.D. and Barber, J.R. (2015) A phantom road experiment reveals traffic noise is an invisible source of habitat degradation. *Proceedings of the National Academy of Sciences of the United States of America* 112(39), 12105–12109. DOI: 10.1073/pnas.1504710112.

Warren, P.S., Katti, M., Ermann, M. and Brazel, A. (2006) Urban bioacoustics: it's not just noise. *Animal Behaviour* 71(3), 491–502. DOI: 10.1016/j.anbehav.2005.07.014.

Warren, P.S., Harlan, S.L., Boone, C., Lerman, S.B., Shochat, E. *et al.* (2010) Urban ecology and human social organization. In: Gaston, K. (ed.) *Urban Ecology*. Cambridge University Press, Cambridge, UK, pp. 172–201.

Warren, P.S., Lerman, S.B., Andrade, R., Larson, K.L. and Bateman, H.L. (2019) The more things change: species losses detected in Phoenix despite stability in bird–socioeconomic relationships. *Ecosphere* 10(3), e02624. DOI: 10.1002/ecs2.2624.

Wenny, D.G., DeVault, T.L., Johnson, M.D., Kelly, D., Sekercioglu, C. *et al.* (2011) The need to quantify ecosystem services provided by birds. *The Auk* 128(1), 1–14. DOI: 10.1525/auk.2011.10248.

Werrell, P.A., Langellotto, G.A., Morath, S.U. and Matteson, K.C. (2009) The influence of garden size and floral cover on pollen deposition in urban community gardens. *Cities and the Environment* 2(1), 1–16. DOI: 10.15365/cate.2162009.

Whelan, C.J., Wenny, D.G. and Marquis, R.J. (2008) Ecosystem services provided by birds. *Annals of the New York Academy of Sciences* 1134(1), 25–60. DOI: 10.1196/annals.1439.003.

Whelan, C.J., Şekercioğlu, Çağan H. and Wenny, D.G. (2015) Why birds matter: from economic ornithology to ecosystem services. *Journal of Ornithology* 156(S1), 227–238. DOI: 10.1007/s10336-015-1229-y.

Wolkovich, E.M. and Cleland, E.E. (2011) The phenology of plant invasions: a community ecology perspective. *Frontiers in Ecology and the Environment* 9(5), 287–294. DOI: 10.1890/100033.

Yanes, M., Velasco, J.M. and Suárez, F. (1995) Permeability of roads and railways to vertebrates: the importance of culverts. *Biological Conservation* 71(3), 217–222. DOI: 10.1016/0006-3207(94)00028-O.

Youngsteadt, E., Henderson, R.C., Savage, A.M., Ernst, A.F., Dunn, R.R. *et al.* (2015) Habitat and species identity, not diversity, predict the extent of refuse consumption by urban arthropods. *Global Change Biology* 21(3), 1103–1115.

Ziter, C. (2016) The biodiversity-ecosystem service relationship in urban areas: a quantitative review. *Oikos* 125(6), 761–768. DOI: 10.1111/oik.02883.

4 Urban Plant Ecophysiology

Nancy Falxa Sonti*
USDA Forest Service, Baltimore, Maryland

Introduction

Plants have long been cultivated to improve quality of life in dense human settlements, mitigating the environmental stresses of urban living. Urban landscape elements include gardens, trees and lawns designed to provide aesthetic and functional benefits to local residents, as well as urban natural areas that reflect the native biome vegetation. Different types of informal green space are typically found in interstitial urban areas wherever plants find space, light, water and nutrients to grow (Rupprecht and Byrne, 2014). A growing body of literature evaluates the health and well-being benefits of these diverse types of intentional and unintentional urban nature, and advocates for their inclusion in sustainable urban design (Konijnendijk *et al.*, 2013; Kowarik, 2018; Threlfall and Kendal, 2018). But how does the urban environment impact plant physiological function, whether cultivated or not, native or introduced species, across the range of habitats found within a metropolitan area? And how might those impacts affect the ability of urban plants to perform the ecosystem services desired by urban residents?

Plant ecophysiology (or physiological plant ecology) is a field of study concerned with the function and performance of plants under constraints imposed by their growing environment. Ecophysiological research is integral to understanding and managing the fluxes of heat, water, gases and nutrients that underlie urban ecosystem science and that help make cities both liveable and sustainable (Alberti, 2005). The past few decades have seen a rise in research on plant community ecology, but ecophysiological studies have lagged behind, possibly due to methodological challenges, or due to the recent popularity of other topics in plant biology (Beyschlag and Ryel, 2007).

A systematic approach to urban plant ecophysiology that is tied to decision making can support efforts to improve both liveability and sustainability of cities via plant physiological function. Plants are the foundation of most nature-based solutions to environmental, social and economic challenges, and physiological function is the engine that drives the provision of associated ecosystem services. Knowledge about variation in plant physiological function across genotypes, species, urban site types or regions that accounts for differences in management intensity or land use history will help inform the continued (and equitable) provision of these services. This chapter presents a comprehensive approach to the study of urban plant ecophysiology and provides a framework for future study, examining existing research, methodological challenges and linkages to nature-based solutions in cities.

*nancy.f.sonti@usda.gov

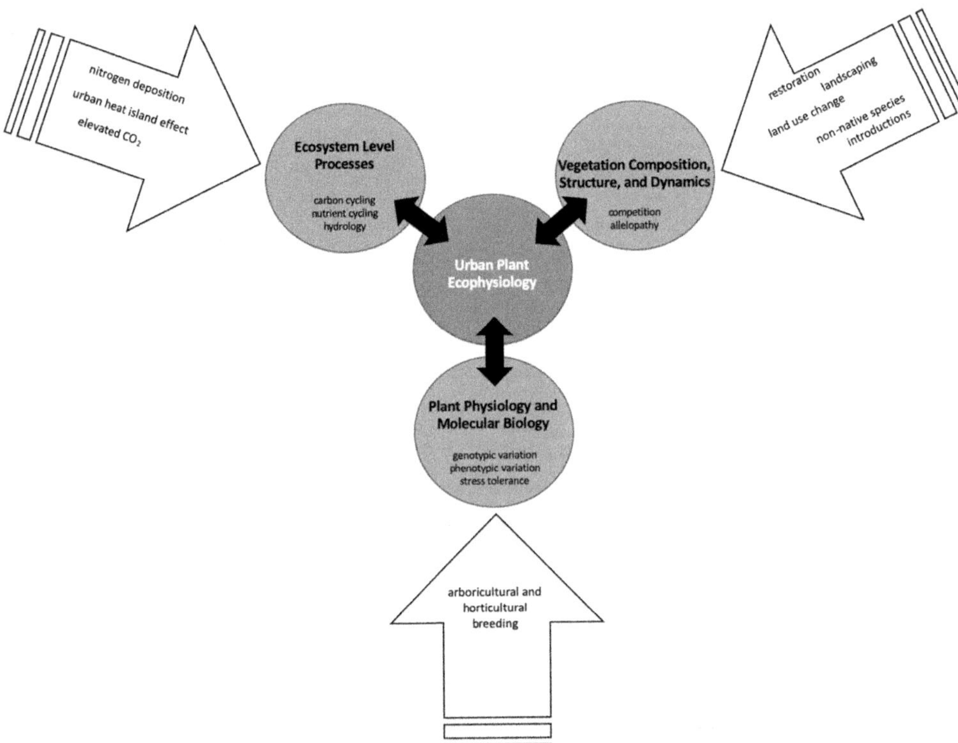

Fig. 4.1. A framework for understanding urban plant ecophysiology in relation to other fields of study and related social-ecological processes. Circles represent the major disciplines in plant biology and include examples of phenomena studied within each discipline. Black arrows represent the feedbacks between urban plant ecophysiology and these ecosystem, community or organismal processes. Clear arrows illustrate examples of social and biophysical processes that impact urban ecosystems at different scales of space and time through direct or indirect human activity. (Adapted from Beyschlag and Ryel, 2007 to include examples of phenomena within each discipline and of social and biophysical processes)

Approaches to the Study of Urban Plant Ecophysiology

The study of plant ecophysiology in any environmental context creates an important link between scales and disciplines in plant science (Beyschlag and Ryel, 2007; Fig. 4.1). Research in plant ecophysiology addresses spatial scales from organelles to ecosystems and corresponding timescales from seconds to millennia across biochemical, physiological, ecological, and evolutionary processes (Osmond *et al.*, 1980). In the urban context, plant ecophysiology provides a mechanistic understanding underlying current advances in the study of urban ecosystem fluxes (e.g. Reisinger *et al.*, 2016; Menzer and McFadden, 2017) and plant community ecology (e.g. Aronson *et al.*, 2016; Kowarik and von der Lippe, 2018).

Predictions of individual plant, community and ecosystem responses to social-ecological urban environmental factors are only possible with knowledge of the underlying physiological mechanisms. Figure 4.1 illustrates examples of social and biophysical processes that impact urban ecosystems at different scales of space and time through direct or indirect human activity. Some processes, such as the urban heat island effect or non-native species introductions have been well studied across many urban areas. In contrast, the processes of natural selection compared to domestic selection through arboricultural and horticultural practices and resulting impacts on plant ecology are an emerging area of research that has received little attention.

Until now, the study of urban plant physiology has been dominated by the paradigm of urban environmental stress as well as dominated by the urban-to-rural gradient approach. The two concepts may be combined, such that an urban–rural gradient is described as a 'stress gradient' (Calfapietra *et al.*, 2015). Although these paradigms have provided an initial approach to urban plant ecophysiology and resulted in important research findings, they are incomplete and reveal the need for a more comprehensive framework.

It is commonly assumed that urban plants must acclimate or adapt to conditions of 'permanent stress' (Calfapietra *et al.*, 2015). Indeed, there are many scenarios in which plants experience stress in urban environments, resulting in less vegetative growth, reduced flowering and seed production, failure to germinate, or mortality. The harsh growing conditions of street trees in particular have been studied for several decades (Bassuk and Whitlow, 1987; Cregg and Dix, 2001; Meineke *et al.*, 2013), leading to a search for stress-tolerant tree species suitable for highly developed urban sites (Sjöman and Busse Nielsen, 2010). Street tree pits surrounded by pavement are known to provide particularly stressful environments due to soil compaction, high soil pH, soil solution chemistry affected by dog urine and de-icing salts, waterlogging, lack of water, air pollution, high summer temperatures, and mechanical damage from people and vehicles (Whitlow and Bassuk, 1988). However, there are many environmental constraints to plant growth that are reduced in urban environments, depending on the plant species and site type (Table 4.1). While highly visible, the street tree pit represents only a fraction of urban tree sites, let alone the habitats of all vegetation contained within a city. Trees planted in lawn strips, residential gardens, parks, institutional grounds, or naturally regenerating across a variety of public and private land uses may have access to more light, nutrients and water than trees in a rural forested condition, and the added heat or heavy metals may not be enough to negatively impact physiological function, depending on the species and site type (Iakovoglou *et al.*, 2001; Pretzsch *et al.*, 2017; Sonti, 2019). Even street trees are likely to experience less competition and reduced herbivory from deer and other mammals compared to forest-grown trees. The variety of plant species and site types found within cities means that various abiotic and biotic aspects of the urban environment may be stressful or beneficial to plant physiological function.

Urban influences on environmental constraints do not occur in isolation, but interact to determine plant growth, physiological function and reproductive success, which in turn influence community composition and ecosystem fluxes. For example, research from the eastern and midwestern USA shows that insect herbivore density increases with elevated urban temperatures (Youngsteadt *et al.*, 2015), and that drought stress can predispose urban trees to insect pest infestation (Cregg and Dix, 2001). Management practices, including irrigation and pruning, reduce water-use efficiency and lead to sunscald on urban trees in the arid south-west USA (Martin and Stabler, 2004; Martin and Stutz, 2013). However, interactions between urban environmental factors can also reduce constraints to plant growth and function. Nitrogen availability in urban conditions may interact with changes in temperature and water availability to either enhance or limit tree growth (Searle *et al.*, 2011; Osone *et al.*, 2014). A nuanced view of environmental constraints across plant species, cities and urban site types will allow for a more accurate picture that reflects the values of urban green spaces for ecosystem function.

A second common approach to urban plant physiology has been the use of urban-to-rural gradients to understand the influence of the urban environment on plant physiology. This approach can help gain insight into future global change conditions (Gregg *et al.*, 2003; Searle *et al.*, 2012; Calfapietra *et al.*, 2015). The gradient approach can be a useful construct for examining the impact of large-scale anthropogenic factors such as CO_2 emissions, nitrogen deposition, changes in ozone concentration and precipitation patterns (Lahr *et al.*, 2018a). Urban–rural gradient experiments conducted using uniform soil conditions (in pots or constructed plots) demonstrate that the elevated temperatures, increased levels of atmospheric

Table 4.1. Urban influences on abiotic and biotic environmental factors affecting plant growth, including direct and indirect human impacts.

Environmental factors	Urban influence
Abiotic	
Radiation	Shade from tall buildings or planted trees Reflected UV radiation from built surfaces Reduced UV radiation due to air pollution
Temperature	Urban heat island effect Soil freezing affected by soil compaction; snow/leaf litter removal
Water	
Humidity	Wind tunnels Urban heat island effect
Soil moisture	Changes in soil organic matter Soil compaction Flashy precipitation and stormwater run-off Engineered stormwater controls Stream channelization
Gas	
Oxygen deficiency	Modified flooding frequency Soil compaction
Ozone exposure	Elevation or depletion
Atmospheric CO_2	Elevation from local point sources Modified seasonal dynamics
Particulate matter	Increased exposure and deposition
Chemicals/minerals	Nutrient excess, deficiency, or imbalance from pollution or fertilizer Altered pH from fertilizer or anthropogenic construction materials Changes in salinity from road salt or altered coastal flooding patterns Increased heavy metals Pesticide or herbicide application
Mechanical effects	
Wind	Wind tunnels or wind breaks from tall buildings
Burial	Construction activities Planting practices
Snow and ice cover	Urban heat island effect Snow clearing
Fire	Fire suppression Arson
Vegetation management	Pruning Weeding Lawn mowing
Biotic	
Plants	Reduced competition through vegetation control Increased competition from invasive species or planting Allelopathy from introduced species

Continued

Table 4.1. Continued

Environmental factors	Urban influence
Microorganisms	Introduced diseases (virus, bacteria, fungi) Loss of mycorrhizal fungi Changes in microorganism biodiversity Altered rates of nitrification
Animals	Herbivory from insect pests (native and introduced) Trampling and soil compaction from pets Herbivory and seed predation by urban mammals (e.g. deer, rodents) Increased faeces and urine from pets

CO_2 and/or reduced ozone found in urban areas can lead to greater plant productivity (Gregg *et al.*, 2003; George *et al.*, 2009; Searle *et al.*, 2012). Mature urban trees growing *in situ* are also more productive than their rural counterparts across both open-grown and forested settings (Pretzsch *et al.*, 2017; Sonti *et al.*, 2019), although tree longevity may be reduced in urban areas (Smith *et al.*, 2019). This increase in productivity has been observed using remote sensing as well (Zhao *et al.*, 2016). Urban conditions can also enhance growth of species that are problematic for human health, including ragweed and poison ivy (Ziska, 2003; Ziska *et al.*, 2007). Urbanization is also known to alter nutrient availability, with implications for foliar nitrogen content and physiological function (Vallano and Sparks, 2013; Falxa-Raymond *et al.*, 2014;; Trammell *et al.*, 2016). However, urbanization gradients are applied inconsistently across studies and can be an oversimplified and misleading concept in many situations (McDonnell and Hahs, 2008; Raciti *et al.*, 2012a; Short Gianotti *et al.*, 2016). Urban–rural gradients are inadequate to capture many of the environmental factors affecting plant physiology that vary on small spatial scales within cities and suburbs. These factors include impervious surfaces, fertilizer or pesticide application, soil disturbance history, and the influence of domestic versus natural selection.

By focusing on terrestrial and aquatic patches within cities and suburbs as analogues of non-urban habitats, the gradient approach tends to examine ecology *in* the city, rather than a more comprehensive ecology *of* the city that treats the entire urban mosaic as a social-ecological system (Pickett *et al.*, 2016). As illustrated in Fig. 4.2, differences in human-management intensity and land-use history both within a city and across urban–rural gradients will impact both plant community composition and physiological function across different site types. This approach provides a comprehensive framework in which to design specific studies of urban plant ecophysiology. For example, in Baltimore, Maryland, we find vacant lots with ruderal plant species and patches of closed canopy forest (Avins, 2013; Johnson *et al.*, 2018); both are examples of sites with little management or cultivation, but their species composition and ecophysiological function vary greatly due to the history of development or disturbance on that site. Even within a particular type of urban green space, there may be a great deal of variation in environmental conditions due to land-use history. For example, urban soils collected from several New York City forest restoration sites supported significantly different levels of tree seedling health and growth (Pregitzer *et al.*, 2016). In addition, we may find particular site types, such as a lawn or a patch of forest, that occur within urban, suburban and rural areas. Rather than frame research questions around an urbanization gradient, it may make more sense to conceptualize and compare sites within cities or metropolitan areas that vary according to specific combinations of environmental parameters (McDonnell and Hahs, 2008).

Within urban areas, there is a variety of formal and informal green spaces that provide habitat for plant life (Kowarik, 2011; Rupprecht and Byrne, 2014). Management practices vary across these sites, leading to differences in physiological function (Fig. 4.2). For example, water use (measured as sap flux) varies between native and non-native tree species, as well as between natural, irrigated, unirrigated and street tree

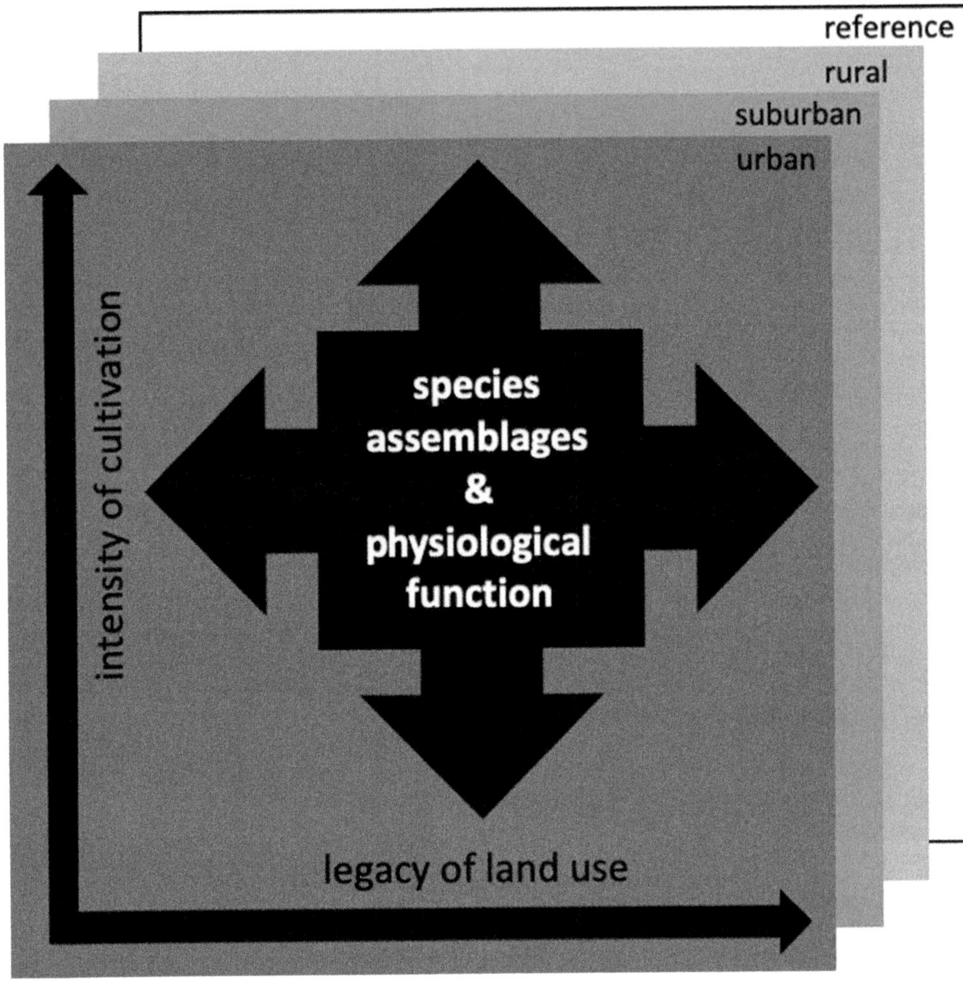

Fig. 4.2. Urban plants grow in a variety of site types that vary along axes of land-use history and intensity of cultivation or management, leading to differences in community composition and physiological function. These site types are found throughout the urban-to-rural gradient and can be utilized to construct or refine experimental studies of plant ecophysiological function.

sites in Los Angeles, California (McCarthy and Pataki, 2010; Pataki *et al.*, 2011). Research has also found that landscaping choices may have socio-economic drivers (Kinzig *et al.*, 2005) and result in different ecophysiological outcomes. Empirical models of turfgrass and tree evapotranspiration combined with remotely sensed estimates of Los Angeles vegetation cover reveal that coniferous and palm tree species contribute little transpiration compared to angiosperms, and that total modelled evapotranspiration is correlated with median household income (Litvak *et al.*, 2017).

Private residential lands are beginning to receive recognition from ecologists as one of the most prevalent land uses in cities, with important consequences for urban biodiversity and ecosystem function (Locke *et al.*, 2018; Pearse *et al.*, 2018). Lawns in particular dominate urban green space land cover across many cities (Ignatieva *et al.*, 2015), and much remains to be learned about their role in urban ecosystem functioning. Microclimate factors such as shading may have a strong influence on turfgrass water use, which has implications for lawn irrigation practices (Litvak and Pataki, 2016). Analysis of residential lawns across the USA suggests that nitrogen additions interact with overall rates of nitrogen cycling and losses to drive foliar nitrogen concentration and isotopic composition (Trammell *et al.*, 2016). Carbon assimilation was found to be greater in arid residential landscapes than in the surrounding native desert across different plant life forms (Martin and Stabler, 2002).

In addition to gradients of urbanization, it is important to consider the impact of latitudinal gradients and range limits of both native and introduced species found in urban areas. For example, elevated urban temperatures may switch from beneficial to harmful in the lower latitudes of a species's range (Ghannoum and Way, 2011). Plant ecophysiology also underlies theory and practice related to the role of cities in plant migration related to climate change (Woodall *et al.*, 2010). The role of urban landscapes in plant migration and species range distributions deserves further study, particularly in light of projected changes in climatic conditions influencing ecophysiological function.

Across biomes and land uses, root function and other below-ground processes are notoriously difficult to measure, and this is even more true in urban systems where soil is often locked under impervious surfaces and destructive sampling is discouraged (Raciti *et al.*, 2012b). However, below-ground processes are a critical link in understanding urban plant productivity and ecosystem function (Högberg and Read, 2006). Research from several cities across the USA suggests that paved surfaces can reduce tree growth due to increased rhizosphere temperatures, among other factors (Martin *et al.*, 2002; McClung and Ibáñez, 2018). Soils beneath pavement may become compacted without the use of structural materials to maintain pore spaces (Grabosky *et al.*, 2009). Soil compaction is considered a major stress to urban plants, potentially restricting root access to oxygen and water (Mullaney *et al.*, 2015). Urban soils may also become waterlogged if drainage is inadequate, causing a lack of oxygen in the root zone. De-icing salts can cause 'chemical drought' to roots; chloride ions taken up by plant roots can accumulate to toxic levels in the leaves and result in leaf necrosis (Bassuk and Whitlow, 1987). Urban fill contains cement with limestone and is often alkaline, which limits availability of nutrients like iron and manganese in some urban sites. However, less heavily disturbed urban soils may provide greater nutrient availability than surrounding native ecosystems (Sonti *et al.*, 2019). Urban soils from a variety of land uses may support lower rates of tree root colonization by mycorrhizal fungi, which provide increased water and nutrient absorption to the host plant (Martin *et al.*, 2002; Karpati *et al.*, 2011; Tyburska *et al.*, 2013). The impacts of urbanization on root herbivory also remain largely unexplored (Rasmann and Agrawal, 2008).

Urban plant populations and communities are shaped by the forces of both natural selection and domestic selection (i.e. artificial selection, propagation and dissemination of plant genotypes with traits deemed advantageous or desirable for human use). Species or cultivars planted in gardens and other highly cultivated sites are selected for specific traits related to their physiological function (shade, water use), aesthetic appeal (flowers, fall foliage) and/or ease of maintenance (male dioecious trees that do not produce seeds or fruit) (Cariñanos and Casares-Porcel, 2011; Pataki *et al.*, 2013; Avolio *et al.*, 2018). In sites with less maintenance, conventional traits related to natural selection may be more relevant to reproductive success (Johnson *et al.*, 2018). The success of ubiquitous urban species such as *Ailanthus altissima* and *Ginkgo biloba* may be explained by ecophysiological traits allowing for high rates of photosynthesis and water-use efficiency in urban growing conditions (Hamerlynck, 2001; Osone *et al.*, 2014). There is little research comparing physiological function of ornamental species compared to the native wildtypes they replace across different plant functional groups (Lahr *et al.*, 2018b) compared to photosynthesis and stomatal conductance rates among red maple urban-planted cultivars and wildtype trees, finding that cultivars had higher maximum rates of photosynthesis but lower water-use efficiency than wildtype trees. Red maple is a common urban tree species that occurs naturally in urban-to-rural landscapes through the eastern USA and has also been used to develop dozens of cultivars available for purchase in nurseries. However, cultivars and varieties of many other native species are available to urban residents and land managers, and it is not clear how these domestic selection processes and the resulting genotypic and phenotypic variation in urban plant populations impact physiological function across the urban landscape.

In addition to below-ground processes and horticultural practices, there are other urban environmental factors affecting plant growth (Table 4.1) that have received little attention in the literature. For example, although it has been suggested that wind tunnels caused by tall buildings may increase leaf desiccation (Bassuk and Whitlow, 1987), there is far more research about the role of trees as wind breaks in urban areas (Trowbridge and Bassuk, 2004). Presumably, if trees are strategically placed to block strong winds, there may be impacts to growth and physiological function. The ability of tall buildings to block and reflect light may also impact plant growth form or phenology.

Social-ecological factors may influence urban plant ecology in completely different ways across humid versus arid cities, necessitating different approaches to ecophysiological study. In temperate regions of North America and Europe, many cities have reduced tree canopy cover compared to the native forest biome they replaced, often leading to general changes such as more sunlight and less soil organic matter. However, arid cities generally have more tree cover than the surrounding native ecosystem, resulting in more shade and soil organic matter. In addition, regional differences in irrigation, fertilizer application and pesticide application practices affect plant physiological function differently across local land uses and site types. Comparative plant ecophysiology research across urban areas will be critical to determine whether these environmental impacts are similar across urban geographies (Sonti, 2019). Urban site conditions of different regions vary according to the unique sets of interacting human and biophysical variables at each location, including latitudinal gradient, physiography and land-use history, requiring careful consideration of each of these factors and their interactions. Traditionally, the study of ecophysiology tends to focus on either model plant systems (e.g. *Arabidopsis thaliana*), natural systems, or crops (Ainsworth *et al.*, 2016). In this context, urban systems may be viewed as natural systems, or may be akin to agricultural systems, depending on the level of cultivation, intensity of continued maintenance, and expected performance of the urban plants in question.

Methodological Considerations for Field Studies of Urban Plant Physiology

The focus of plant ecophysiology is often on the scale of a single leaf and extends up to a whole plant, although the discipline includes techniques for examining processes from molecular to global scales (Pearcy *et al.*, 1989). Field methods include assessments of all of the environmental factors listed in Table 4.1 and their resulting impacts on myriad physiological processes including gas exchange, nutrient uptake, water use, carbon allocation, growth and reproduction. Field methods in plant ecophysiology have been adapted to urban conditions with varying levels of success, and there may be opportunities to develop new directions and research techniques suited to urban ecology (Fig. 4.3).

As the study of urban ecology has advanced, field researchers are recognizing the need to include sites beyond formal protected areas such as parks or other institutional grounds. However, fieldwork on private urban land or informal green spaces requires careful precautions to ensure security of equipment, researchers and the local community. Studies of urban plant biodiversity and ecosystem fluxes have outpaced research on urban plant physiology, which may be partly due to the challenges associated with plant ecophysiology methods and their reliance on expensive instrumentation. A study of common problems experienced by urban ecology researchers includes several issues of particular relevance to plant physiological ecologists, who often need to destructively sample plants and/or leave expensive equipment on site that is not easy to hide (Dyson *et al.*, 2019). Though rare, property damage may occur during collection of tree cores or other plant tissues, as well as during soil sampling or root studies, making it particularly difficult to receive permission to conduct

Urban Tree Ecophysiology in Action

Assessing the tree health impacts of salt water flooding in coastal cities: a case study in New York City

Urban tree specific gravity and ash content: a case study from Baltimore, Maryland USA

In this study, researchers evaluated the health of two common tree species (*Acer rubrum* and *Platanus × acerifolia*) after saltwater flooding associated with Hurricane Sandy in New York City. A complete street tree inventory allowed for selection of target tree species meeting the study's minimum diameter requirements located adjacent to certain land uses both inside and outside of the Hurricane Sandy inundation zone. Tree health and physiology measurements included visual health assessments and collection of five sun leaves from each tree canopy. These data collection procedures caused minimal impact to the tree and did not pose substantial safety risks to the researchers or surrounding community, and so permission was easily granted from the NYC Department of Parks & Recreation. Sun leaves were easily accessed on these open grown street trees, with an arborist throw bag used to collect foliage from particularly tall trees. However, the assessment of dynamic physiological traits, including leaf discoloration, chlorophyll fluorescence, and foliar nutrients necessitated data collection limited to a two-week period during the growing season in order to ensure that results were comparable across the entire study population for a given sampling year.

Hallett R, Johnson ML, Sonti NF. 2018. Assessing the tree health impacts of salt water flooding in coastal cities: A case study in New York City. Landscape and Urban Planning 177: 171-177.

To examine potential differences in wood properties (specific gravity and ash content) between trees grown in forest and urban settings, wood samples were collected from various tree species in the city of Baltimore, MD and compared with published values from studies based on forest-grown trees. In this case, the lack of a street tree inventory necessitated a convenience sample that reduced generalizability across site types . Tree removal logs provided by the Baltimore City Forestry Division allowed researchers to identify candidate stumps for wood sample collection within a 2-8 week period after the tree was removed. However, the lack of detailed species and location information in the tree removal work orders meant that more field work was required to locate appropriate tree stumps. In addition, the need for chainsaw operation to obtain wood samples required extra safety precautions and proof of certification before permission was granted to begin field work. However, the lag time between tree removal and stump grinding combined with the stable nature of wood samples (compared to foliage) provided a flexible timeframe for field work, and analysis of wood properties was conducted in the lab without time constraints.

Westfall JA, Sonti NF, Wiemann MC, Eberhardt TL, So C-L. 2020. Urban tree specific gravity and ash content: a case study from Baltimore, Maryland USA. *Urban Forestry & Urban Greening* 48: 126556.

Fig. 4.3. Two street tree studies conducted by scientists from the USDA Forest Service Northern Research Station illustrate the challenges and feasibility associated with design and implementation of ecophysiological research methods in urban contexts (Hallett *et al.*, 2018; Westfall *et al.*, 2020)

these types of studies on some urban land uses or ownerships. Because small-scale management practices can impact individual plant function, there is also a need to control or document these types of activities during the course of a research project. For example, a study of residential tree water use or lawn nutrient dynamics will be impacted by irrigation or fertilizer application, so it is important to communicate with property owners about the frequency of these activities or request that they suspend management for the duration of the research. This is not a unique consideration for urban research, but the large number of property owners involved in a given urban ecology study makes these types of communications and management activities more complex.

Traditional ecophysiology methods include instrumentation of individual plants to measure physical and chemical parameters (Pearcy *et al.*, 1989). For example, measurements of tree sap flux or leaf-level gas exchange require expensive equipment to be attached to plants *in situ* and left in place anywhere from hours to months at a time. As a result, many of these studies have been conducted in protected urban sites such as an arboretum or university campus (but see, for example, Osone *et al.*, 2014; Lahr *et al.*, 2018b). Access to tree canopies is a challenge in many ecophysiological studies, but the urban setting may preclude the use of techniques like scaffolding to reach the canopy or shotgun use to collect foliage. Pre-dawn measurements of chlorophyll fluorescence or leaf-water potential may be challenging if sites are not considered safe in the dark. At the same time, urban sites are generally more accessible and less remote than wilderness areas, making it easier to transport heavy equipment or find supplies to make repairs.

Methods that require collection of a minimal amount of foliage can be successfully employed in urban areas, including assessments of foliar nutrients, isotopic composition, chlorophyll fluorescence, or chlorophyll content.

For example, foliar nitrogen concentration and isotopic composition of residential lawns may be associated with fertilization practices and housing age and may differ from associated native ecosystems in many cities (Trammell *et al.*, 2016). Nitrogen and oxygen isotope ratios of winter annual plants in Los Angeles, CAalifornia, vary with the spatial distribution of atmospheric NO_2 and ozone, suggesting that these pollutants influence stomatal conductance (Wang and Pataki, 2010). Foliar nitrogen concentrations and isotope signatures may also reveal differences in species responses to altered nitrogen availability in urban green spaces, with implications for differential growth and survival (Falxa-Raymond *et al.*, 2014).

Growth rates of woody plants across urban site types and management regimes may be assessed easily using repeated measurements of stem diameter, or sometimes through increment cores when permission is granted (e.g. Pretzsch *et al.*, 2017; Bialecki *et al.*, 2018). Because of the difficulty in assessing tree root growth and function in urban sites, some researchers have made use of ground-penetrating radar to assess rooting depth in different site conditions (Bassuk *et al.*, 2011; Grabosky and Bassuk, 2016).

One approach to studying ecophysiology along an urbanization gradient is to use potted plants that may be grown and harvested without harming existing vegetation. These types of studies measure plant physiological responses to atmospheric factors such as ozone, carbon dioxide, temperature and nitrogen deposition, but do not account for below-ground influences of the urban environment (e.g. Gregg *et al.*, 2003; Searle *et al.*, 2012; Vallano and Sparks, 2013). Of course, greenhouses, growth chambers and experimental plots may also be used to simulate urban environmental conditions, avoiding many of the difficulties of urban ecophysiology field research (e.g. Mueller and Day, 2005; Bartens *et al.*, 2008; Searle *et al.*, 2011, 2012; Rahman *et al.*, 2014; Pregitzer *et al.*, 2016). These methods also have the advantage of direct experimental manipulation and subsequent determination of causal relationships between environment and plant physiological response, which is difficult to achieve in the urban setting. However, the applicability of these results to urban field conditions may be more limited.

Flux towers can be used to examine the influence of urban vegetation on atmospheric CO_2 fluxes (Briber *et al.*, 2013), particularly when combined with empirical data such as leaf-level gas exchange (Ng *et al.*, 2015) or tree-level carbon sequestration estimates (Velasco *et al.*, 2016). Flux tower data has also shown that tree canopy can augment atmospheric elemental inputs to urban ground surfaces (Decina *et al.*, 2018). Although flux towers are large installations that may be difficult to establish initially, they can be used to collect continuous long-term data unobtrusively once they are installed.

High spatial resolution remote sensing may also provide insight into urban plant ecophysiology. For example, evapotranspiration rates of different types of urban vegetation cover may be associated with Normalized Difference Vegetation Index (NDVI) values derived from high spatial resolution imagery (Nouri *et al.*, 2013); although, at a lower spatial resolution, Moderate Resolution Imaging Spectroradiometer (MODIS) data have been used to demonstrate the effect of the urban heat island on vegetation phenology (Zhang *et al.*, 2004) as well as to estimate net primary productivity in urban areas (Lu *et al.*, 2010; Yan *et al.*, 2018). High-resolution hyperspectral imagery and LiDAR data may be used to map tree species, health status, leaf area index (LAI) and carbon storage (Alonzo *et al.*, 2016; Pontius *et al.*, 2017). Ground-based LiDAR can also be used to model urban tree biomass in a non-destructive manner (McHale *et al.*, 2009). Unmanned aerial vehicle (UAV) remote sensing also has great potential for ultra-high-resolution (both spatial and temporal) mapping of vegetation structure and function in complex urban landscapes provided that safety and privacy concerns of this emerging technology are addressed (Feng *et al.*, 2015; Gallacher, 2016). The increased availability of high-resolution data about the structure and arrangement of urban vegetation in relation to the built environment also allows for improved design of plant ecophysiology studies. For example, mapping of woody and herbaceous vegetation in front versus backyards in Boston, Massachusetts, would allow for investigation of physiological function of these different vegetation types across residential management contexts (Ossola *et al.*, 2019).

Urban Plant Ecophysiology as the Foundation of Nature-based Solutions

The field of plant ecophysiology has long been central to the mitigation of anthropogenic global change factors, as well as maintaining humankind's supply of goods and service in the face of increasing environmental stress (Ainsworth *et al.*, 2016). Thus, the discipline is a natural fit with urban ecology, which also seeks to address issues of sustainability, environmental quality and human well-being (McDonnell and MacGregor-Fors, 2016). Using an approach which has been termed ecology *for* the city, interdisciplinary scientists seek to inform the design, planning, construction and management of cities and towns through the study of the structure and function of urban ecosystems (Childers *et al.*, 2015). The provision of such evidence-based knowledge is essential for the success of 'nature-based solutions' that aim to promote public health and safety, enhance quality of life, and restore natural hydrologic and ecological processes in urban areas (Nesshöver *et al.*, 2017; Keeler *et al.*, 2019). Cost-effective, nature-based solutions are often presented as 'green infrastructure' that provide additional social-ecological co-benefits, unlike costly technological 'grey infrastructure'. Plants provide the foundation of most nature-based solutions, and a comprehensive understanding of their physiological functioning in relation to current and future urban environmental conditions will allow for more effective design, implementation and maintenance of these ecological features. Of course, it is important to consider the ways in which interactions among social, ecological and technological factors (including climate change) will impact the efficacy with which plants provide urban ecosystem services and disservices (Keeler *et al.*, 2019).

As the use of nature-based solutions becomes more widespread in urban areas, there is a need to evaluate the current physiological performance of plants within these installations as well as the suitability of species and cultivars or varieties of plants for future installations. To this end, Ordóñez *et al.* (2019) have developed a performance assessment framework, using street trees as a model nature-based solution. Unsurprisingly, tree ecophysiology is central to the conceptual and methodological framework, as the performance indicators centre around measures of tree physiological performance and the abiotic and biotic environmental conditions that influence it (Ordóñez *et al.*, 2019). Research by Ballinas and Barradas (2015,2016) illustrates the utility of this approach, showing that measurements of transpiration and stomatal conductance may be used to inform urban tree-planting arrangements that will dissipate the greatest possible amount of heat, mitigating the urban heat island. Similarly, water-use efficiency can be used as a metric of urban ecosystem service provision by different species in arid cities, given that it captures the trade-off between plant water use and growth (McCarthy *et al.*, 2011).

Ecophysiological function of certain nature-based solutions has been studied more thoroughly than others, including green roofs. In these highly designed settings, species selection and management practices are extremely important in order to achieve the desired physiological performance and environmental outcomes. For example, research shows that green roof substrate design, vegetation type and species have been found to affect plant performance (e.g. photosynthesis or water use) and/or green roof performance (water run-off) (VanWoert *et al.*, 2005; Nagase and Dunnett, 2012; Starry *et al.*, 2014). Green roof plant functional traits may also impact provision of ecosystem services such as canopy density, substrate organic matter, substrate temperature index, and substrate nutrient retention (Xie *et al.*, 2018). Selection of green roof plant species based on evapotranspiration rates provides a tangible example of the use of ecophysiological properties to inform design of nature-based solutions (Kemp *et al.*, 2019). Interactions between species are also important in these systems. For example, mycorrhizae may be introduced into green roofs or other nature-based solutions to enhance productivity and performance (John *et al.*, 2017). Sedum groundcover may be used strategically to enhance performance and phenolic concentrations in edible plants used as part of a sustainable urban agriculture green roof programme (Ahmed *et al.*, 2017).

Phytoremediation systems are another example of nature-based solutions that use more advanced species and genotype selection processes in order to achieve specific ecosystem service outcomes. Dozens of plant species have been identified as hyperaccumulators of heavy metals and at least ten uptake mechanisms have been identified as part of a comprehensive model of phytoremediation technology (Mahar *et al.*, 2016). Salicaceae species are particularly well studied, with a combination of field and laboratory trials exploring the growth and physiology of *Populus* and *Salix* spp. genotypes used in phytoremediation of heavy metals and organics (Marmiroli *et al.*, 2011). Further research is needed on both intraspecific and interspecific variation in productivity across a range of urban conditions in order to maximize plant function and associated performance of nature-based solutions.

Conclusions: Future Directions for Urban Plant Ecophysiology Research

In 1988, Whitlow and Bassuk published 'An Agenda for Urban Ecophysiological Research', focused on street trees, which are highly visible and managed components of many cities, and which provide important benefits to urban residents (Whitlow and Bassuk, 1988). Many of the questions they identified remain to be answered, both for street trees and other types of urban vegetation. However, our increased recognition of the value of urban plants across a diverse range of habitats, and their central role in the function of nature-based solutions, requires an expansion of this research agenda. Over half of the world's population already lives in urban areas and it is projected that 60% will live in cities by 2030 (United Nations *et al.*, 2018). Ambitious urban greening agendas worldwide recognize the value of urban vegetation in sustaining and enhancing human and ecosystem health and well-being, particularly in the face of changing climate conditions (Tan *et al.*, 2013; Anguelovski *et al.*, 2018). Successful implementation of these greening initiatives will require a mechanistic understanding of plant productivity that is critical to the design and management of urban green spaces.

Several areas of urban ecophysiological research have received little attention but have particular potential to inform the practice of urban natural resource management. There is a need to elucidate the impacts of domestic selection on genotypic and species diversity of urban plants and on associated ecosystem function. Continued research on both inter- and intraspecific variation in physiological function of native and introduced urban plants can help inform planning and design of urban landscapes. The ability to match plant species and genotypes with specific urban site conditions can help reduce the resources needed to sustain urban landscapes that provide a high quality of life to residents. There is also a need to quantify plant performance and physiological function across a range of urban site types, from the most intensively managed to urban wilderness areas. When informed by an ecophysiological approach, patch analysis of the urban mosaic using remotely sensed data may be used to estimate the extent of urban vegetation site types, with implications for ecosystem function at a metropolitan or regional scale (e.g. Cadenasso *et al.*, 2007). In addition, plant growth in each of these social-ecological contexts will lead to different positive and negative outcomes for human health and well-being. Research from landscape architecture and social science perspectives can provide complementary information about human values and perceptions of urban vegetation form and function (e.g. Heynen *et al.*, 2006; Jansson *et al.*, 2013). Together, social and ecophysiological research can inform optimization of the trade-offs between resources (carbon emissions, water) and ecosystem services or disservices (biophysical and socio-cultural).

Cities are dynamic ecosystems made up of a mosaic of heterogeneous sites that may be heavily managed or not managed at all. Plants may be integrated into urban landscapes in highly intentional ways with expected or unanticipated functional outcomes (Pataki, 2015), or they may exist and function in ways that are not intentional at all. As large-scale urban greening initiatives become increasingly widespread and the impact of their changes to the urban landscape are borne out, a greater understanding of plant ecophysiology will provide insight into carbon, water and nutrient dynamics essential to sustaining life in cities. Urban ecology and plant ecophysiology are both areas of study that cross scales and disciplines in order to describe mechanisms underlying observed ecological patterns, but also to inform desired ecosystem function and related impacts on human health and well-being. Advancing an urban plant ecophysiology research agenda can help support the design and implementation of more sustainable and resilient cities in the face of complex and uncertain future social-ecological conditions.

References

Ahmed, S., Buckley, S., Stratton, A.E., Asefaha, F., Butler, C. *et al.* (2017) Sedum groundcover variably enhances performance and phenolic concentrations of perennial culinary herbs in an urban edible green roof. *Agroecology and Sustainable Food Systems* 41, 487–504.

Ainsworth, E.A., Bernacchi, C.J. and Dohleman, F.G. (2016) Focus on ecophysiology. *Plant Physiology* 172, 619–621.

Alberti, M. (2005) The effects of urban patterns on ecosystem function. *International Regional Science Review* 28, 168–192.

Alonzo, M., McFadden, J.P., Nowak, D.J. and Roberts, D.A. (2016) Mapping urban forest structure and function using hyperspectral imagery and lidar data. *Urban Forestry and Urban Greening* 17, 135–147.

Anguelovski, I., Argüelles, L., Baró, F., Cole, H.V.S., Connolly, J.J.T. *et al.* (2018) *Green trajectories: Municipal policy trends and strategies for greening in Europe, Canada and United States (1990–2016).* Barcelona Laboratory for Urban Environmental Justice and Sustainability, Barcelona, p. 260.

Aronson, M.F.J., Nilon, C.H., Lepczyk, C.A., Parker, T.S., Warren, P.S. *et al.* (2016) Hierarchical filters determine community assembly of urban species pools. *Ecology* 97, 2952–2963.

Avins, M. (2013) *Baltimore's Forest Patches: Emerald Assets for Ecosystem Services*. Baltimore Green Space.

Avolio, M.L., Pataki, D.E., Trammell, T.L. and Endter-Wada, J. (2018) Biodiverse cities: the nursery industry, homeowners, and neighborhood differences drive urban tree composition. *Ecological Monographs* 88, 259–276.

Ballinas, M. and Barradas, V.L. (2015) The urban tree as a tool to mitigate the urban heat island in Mexico City: a simple phenomenological model. *Journal of Environment Quality* 45, 157.

Ballinas, M. and Barradas, V.L. (2016) Transpiration and stomatal conductance as potential mechanisms to mitigate the heat load in Mexico City. *Urban Forestry and Urban Greening* 20, 152–159.

Bartens, J., Day, S.D., Harris, J.R., Dove, J.E. and Wynn, T.M. (2008) Can urban tree roots improve infiltration through compacted subsoils for stormwater management? *Journal of Environment Quality* 37, 2048–2057.

Bassuk, N., Grabosky, J., Mucciardi, A. and Raffel, G. (2011) Ground-penetrating radar accurately locates tree roots in two soil media under pavement. *Arboriculture and Urban Forestry* 37, 160–166.

Bassuk, N.L. and Whitlow, T.H. (1987) Environmental stress in street trees. *Acta Horticulturae* 195, 49–57.

Beyschlag, W. and Ryel, R.J. (2007) Plant physiological ecology: an essential link for integrating across disciplines and scales in plant ecology. *Flora: Morphology, Distribution, Functional Ecology of Plants* 202, 608–623.

Bialecki, M.B., Fahey, R.T. and Scharenbroch, B. (2018) Variation in urban forest productivity and response to extreme drought across a large metropolitan region. *Urban Ecosystems* 21, 157–169.

Briber, B., Hutyra, L., Dunn, A., Raciti, S. and Munger, J. (2013) Variations in atmospheric CO_2 mixing ratios across a Boston, MA urban to rural gradient. *Land* 2, 304–327.

Cadenasso, M.L., Pickett, S.T. and Schwarz, K. (2007) Spatial heterogeneity in urban ecosystems: reconceptualizing land cover and a framework for classification. *Frontiers in Ecology and the Environment* 5, 80–88.

Calfapietra, C., Peñuelas, J. and Niinemets, Ü. (2015) Urban plant physiology: Adaptation-mitigation strategies under permanent stress. *Trends in Plant Science* 20, 72–75.

Cariñanos, P. and Casares-Porcel, M. (2011) Urban green zones and related pollen allergy: a review. Some guidelines for designing spaces with low allergy impact. *Landscape and Urban Planning* 101, 205–214.

Childers, D.L., Cadenasso, M.L., Grove, J.M., Marshall, V., McGrath, B. *et al.* (2015) An ecology for cities: a transformational nexus of design and ecology to advance climate change resilience and urban sustainability. *Sustainability* 7, 3774–3791.

Cregg, B.M. and Dix, M.E. (2001) Tree moisture stress and insect damage in urban areas in relation to heat island effects. *Journal of Arboriculture* 27, 8–17.

Decina, S.M., Templer, P.H. and Hutyra, L.R. (2018) Atmospheric inputs of nitrogen, carbon, and phosphorus across an urban area: unaccounted fluxes and canopy influences. *Earth's Future* 6, 134–148.

Dyson, K., Ziter, C., Fuentes, T.L. and Patterson, M.S. (2019) Conducting urban ecology research on private property: advice for new urban ecologists. *Journal of Urban Ecology* 5, 1–10.

Falxa-Raymond, N., Palmer, M. I., McPhearson, T. and Griffin, K.L. (2014) Foliar nitrogen characteristics of four tree species planted in New York City reforestation sites. *Urban Ecosystems* 17, 807–824.

Feng, Q., Liu, J. and Gong, J. (2015) UAV remote sensing for urban vegetation mapping using random forest and texture analysis. *Remote Sensing* 7, 1074–1094.

Gallacher, D. (2016) Drones to manage the urban environment: risks, rewards, alternatives. *Journal of Unmanned Vehicle Systems* 4, 115–124.

George, K., Ziska, L.H., Bunce, J.A., Quebedeaux, B., Horn, J.L. *et al.* (2009) Macroclimate associated with urbanization increases the rate of secondary succession from fallow soil. *Oecologia* 159, 637–647.

Ghannoum, O. and Way, D.A. (2011) On the role of ecological adaptation and geographic distribution in the response of trees to climate change. *Tree Physiology* 31, 1273–1276.

Grabosky, J., Haffner, E. and Bassuk, N. (2009) Plant available moisture in stone-soil media for use under pavement while allowing urban tree root growth. *Arboriculture and Urban Forestry* 35, 271–278.

Grabosky, J. and Bassuk, N. (2016) Seventeen years' growth of street trees in structural soil compared with a tree lawn in New York City. *Urban Forestry and Urban Greening* 16, 103–109.

Gregg, J.W., Jones, C.G. and Dawson, T.E. (2003) Urban ozone depletion: why a tree grows better in New York City. *Nature* 424, 183–187.

Hallett, R., Johnson, M.L. and Sonti, N.F. (2018) Assessing the tree health impacts of salt water flooding in coastal cities: a case study in New York City. *Landscape and Urban Planning* 177, 171–177.

Hamerlynck, E. (2001) Chlorophyll fluorescence and photosynthetic gas exchange responses to irrandiance of Tree of Heaven (*Ailanthus altissima*) in contrasting urban environments. *Photosynthetica* 39, 79–86.

Heynen, N., Perkins, H.A. and Roy, P. (2006) The political ecology of uneven urban green space: the impact of political economy on race and ethnicity in producing environmental inequality in Milwaukee. *Urban Affairs Review* 42, 3–25.

Högberg, P. and Read, D.J. (2006) Towards a more plant physiological perspective on soil ecology. *Trends in Ecology and Evolution* 21, 548–554.

Iakovoglou, V., Thompson, J., Burras, L. and Kipper, R. (2001) Factors related to tree growth across urban-rural gradients in the Midwest, USA. *Urban Ecosystems* 5, 71–85.

Ignatieva, M., Ahrné, K., Wissman, J., Eriksson, T., Tidåker, P. *et al.* (2015) Lawn as a cultural and ecological phenomenon: a conceptual framework for transdisciplinary research. *Urban Forestry and Urban Greening* 14, 383–387.

Jansson, M., Fors, H., Lindgren, T. and Wiström, B. (2013) Perceived personal safety in relation to urban woodland vegetation: a review. *Urban Forestry and Urban Greening* 12, 127–133.

John, J., Kernaghan, G. and Lundholm, J. (2017) The potential for mycorrhizae to improve green roof function. *Urban Ecosystems* 20, 113–127.

Johnson, A.L., Borowy, D. and Swan, C.M. (2018) Land use history and seed dispersal drive divergent plant community assembly patterns in urban vacant lots. *Journal of Applied Ecology* 55, 451–460.

Karpati, A.S., Handel, S.N., Dighton, J. and Horton, T.R. (2011) *Quercus rubra*-associated ectomycorrhizal fungal communities of disturbed urban sites and mature forests. *Mycorrhiza* 21, 537–547.

Keeler, B.L., Hamel, P., McPhearson, T., Hamann, M.H., Donahue, M.L. *et al.* (2019) Social-ecological and technological factors moderate the value of urban nature. *Nature Sustainability* 2, 29–38.

Kemp, S., Hadley, P. and Blanuša, T. (2019) The influence of plant type on green roof rainfall retention. *Urban Ecosystems* 22, 355–366.

Kinzig, A.P., Warren, P., Martin, C., Hope, D. and Katti, M. (2005) The effects of human socioeconomic status and cultural characteristics on urban patterns of biodiversity. *Ecology and Society* 10, 23–35.

Konijnendijk, C.C., Annerstedt, M., Nielsen, A.B. and Maruthaveeran, S. (2013) Benefits of urban parks: a systematic review. A report for IPFRA.

Kowarik, I. (2011) Novel urban ecosystems, biodiversity, and conservation. *Environmental Pollution* 159, 1974–1983.

Kowarik, I. (2018) Urban wilderness: supply, demand, and access. *Urban Forestry and Urban Greening* 29, 336–347.

Kowarik, I. and von der Lippe, M. (2018) Plant population success across urban ecosystems: a framework to inform biodiversity conservation in cities. *The Journal of Applied Ecology* 55, 2354–2361.

Lahr, E.C., Dunn, R.R. and Frank, S.D. (2018a) Getting ahead of the curve: cities as surrogates for global change. *Proceedings of the Royal Society B: Biological Sciences* 285, 20180643.

Litvak, E., Manago, K.F., Hogue, T.S. and Pataki, D.E. (2017) Evapotranspiration of urban landscapes in Los Angeles, California at the municipal scale. *Water Resources Research* 53, 4236–4252.

Lahr, E.C., Dunn, R.R. and Frank, S.D. (2018b) Variation in photosynthesis and stomatal conductance among red maple (*Acer rubrum*) urban planted cultivars and wildtype trees in the southeastern United States. *PLOS ONE* 13, 1–18.

Litvak, E. and Pataki, D.E. (2016) Evapotranspiration of urban lawns in a semi-arid environment: An *in situ* evaluation of microclimatic conditions and watering recommendations. *Journal of Arid Environments* 134, 87–96.

Locke, D.H., Avolio, M., Trammel, T., Chowdhury, R.R., Grove, J.M. *et al.* (2018) A multi-city comparison of front and backyard differences in plant species diversity and nitrogen cycling in residential landscapes. *Landscape and Urban Planning* 178, 102–111.

Lu, D., Xu, X., Tian, H., Moran, E., Zhao, M. *et al.* (2010) The effects of urbanization on net primary productivity in southeastern China. *Environmental Management* 46, 404–410.

Mahar, A., Wang, P., Ali, A., Awasthi, M.K., Lahori, A.H. *et al.* (2016) Challenges and opportunities in the phytoremediation of heavy metals contaminated soils: a review. *Ecotoxicology and Environmental Safety* 126, 111–121.

Marmiroli, M., Pietrini, F., Maestri, E., Zacchini, M., Marmiroli, N. *et al.* (2011) Growth, physiological and molecular traits in Salicaceae trees investigated for phytoremediation of heavy metals and organics. *Tree Physiology* 31, 1319–1334.

Martin, C.A. and Stabler, L.B. (2002) Plant gas exchange and water status in urban desert landscapes. *Journal of Arid Environments* 51, 235–254.

Martin, C.A. and Stabler, L.B. (2004) Urban horticultural ecology: interactions between plants, people and the physical environment. *Acta Horticulturae* 639, 97–101.

Martin, C.A. and Stutz, J.C. (2013) Tree health in Phoenix, Arizona, US. *Arboriculture and Urban Forestry* 39, 286–291.

Martin, C.A., Stabler, L.B., Celestian, S.B. and Stutz, C. (2002) Urban plant ecology: a horticultural perspective. Initial results from a Long Term Ecological Research (LTER) Site. *Proceedings of the 12th METRIA Conference.*

McCarthy, H.R. and Pataki, D.E. (2010) Drivers of variability in water use of native and non-native urban trees in the greater Los Angeles area. *Urban Ecosystems* 13, 393–414.

McCarthy, H.R., Pataki, D.E. and Jenerette, G.D. (2011) Plant water-use efficiency as a metric of urban ecosystem services. *Ecological Applications* 21, 3115–3127.

McClung, T. and Ibáñez, I. (2018) Quantifying the synergistic effects of impervious surface and drought on radial tree growth. *Urban Ecosystems* 21, 147–155.

McDonnell, M.J. and Hahs, A.K. (2008) The use of gradient analysis studies in advancing our understanding of the ecology of urbanizing landscapes: current status and future directions. *Landscape Ecology* 23, 1143–1155.

McDonnell, M.J. and MacGregor-Fors, I. (2016) The ecological future of cities. *Science* 352, 936–938.

McHale, M.R., Burke, I.C., Lefsky, M.A., Peper, P.J. and McPherson, E.G. (2009) Urban forest biomass estimates: is it important to use allometric relationships developed specifically for urban trees? *Urban Ecosystems* 12, 95–113.

Meineke, E.K., Dunn, R.R., Sexton, J.O. and Frank, S.D. (2013) Urban warming drives insect pest abundance on street trees 8, 2–8.

Menzer, O. and McFadden, J.P. (2017) Statistical partitioning of a three-year time series of direct urban net CO_2 flux measurements into biogenic and anthropogenic components. *Atmospheric Environment* 170, 319–333.

Mueller, E.C. and Day, T.A. (2005) The effect of urban ground cover on microclimate, growth and leaf gas exchange of oleander in Phoenix, Arizona. *International Journal of Biometeorology* 49, 244–255.

Mullaney, J., Lucke, T. and Trueman, S.J. (2015) A review of benefits and challenges in growing street trees in paved urban environments. *Landscape and Urban Planning* 134, 157–166.

Nagase, A. and Dunnett, N. (2012) Amount of water runoff from different vegetation types on extensive green roofs: effects of plant species, diversity and plant structure. *Landscape and Urban Planning* 104, 356–363.

Nesshöver, C., Assmuth, T., Irvine, K.N., Rusch, G.M., Waylen, K.A. *et al.* (2017) The science, policy and practice of nature-based solutions: an interdisciplinary perspective. *Science of the Total Environment* 579, 1215–1227.

Ng, B.J.L., Hutyra, L.R., Nguyen, H., Cobb, A.R., Kai, F.M. *et al.* (2015) Carbon fluxes from an urban tropical grassland. *Environmental Pollution* 203, 227–234.

Nouri, H., Beecham, S., Kazemi, F. and Hassanli, A.M. (2013) A review of ET measurement techniques for estimating the water requirements of urban landscape vegetation. *Urban Water Journal* 10, 247–259.

Ordóñez, C., Grant, A., Millward, A.A., Steenberg, J. and Sabetski, V. (2019) Developing performance indicators for nature-based solution projects in urban areas: the case of trees in revitalized commercial spaces. *Cities and the Environment* 12, Article 1.

Osmond, C.B., Björkman, O. and Anderson, D.J. (1980) *Physiological Processes in Plant Ecology: Towards a Synthesis with Atriplex*. Springer-Verlag, Berlin.

Osone, Y., Kawarasaki, S., Ishida, A., Kikuchi, S., Shimizu, A. *et al.* (2014) Responses of gas-exchange rates and water relations to annual fluctuations of weather in three species of urban street trees. *Tree Physiology* 34, 1056–1068.

Ossola, A., Locke, D., Lin, B. and Minor, E. (2019) Greening in style: urban form, architecture and the structure of front and backyard vegetation. *Landscape and Urban Planning* 185, 141–157.

Pataki, D.E. (2015) Grand challenges in urban ecology. *Frontiers in Ecology and Evolution* 3, 1–6.

Pataki, D.E., McCarthy, H.R., Litvak, E. and Pincetl, S. (2011) Transpiration of urban forests in the Los Angeles metropolitan area. *Ecological Applications* 21, 661–677.

Pataki, D.E., McCarthy, H.R., Gillespie, T., Jenerette, G.D. and Pincetl, S. (2013) A trait-based ecology of the Los Angeles urban forest. *Ecosphere* 4, 1–20.

Pearcy, R.W., Ehleringer, J.R., Mooney, H. and Rundel, P.W. (1989) Introduction. In: *Plant Physiological Ecology: Field Methods and Instrumentation*. Springer Science and Business Media, Dordrecht, The Netherlands, pp. xv–xix.

Pearse, W.D., Cavender-Bares, J., Hobbie, S.E., Avolio, M.L., Bettez, N. *et al.* (2018) Homogenization of plant diversity, composition, and structure in North American urban yards. *Ecosphere* 9, 1–17.

Pickett, S.T.A., Cadenasso, M.L., Childers, D.L., Mcdonnell, M.J. and Zhou, W. (2016) Evolution and future of urban ecological science: ecology in, of, and for the city. *Ecosystem Health and Sustainability* 2, e01229.

Pontius, J., Hanavan, R.P., Hallett, R.A., Cook, B.D. and Corp, L.A. (2017) High spatial resolution spectral unmixing for mapping ash species across a complex urban environment. *Remote Sensing of Environment* 199, 360–369.

Pregitzer, C.C., Sonti, N.F. and Hallett, R.A. (2016) Variability in urban soils influences the health and growth of native tree seedlings. *Ecological Restoration* 34, 106–116.

Pretzsch, H., Biber, P., Uhl, E., Dahlhausen, J., Schütze, G. *et al.* (2017) Climate change accelerates growth of urban trees in metropolises worldwide. *Scientific Reports* 7, 1–10.

Raciti, S.M., Hutyra, L.R., Rao, P. and Finzi, A.C. (2012a) Inconsistent definitions of "urban" result in different conclusions about the size of urban carbon and nitrogen stocks. *Ecological Applications* 22, 1015–1035.

Raciti, S.M., Hutyra, L.R. and Finzi, A.C. (2012b) Depleted soil carbon and nitrogen pools beneath impervious surfaces. *Environmental Pollution* 164, 248–251.

Rahman, M.A., Armson, D. and Ennos, A.R. (2014) Effect of urbanization and climate change in the rooting zone on the growth and physiology of *Pyrus calleryana*. *Urban Forestry and Urban Greening* 13, 325–335.

Rasmann, S. and Agrawal, A.A. (2008) In defense of roots: a research agenda for studying plant resistance to belowground herbivory. *Plant Physiology* 146, 875–880.

Reisinger, A.J., Groffman, P.M. and Rosi-Marshall, E.J. (2016) Nitrogen cycling process rates across urban ecosystems. *FEMS Microbiology Ecology* 92, 1–11.

Rupprecht, C.D.D. and Byrne, J.A. (2014) Informal urban greenspace: a typology and trilingual systematic review of its role for urban residents and trends in the literature. *Urban Forestry and Urban Greening* 13, 597–611.

Searle, S.Y., Bitterman, D.S., Thomas, S., Griffin, K.L., Atkin, O.K. *et al.* (2011) Respiratory alternative oxidase responds to both low- and high-temperature stress in *Quercus rubra* leaves along an urban-rural gradient in New York. *Functional Ecology* 25, 1007–1017.

Searle, S.Y., Turnbull, M.H., Boelman, N.T., Schuster, W.S.F., Yakir, D. *et al.* (2012) Urban environment of New York City promotes growth in northern red oak seedlings. *Tree Physiology* 32, 389–400.

Short Gianotti, A.G., Getson, J.M., Hutyra, L.R. and Kittredge, D.B. (2016) Defining urban, suburban, and rural: a method to link perceptual definitions with geospatial measures of urbanization in central and eastern Massachusetts. *Urban Ecosystems* 19, 823–833.

Sjöman, H. and Busse Nielsen, A. (2010) Selecting trees for urban paved sites in Scandinavia – a review of information on stress tolerance and its relation to the requirements of tree planners. *Urban Forestry and Urban Greening* 9, 281–293.

Smith, I.A., Dearborn, V.K. and Hutyra, L.R. (2019) Live fast, die young: accelerated growth, mortality, and turnover in street trees. *PLOS ONE* 14, e0215846.

Sonti, N.F. (2019) Ecophysiological and social functions of urban forest patches. Ph.D. dissertation. University of Maryland, College Park, MD. 166 p.

Sonti, N.F., Hallett, R.A., Griffin, K.L. and Sullivan, J.H. (2019) White oak and red maple tree ring analysis reveals enhanced productivity in urban forest patches. *Forest Ecology and Management* 453, 117626.

Starry, O., Lea-Cox, J.D., Kim, J. and van Iersel, M.W. (2014) Photosynthesis and water use by two Sedum species in green roof substrate. *Environmental and Experimental Botany* 107, 105–112. DOI: 10.1016/j.envexpbot.2014.05.014.

Tan, P.Y., Wang, J. and Sia, A. (2013) Perspectives on five decades of the urban greening of Singapore. *Cities* 32, 24–32. DOI: 10.1016/j.cities.2013.02.001.

Threlfall, C.G. and Kendal, D. (2018) The distinct ecological and social roles that wild spaces play in urban ecosystems. *Urban Forestry & Urban Greening* 29, 348–356. DOI: 10.1016/j.ufug.2017.05.012.

Trammell, T.L.E., Pataki, D.E., Cavender-Bares, J., Groffman, P.M., Hall, S.J. *et al.* (2016) Plant nitrogen concentration and isotopic composition in residential lawns across seven US cities. *Oecologia* 181(1), 271–285. DOI: 10.1007/s00442-016-3566-9.

Trowbridge, P.J. and Bassuk, N.L. (2004) *Trees in the Urban Landscape: Site Assessment, Design, and Installation*. John Wiley and Sons, Hoboken, New Jersey.

Tyburska, J., Frymark-Szymkowiak, A., Kulczyk-Skrzeszewska, M. and Kieliszewska-Rokicka, B. (2013) Mycorrhizal status of forest trees grown in urban and rural environments in Poland. *Ecological Questions* 18, 49–57. DOI: 10.12775/ecoq-2013-0005.

United Nations, Department of Economic and Social Affairs, Population Division (2018) *The World's Cities in 2018*—Data Booklet (ST/ESA/ SER.A/417). 29 p.

Vallano, D.M. and Sparks, J.P. (2013) Foliar ^{15}N is affected by foliar nitrogen uptake, soil nitrogen, and mycorrhizae along a nitrogen deposition gradient. *Oecologia* 172(1), 47–58. DOI: 10.1007/s00442-012-2489-3.

VanWoert, N.D., Rowe, D.B., Andresen, J.A., Rugh, C.L. and Xiao, L. (2005) Watering regime and green roof substrate design affect Sedum plant growth. *HortScience* 40(3), 659–664. DOI: 10.21273/HORTSCI.40.3.659.

Velasco, E., Roth, M., Norford, L. and Molina, L.T. (2016) Does urban vegetation enhance carbon sequestration? *Landscape and Urban Planning* 148, 99–107. DOI: 10.1016/j.landurbplan.2015.12.003.

Wang, W. and Pataki, D.E. (2010) Spatial patterns of plant isotope tracers in the Los Angeles urban region. *Landscape Ecology* 25(1), 35–52. DOI: 10.1007/s10980-009-9401-5.

Westfall, J.A., Sonti, N.F., Wiemann, M.C., Eberhardt, T.L. and So, C.-L. (2020) Urban tree specific gravity and ash content: a case study from Baltimore, Maryland USA. *Urban Forestry & Urban Greening* 48, 126556. DOI: 10.1016/j.ufug.2019.126556.

Whitlow, T.H. and Bassuk, N.L. (1988) Ecophysiology of urban trees and their management—the North American experience. *HortScience: a publication of the American Society for Horticultural Science* 23, 542–546.

Woodall, C.W., Nowak, D.J., Liknes, G.C. and Westfall, J.A. (2010) Assessing the potential for urban trees to facilitate forest tree migration in the eastern United States. *Forest Ecology and Management* 259(8), 1447–1454. DOI: 10.1016/j.foreco.2010.01.018.

Xie, G., Lundholm, J.T. and Scott MacIvor, J. (2018) Phylogenetic diversity and plant trait composition predict multiple ecosystem functions in green roofs. *Science of the Total Environment* 628-629, 1017–1026. DOI: 10.1016/j.scitotenv.2018.02.093.

Yan, Y., Liu, X., Wang, F., Li, X., Ou, J. *et al.* (2018) Assessing the impacts of urban sprawl on net primary productivity using fusion of Landsat and MODIS data. *Science of the Total Environment* 613-614, 1417–1429. DOI: 10.1016/j.scitotenv.2017.09.139.

Youngsteadt, E., Dale, A.G., Terando, A.J., Dunn, R.R. and Frank, S.D. (2015) Do cities simulate climate change? A comparison of herbivore response to urban and global warming. *Global Change Biology* 21(1), 97–105. DOI: 10.1111/gcb.12692.

Zhang, X., Friedl, M.A., Schaaf, C.B., Strahler, A.H. and Schneider, A. (2004) The footprint of urban climates on vegetation phenology. *Geophysical Research Letters* 31(12), L12209. DOI: 10.1029/2004GL020137.

Zhao, S., Liu, S. and Zhou, D. (2016) Prevalent vegetation growth enhancement in urban environment. *Proceedings of the National Academy of Sciences* 113(22), 6313–6318. DOI: 10.1073/pnas.1602312113.

Ziska, L.H. (2003) Evaluation of the growth response of six invasive species to past, present and future atmospheric carbon dioxide. *Journal of Experimental Botany* 54(381), 395–404. DOI: 10.1093/jxb/erg027.

Ziska, L.H., George, K. and Frenz, D.A. (2007) Establishment and persistence of common ragweed (*Ambrosia artemisiifolia* L.) in disturbed soil as a function of an urban-rural macro-environment. *Global Change Biology* 13(1), 266–274. DOI: 10.1111/j.1365-2486.2006.01264.x.

5 Urban Biodiversity: Between Elsewhere-Nature and the Post-Wild World

Lea R. Johnson*
Research and Conservation Division, Longwood Gardens, Kennett Square, PA

Introduction

Most of humanity now lives in cities and towns, and human experiences of biodiversity are more and more often urban ones (Fig. 5.1). These encounters may evoke in us an array of responses: delight and fear, repulsion and curiosity, annoyance and wonder. We build understanding from experience, and the opinions and perceptions we develop from these encounters with biodiversity influence our actions. Urban biodiversity is therefore increasingly central to the way humanity perceives and acts toward the diversity of life on earth.

Over the past half-century, the field of urban ecology has revealed diverse species, regionally rare habitats and complex relationships among urban biodiversity, the built environment and human choices (Pickett *et al.*, 2011). At the same time, urbanization destroys habitats and replaces ecosystems,and proposals for urban biodiversity conservation are often met with surprise or skepticism (Soanes *et al.*, 2019). There is broad concern about the potential effects of 'nature deficit disorder' and 'extinction of experience' of nature in human life (Miller, 2005; Louv, 2011; Palomino *et al.*, 2016). What can be done to enrich and connect experiences of urban nature to larger environmental systems, and to stewardship of biodiversity?

To explore this question, we will first visit two poles in thinking about urban biodiversity, 'elsewhere-nature' and the 'post-wild world', and the middle ground between them. We will then examine relationships between these tendencies and our current understanding of urban biodiversity, and implications of these tendencies for perceptions and actions. We will conclude with an exploration of some ways forward: embracing uncertainty and inquiry, possibility-based management, and experiences of wonder and beauty.

What Do We Mean When We Talk about Biodiversity?

At its broadest, biodiversity encompasses the variety of life on earth; from genes to species, and from communities to ecosystems. It can be viewed as accumulated knowledge about how to survive the changing environment of this planet, stored in evolving species or expressed as an equation (Harper *et al.*, 1994). For people, biodiversity is the source of products and processes on which our lives depend, though the relationship may not be obvious in urban life (TEEB, 2011). Biodiversity can be a source of solace, recreation, recharge and relaxation (van den Bosch and Ode Sang, 2017). As a measurement,

*ljohnson@longwoodgardens.org

Fig. 5.1. A few urban experiences of biodiversity in temperate-zone cities. Top, left to right: pigeon (*Columba livia*)-shaped bench at a bus stop; willow oak (*Quercus phellos*) sprouting from an acorn on a paved walkway; sharing lunch with an Eastern gray squirrel (*Sciurus carolinensis*). Bottom: monarch butterfly (*Danaus plexippus*) on a butterfly bush (*Buddleja davidii*); slug on a sandwich bag; white-tailed deer (*Odocoileus virginianus*) browsing in a residential garden. (All photos by L. R. Johnson)

biodiversity's power is that it allows us to understand similarities and differences among places, across space and time – from arctic tundra to tropical rainforest, and through changes in an abandoned field as it grows into an old forest. The numbers we use to calculate biodiversity are generated by identifying and counting how many kinds of things there are, and how many of each of them are found in an area of a certain size. From these counts, we can calculate richness and eveness (the number of different kinds found in an area and whether there are few or many of each kind there). From these measures, we can evaluate indicators of habitat quality, such as diversity of floral resources for pollinators, or understand how unique a place is, in a global context. The importance of the increasingly rare human skill of identifying diverse life cannot be over-emphasized here (Drew, 2011).

Whether we are interested in trees in a park, microbes in a teaspoon of soil, or birds migrating through an urban region, the scale of our focus reveals different dimensions of biodiversity. Considering urban biodiversity, we might be interested in knowing how the diversity of plants in an urban green space varies across a city (Pregitzer *et al.*, 2019), or how urban regions on different continents compare to their surroundings (Aronson *et al.*, 2014). At each level of scale, we may find different patterns (Hulme, 2008). Biodiversity helps us to understand the distribution of life on the planet and to see regional uniqueness. It is a measure that shows us the mind-boggling array of ways that life has evolved, and the effects of human actions and decisions on ecosystems.

'Elsewhere-nature' and the 'Post-wild World': How We Think about Biodiversity in Cities

For some people, 'urban biodiversity' would seem to be a contradiction in terms, while for others true wilderness is found along train tracks and in abandoned lots. A spectrum in popular thinking about urban biodiversity ranges between these two poles (Fig. 5.2). What formative experiences have informed these divergent perceptions, and what implications do they have for decision making about the environment?

One end of this spectrum is informed by a tendency to think of nature as something you

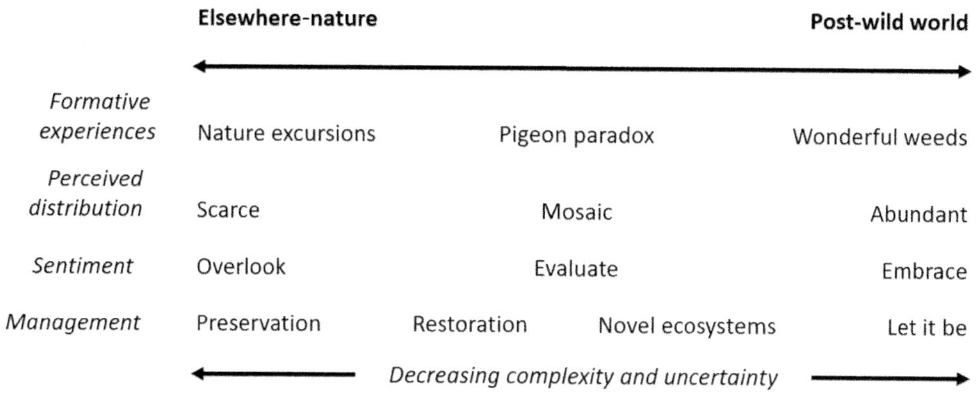

Fig. 5.2. A spectrum of tendencies in popular thinking about urban biodiversity. Human perceptions and opinions based in experiences influence approaches to management of urban biodiversity.

find far away from cities: deep forest, desert, mountain peaks. Although it is beyond the scope of this chapter to include a review of the many definitions of 'nature', this view aligns with a current dictionary definition: *existing or happening independently of people* (Cambridge English Dictionary, 2019). This 'elsewhere' nature is real but distant from everyday urban life. We learn about it via nature excursions to these faraway places that we, or others, make. Elsewhere-nature is made vivid by photographers who sit still among mosquitoes for days to capture a single image, and by a sunrise bird chorus on a weekend camping trip. Elsewhere-nature can be far away in time, as in a landscape that has been destroyed. It may evoke feelings of longing or nostalgia, or anxiety arising from an awareness that it is increasingly scarce. Elsewhere-nature is a source of joy and beauty, and motivation for preservation of remaining large wild places. It inspires the funding of organizations that do important work in places and on behalf of species that most people will never see with their own eyes. However, this long-distance view, as it focuses on the horizon, overlooks local biodiversity.

A contrasting view has arisen in cities. Resisting the definition of the city as anti-nature, a new enthusiasm has grown for overlooked plants and animals in urban everyday life. Murals of weeds sprout larger than life (Caron, 2019). Foragers prowl parks in search of fruits and vegetables (Foderaro, 2011; McLain *et al.*, 2014; Shackleton *et al.*, 2017). Weeds, viewed in another light, turn out to be herbs, spices and flowers (Seiter, 2016). Some go as far as to declare the entire world 'post-wild' (Marris, 2013), embracing novelty as the new status quo and accepting human dominance of all ecosystems. Some proponents of the post-wild world view it from a conservation perspective, advocating for more naturalistic approaches to landscape design of the built environment (Rainer and West, 2015). Applied to cities, this tendency inspires joy in, and recognition of, the plants and animals in everyday urban life (Cazabon, 2019). This perspective is refreshing in its ability to see and embrace life in places where it has been overlooked and has potential for positive effects on an ecological sense of place in urban ecosystems. This enthusiasm would seem to bode well for conservation in the broader landscape. Experiences of biodiversity in childhood inform our adult perceptions of quality and change in the environment, and awareness of urban biodiversity could counteract 'extinction of experience' of nature in urban life, particularly for children. However, it remains to be seen whether effects of 'shifting baselines' (Turvey *et al.*, 2010) or 'environmental generational amnesia' (Kahn and Weiss, 2017) could result in lowered expectations for both local and global biodiversity. As species are lost, children perceive the environment into which they are born as normal, leading to a downward-shifting baseline in expectations of species diversity.

These perspectives on urban biodiversity share a limitation in their simplicity. Simpler, or seemingly simpler, solutions are more quickly grasped, but at the extreme of both positions, the potential for urban environments to contribute to global biodiversity could be oversimplified and underestimated. On the elsewhere-nature end of the spectrum, urban biodiversity is overlooked entirely. At the other extreme, the post-wild world accepts a shifted baseline. In both cases, opportunities to preserve and enhance biodiversity could be missed.

Understanding Urban Biodiversity

Patterns in urban biodiversity have been revealed by ecological research in cities across the globe (e.g., Shochat *et al.*, 2006, 2010; Wittig, 2010; Faeth *et al.*, 2011; Swan *et al.*, 2011; Aronson *et al.*, 2014, 2016; McDonnell and Hahs, 2015; McPhearson *et al.*, 2016; Lepczyk *et al.*, 2017). This growing body of knowledge lends support a spectrum of views of urban biodiversity.

On one hand, cities clearly destroy habitats as they expand. Habitat loss is the primary driver of species extinctions at a global scale (Pimm and Raven, 2000). If current trends in human population density continue, the proportion of the planet's surface covered by urban land will nearly triple between 2000 and 2030, increasing by more than a million square kilometres (Seto *et al.*, 2012). As a city expands, native ecosystems are destroyed, fragmented and transformed. Pavement, roofing and pipes alter the flow of water (Hughes *et al.*, 2014), cars and factories add pollution to the air, and heat trapped by buildings and roads warm the local climate (Alcoforado and Andrade, 2008). Soils are sealed, deposited on, turned over and built upon (Effland and Pouyat, 1997; Craul, 1999; Wessolek, 2008). People introduce new species both intentionally and by accident, leading to new interactions among competitors, herbivores, predators and prey (Aronson and Handel, 2011; Johnson *et al.*, 2019a). Remnant habitats are subject to species loss due to reduced area, while at the same time introduced species reduce native diversity via competitive exclusion (Shochat *et al.*, 2010). These changes make cities unsuitable for many species, filtering out a great deal of the biodiversity of the region (Aronson *et al.*, 2016).

However, biodiversity loss is not uniform across an urbanized area, and urban environments create new opportunities for some species. Studies of urban landscape ecology reveal that the fabric of cities is a heterogeneous mosaic of very different patches at a very fine scale (Band *et al.*, 2005; Pickett and Cadenasso, 2009; Pickett *et al.*, 2017). Fragments of original woodlands continue growing (Loeb, 2011; Atha *et al.*, 2016). Processes of soil formation continue as they have for millennia, breaking down new urban parent materials (Effland and Pouyat, 1997). Cities can support populations of regionally rare and threatened plants and animals (Güneralp and Seto, 2013; Ives *et al.*, 2016; Rebelo *et al.*, 2011; Kantsa *et al.*, 2013; Kowarik *et al.*, 2019). Plants and animals that thrive in cities can be more abundant than they were in their original habitats (Faeth *et al.*, 2011). Novel assemblages of native and introduced species emerge where spontaneous vegetation is allowed to grow, providing refugia for native species and nature experiences for urban residents (Kowarik and Körner, 2005; Kowarik *et al.*, 2019).

The region in which a city develops is the source of the native species found in that city, leading to diversity in urban native species across the globe. For example, your lunch may be stolen by a squirrel in Chicago (ABC7, 2019), or by a monkey in Jaipur (Singh, 1969). Similarly, the history of environmental change surroundings a city affects patterns of urban biodiversity. Cities surrounded by large-scale agriculture, deserts, or forests vary in the plants and animals that are available. Species also vary in their ability to use and move across an urban mosaic (Graham *et al.*, 2019). Some species are able to take advantage of 'habitat analogues' (Lundholm and Richardson, 2010), making use of human-created structures and resources that are similar to elements of their original habitat; a classic example is the Eurasian rock dove (*Columba livia*), which finds building ledges in cities all over the world a fine substitution for the cliffs where it originally nested.

Mobile organisms like blue jays (Lundberg and Moberg, 2003) or plants with wind-blown seeds and pollen may be able to use small patches as stepping stones to travel across a city (Ignatieva *et al.*, 2011; Van Rossum and Triest, 2012). However, those that travel via ant dispersal are not able to move if conditions become unfavorable, or if a parcel is cleared for development. For species that can't make use of paved and built surfaces, patches of suitable habitat are surrounded by a matrix of other urban land uses, as in urban parks or vacant lots. There is now evidence that isolation of plant and animal populations in cities is having evolutionary effects on genetic diversity; in New York City, uptown rats are genetically distinct from downtown ones (Johnson and Munshi-South, 2017). In Paris, tiny plants in sidewalk cracks disperse their seeds differently from members of the same species growing in agricultural fields outside the city (Cheptou *et al.*, 2008).

Although animal species richness generally declines in cities, the abundance of some groups, especially birds and arthropods, often increase (Faeth *et al.*, 2011). While large predators are often extirpated, smaller predators such as domestic cats (*Felis catus*), raccoons (*Procyon lotor*), skunks (*Mephitis mephitis*) and foxes (*Vulpes vulpes*), whose prey include urban rodents and birds, increase in number (Prange and Gehrt, 2004; Rodewald and Kearns, 2011). While smaller and isolated patches support fewer species, urban habitat diversity can increase the number of species that are found in a city if patches are big enough to support viable populations (Lepczyk *et al.*, 2017). Where patches are large enough or connect to larger areas on the urban fringe, they can be suitable for animals requiring larger territories, as in Berlin where sightings of wild boars are a common occurrence in some neighborhoods (Stillfried *et al.*, 2017).

Species that thrive in human-created habitats have been called synanthropes, synurbic (Francis and Chadwick, 2012), urbanophiles (Wittig *et al.*, 1985), and urban adapters (McKinney, 2006; Wrzesień *et al.*, 2016). Whether called by these or other titles, these species benefit from resource subsidies and structures provided by the built environment and human activities. These include species that would be familiar to people in many cities, such as house sparrows (*Passer domesticus*), brown rats (*Rattus norvegicus*), dandelions (*Taraxacum officinale*), house mice (*Mus musculus*), feral cats (*Felis catus*) and cockroaches (Blattodea). Species that thrive in cities can have much larger populations inside cities than outside of them (McDonnell and Hahs, 2015). The abundance of these species can sometimes lead to human–wildlife conflict (Morzillo *et al.*, 2014).

Cities Are both Rich and Poor in Species

In terms of species richness (the total number of species, not considering how many individuals there are of each) cities are both rich and poor. This pattern is underlain by the distribution of species and resources across urban landscapes as a result of human activities. Cities are rich in plant species because land uses and commercial transactions leave traces in the urban flora. Many cities are built in places that were once highly productive ecosystems, and are global hotspots of species introduction because they lie where trade and transportation converge (Imhoff *et al.*, 2004; Kühn *et al.*, 2004). Members of original plant communities of a place are joined by agricultural introductions – crops and weeds – from early farms engulfed by urban expansion. To these are added the delights and mistakes of ornamental horticulture, some of which escape and spread across any bit of open ground, and by plants imported for a variety of utilitarian purposes along with many hitch-hikers. As a result of these many introductions, cities can contain more species than surrounding rural areas (Kowarik, 2011; Aronson *et al.*, 2014).

Because they are the driver of primary productivity and a major component of habitat structure, plants are both a component of biodiversity and a key element of the conditions for survival of other species in cities. Much of the intentionally introduced plant richness of cities has been selected to be resistant to herbivores and pests; many species are intentionally imported from other continents without herbivores and diseases that have evolved the ability to eat or infect them despite their defences. Patches of spontaneous vegetation with native plants are generally much richer in species than most human-designed plantings, and include more plants that support coevolved relationships with native insects and animals (Burghardt *et al.*, 2009).

Local Uniqueness and Global Homogenization

Because cities contain many elements that are repeated across the globe, particularly the dominance of buildings and pavement, they create similar environments. This observation has led to the hypothesis that urbanization leads to homogenization of species at a global scale (McKinney, 2006). Evidence testing this hypothesis is inconclusive, due in part to both the influence of native diversity on assemblages at continental scales and the range of heterogeneous environments found in cities (Kühn and Klotz, 2006; La Sorte *et al.*, 2014). The homogenization hypothesis seems to be best supported when focused on similar repeating elements of urban form across cities on the same continent (Hall *et al.*, 2016; Pearse *et al.*, 2018) and excluding areas of remnant and regenerating native vegetation (Wittig, 2010; Wittig and Becker, 2010). Where homogenization has been documented, evidence suggests that introduced species and human-changed conditions of the abiotic environment are important interacting drivers (Trentanovi *et al.*, 2013). In Europe, where species have been introduced in great numbers over a long period of time, plants present prior to the year 1500 differ from those introduced more recently in patterns of urban homogenization (Lososová *et al.*, 2012). When considered in the light of time lags of hundreds of years that have been documented in invasive plant species (Kowarik *et al.*, 1995), it is clear that the long-term effects of humanity's great experiment in moving species around the planet remain to be seen.

The role of expanding cities in global biodiversity conservation will become increasingly important. The regions of the world currently projected to urbanize fastest include areas of globally rare biodiversity with currently low urbanization (United Nations, 2014), such as the Eastern Afromontane, West Africa, India and Sri Lanka, with the greatest urban expansion in biodiversity hotspots predicted in South America (Güneralp and Seto, 2013).

Visions and Actions for the Future of Urban Biodiversity

What obligation do we as human beings, members of the species that is changing the planet at an unprecedented rate and scale, have toward the rest of life on earth? How much of it will we bring with us into the future? The planet's biodiversity is supported by a web of complex, long-term, coevolved relationships that we do not yet fully understand. If we do have a responsibility to take these species and their relationships forward in time with us, then urban biodiversity is important to this effort.

Although urbanization is often thought of as a local issue, the combined global impacts of local urban expansion will require significant policy changes to minimize losses of global biodiversity (Seto *et al.*, 2012). Biodiversity can be maintained worldwide only if each country and each region takes care of its unique species and characteristic ecosystem types (Wittig, 2010).

To address this challenge, we need a broad array of perspectives and approaches. Elsewhere-nature and the post-wild world frame a spectrum of views, information, and experience that can help humans as an urban species to see the places where we live and work. Three complementary ways forward have emerged from communities seeking to understand and support urban biodiversity: rigorous curiosity, possibility-focused management and experiences of wonder and beauty.

Rigorous Curiosity

Cities are neither the only problem nor the only answer when considering how to conserve the planet's biodiversity. Researchers in numerous fields have pointed to areas of inquiry that could advance our understanding of cities as ecosystems and our progress toward human societies that make room for other species. As our understanding of urban ecosystems has advanced, there have been surprises and counterintuitive findings (Pickett *et al.*, 2011), and there is still much to learn (McPhearson *et al.*, 2016). To find answers to pressing problems facing biodiversity, we need rigorous curiosity.

Social-ecological systems thinking (Palmer *et al.*, 2016) has emerged as an important tool for grappling with relationships between societies and their environments. Ecologists have pointed to the need to better understand social drivers of ecological processes (Pickett *et al.*, 1997; Niemelä, 1999) and new insights and frameworks for problem-solving in these complex systems have arisen from this approach (Folke *et al.*, 2005; Ostrom, 2009; Johnson *et al.*, 2019b). These approaches require collaboration within and across boundaries of expertise and experience.

New methods for understanding urban biodiversity are being developed as an increasing number of disciplines see the human effects on ecosystems as an important focus for problem-solving Direct observation can now be combined with an array of other tools. As new technologies and approaches develop, they allow for understanding of spatial patterns of urban biodiversity at ever finer scales. For example, landscape-scale patterns in urban vegetation can now be detected more precisely than before using remote sensing (Ossola *et al.*, 2019), and molecular techniques allow detection of evolution in genetically distinct populations of urban animals (Johnson and Munshi-South, 2017).

The spatial scale of these inquiries is important, and there is more to be learned about how findings from one scale apply to social-ecological problems at other scales (McPhearson *et al.*, 2016). Conceptual scale mismatch may be responsible for some of the distance between elsewhere-nature and the post-wild world; at continental scales, urban biodiversity can seem insignificantly small, while at the scale of human experience it is highly important. Comparisons across multiple cities are increasingly being used to understand broad relationships between the structure and function of urban ecosystems and biodiversity (McDonnell and Hahs, 2013; Padullés Cubino *et al.*, 2018). Most studies in urban ecology so far have been conducted in temperate climate zones (McHale *et al.*, 2013, 2015). To maximize biodiversity in the world's most rapidly urbanizing regions, more research is needed in other climates (Seto *et al.*, 2012).

Possibility-focused Management

What biodiversity conservation goals are attainable in cities? Both public and civic stewards of urban nature are in need of guidance, goals, and techniques to address this question. ,Cities across the globe are developing plans for future biodiversity, but many of these plans lack specific targets (Nilon *et al.*, 2017). It is important to specify goals for these at the outset so that progress can be measured and effects can be learned from, whether the aim is preserving habitats, improving connectivity, providing ecosystem services, responding to environmental change, or fulfilling ethical responsibilities (Dearborn and Kark, 2010).

This is the territory in which the elsewhere-nature and post-wild world perspectives meet to offer useful insight: 'possibility-based' approaches to urban ecosystem management that engage with both social and environmental context. The practice of ecological restoration was conceived outside of the urban context. In its simplest form, ecological restoration addresses damage to ecosystems by ending human actions causing harm and allowing ecosystem processes to unfold. Informed by this approach, early efforts at urban biodiversity conservation aimed at pristine target conditions. Urban practitioners of ecological restoration have encountered constraints of the urban social-ecological context on reaching this type of goal. Like the ecological restoration community as a whole, emphasis in urban ecological restoration has shifted from attaining pristine conditions or historic species assemblages to focus on restoration of ecosystem processes and trajectories (McDonald *et al.*, 2016). Introduced species are now expected to be part of urban ecosystems in the future (Davis *et al.*, 2011).

More work is needed to understand interactions among species in these novel assemblages. New and complex relationships between species and human-controlled conditions result in many surprises and make it difficult to predict long-term results of efforts to conserve or enhance urban biodiversity. Although conservation attitudes tuned to elsewhere-nature may find it a stretch to embrace the novelty of urban ecosystems with many non-native species, these emerging urban communities can provide ecosystem services and social benefits, and contribute to biodiversity conservation (Kowarik, 2011).

One such type of community is the early-successional habitat. There is increasing interest in the potential for communities like meadows and grasslands to contribute to urban biodiversity (Kühn, 2006; Fischer *et al.*, 2013; Williams *et al.*, 2014). In some areas, early-successional habitats have become rare, and the high frequency and intensity of ecological disturbance in cities can therefore provide a regionally scarce refuge for species. Abandoned land may have temporary value for urban biodiversity by making space for communities of varied ages (Kattwinkel *et al.*, 2009, 2011), or over longer periods of time grow to be dominated by species found later in ecological succession (Kowarik and Körner, 2005).

Where regionally unique habitats and species are found in urban regions, they require special protection and management (Kühn and Klotz, 2006). High-quality, connected patches of 50 hectares or more are key to maximizing species richness in cities (Beninde *et al.*, 2015). However, the clear value of large and connected patches does not reciprocally mean that small and fragmented patches are without value (Guldin *et al.*, 1990; Godefroid and Koedam, 2003; Bodin *et al.*, 2006; Fahrig, 2017). Small patches of spontaneous vegetation make up an important proportion of the habitat in many cities (e.g.,Avins, 2013). This has implications for land-sharing versus land-sparing approaches to urban planning for biodiversity (Lin and Fuller, 2013; Soga *et al.*, 2014). Although urban habitats can harbor self-sustaining populations of common, rare and endangered native species, they cannot replace the complete functionality of less-disturbed places (Kowarik, 2011). Multiple complementary approaches are needed.

Growing interest in residential areas as potential habitat may point toward new solutions (Padullés Cubino *et al.*, 2018). A major challenge for efforts to support and enhance biodiversity remains the matrix surrounding patches of habitat (Fischer *et al.*, 2005; Hostetler *et al.*, 2011). In cities, this means scaling up conservation planning to include privately held properties (Goddard *et al.*, 2010) and considering the broad range of urban ecosystems, with diversity in many dimensions of pattern and form. Yards and gardens can create corridors of connectivity and provide resources for resident and migratory species (Rudd *et al.*, 2002; Gaston *et al.*, 2005). Homeowner association rules can be designed to invite biodiverse plantings that support local wildlife (Lerman *et al.*, 2012). These approaches to 'bringing nature home' (Tallamy and Darke, 2009) shorten the distance between people and the experience of urban biodiversity.

Experiences of Wonder and Beauty

If 'in the end we will conserve only what we love, we will love only what we understand, and we will understand only what we are taught' (Dioum, 1969), it is essential to biodiversity at a global scale that people in cities can experience biodiversity where they live their everyday lives (Nilon, 2011). This has been called the 'pigeon paradox'; people are most likely to be interested in learning about and conserving nature when they experience it personally, especially as children, but as humanity becomes more urban, most people will experience only species that live in cities (Dunn *et al.*, 2006).

More effort needs to be made to broaden and deepen urban children's interactions with nature, and, whenever possible, with 'big nature' (Kahn and Weiss, 2017). Fortunately, a city park can harbor a wild place big enough to get lost in, and 'outside lies magic' (Stilgoe, 1999) for all those who look. Growing interest in urban nature has opened many new avenues for engagement with biodiversity, from nature play schools to foraging lessons. There is a growing global enthusiasm for forest bathing to decrease stress and improve feelings of well-being (Miyazaki, 2018; Antonelli *et al.*, 2019), and remedies for nature-deficit disorder now include medical prescriptions for outdoor time (Crnic and Kondo, 2019; Koselka *et al.*, 2019). People of all ages are observing urban species as participants in citizen science programes (Schuttler *et al.*, 2018); while urban environmental stewardship networks span and interconnect across cities (Svendsen and Campbell, 2008; Svendsen *et al.*, 2016). Effects of these activities reach beyond the addition of a planted tree; participation in stewardship is often motivated by interest in helping the community in which people live (Sorensen *et al.*, 2018) and can increase civic engagement (Fisher *et al.*, 2015).

Care may be a way to engage people in stewardship of biodiversity by connecting responses to species they notice in everyday life to their effect on larger environmental systems (Nassauer, 2011). Urban ecosystems are the baseline from which future generations will extend their understanding of biodiversity. It is here that the post-wild world perspective, with its ability to find wonder in weeds, can create a bridge to elsewhere-nature by inspiring curiosity and care.

Acknowledgements

I would like to thank Steward Pickett for helpful comments on the draft of this chapter, and Maile Neel for a helpful nudge to write.

References

ABC7 (2019) Squirrels gone wild, help themselves to people food [WWW document]. ABC7 Chicago. Available at: https://abc7chicago.com/3593849/ (accessed 11 August 2019).

Alcoforado, M. and Andrade, H. (2008) Global warming and the urban heat island. *Urban Ecology*, 249–511.

Antonelli, M., Barbieri, G. and Donelli, D. (2019) Effects of forest bathing (shinrin-yoku) on levels of cortisol as a stress biomarker: a systematic review and meta-analysis. *International Journal of Biometeorology* 63(8), 1117–1134. DOI: 10.1007/s00484-019-01717-x.

Aronson, M.F.J. and Handel, S.N. (2011) Deer and invasive plant species suppress forest herbaceous communities and canopy tree regeneration. *Natural Areas Journal* 31(4), 400–407. DOI: 10.3375/043.031.0410.

Aronson, M.F.J., La Sorte, F.A., Nilon, C.H., Katti, M., Goddard, M.A. *et al.* (2014) A global analysis of the impacts of urbanization on bird and plant diversity reveals key anthropogenic drivers. *Proceedings of the Royal Society B: Biological Sciences* 281(1780), 20133330. DOI: 10.1098/rspb.2013.3330.

Aronson, M.F.J., Nilon, C.H., Lepczyk, C.A., Parker, T.S., Warren, P.S. *et al.* (2016) Hierarchical filters determine community assembly of urban species pools. *Ecology* 97(11), 2952–2963. DOI: 10.1002/ecy.1535.

Atha, D.E., Forrest, T., Naczi, R.F.C., Pace, M.C., Rubin, M. *et al.* (2016) The historic and extant spontaneous vascular flora of the New York botanical garden. *Brittonia* 68(3), 245–277. DOI: 10.1007/s12228-016-9417-5.

Avins, M. (2013) *Baltimore's Forest Patches: Emerald Assets for Ecosystem Services*. Available at: chrome-extension://oemmndcbldboiebfnladdacbdfmadadm/https://baltimoregreenspace.org/wp-content/downloads/ForestPatchesWeb.pdf

Band, L.E., Cadenasso, M.L., Grimmond, C.S., Grove, J.M. and Pickett, S.T.A. (2005) Heterogeneity in urban ecosystems: patterns and process. In: Lovett, G.M., Turner, M.G., Jones, C.G. and Weathers, K.C. (eds) *Ecosystem Function in Heterogeneous Landscapes*. Springer New York, New York, pp. 257–278.

Beninde, J., Veith, M. and Hochkirch, A. (2015) Biodiversity in cities needs space: a meta-analysis of factors determining intra-urban biodiversity variation. *Ecology Letters* 18(6), 581–592. DOI: 10.1111/ele.12427.

Bodin, O., Tengö, M., Norman, A., Lundberg, J. and Elmqvist, T. (2006) The value of small size: loss of forest patches and ecological thresholds in southern Madagascar. *Ecological Applications* 16(2), 440–451. DOI: 10.1890/1051-0761(2006)016[0440:TVOSSL]2.0.CO;2.

Burghardt, K.T., Tallamy, D.W. and Gregory Shriver, W. (2009) Impact of native plants on bird and butterfly biodiversity in suburban landscapes. *Conservation Biology* 23(1), 219–224. DOI: 10.1111/j.1523-1739.2008.01076.x.

Cambridge English Dictionary (2019) *Nature*. Cambridge University Press.

Caron, M. (2019) Weeds [WWW document]. Mona Caron. Available at: https://www.monacaron.com/weeds (accessed 11 December 2019).

Cazabon, L. (2019) Uncultivated (2010–present) [WWW document]. Lynn Cazabon. Available at: https://www.lynncazabon.com/uncultivated (accessed 12 May 2019).

Cheptou, P.-O., Carrue, O., Rouifed, S. and Cantarel, A. (2008) Rapid evolution of seed dispersal in an urban environment in the weed *Crepis sancta*. *Proceedings of the National Academy of Sciences* 105(10), 3796–3799. DOI: 10.1073/pnas.0708446105.

Craul, P.J. (1999) *Urban Soils: Applications and Practices*. John Wiley & Sons.

Crnic, M. and Kondo, M.C. (2019) Nature RX: reemergence of pediatric nature-based therapeutic programs from the late 19th and early 20th centuries. *American Journal of Public Health* 109(10), 1371–1378. DOI: 10.2105/AJPH.2019.305204.

Davis, M.A., Chew, M.K., Hobbs, R.J., Lugo, A.E., Ewel, J.J. *et al*. (2011) Don't judge species on their origins. *Nature* 474(7350), 153–154. DOI: 10.1038/474153a.

Dearborn, D.C. and Kark, S. (2010) Motivations for conserving urban biodiversity. *Conservation Biology* 24(2), 432–440. DOI: 10.1111/j.1523-1739.2009.01328.x.

Dioum, B. (1969) Paper presented at the General Assembly of the IUCN. *Presented at the 10th General Assembly of the International Union for the Conservation of Nature and Natural Resources*, New Delhi, India.

Drew, L.W. (2011) Are we losing the science of taxonomy? As need grows, numbers and training are failing to keep up. *BioScience* 61, 942–946.

Dunn, R.R., Gavin, M.C., Sanchez, M.C. and Solomon, J.N. (2006) The pigeon paradox: dependence of global conservation on urban nature. *Conservation Biology* 20(6), 1814–1816. DOI: 10.1111/j.1523-1739.2006.00533.x.

Effland, W.R. and Pouyat, R.V. (1997) The genesis, classification, and mapping of soils in urban areas. *Urban Ecosystems* 1(4), 217–228. DOI: 10.1023/A:1018535813797.

Faeth, S.H., Bang, C. and Saari, S. (2011) Urban biodiversity: patterns and mechanisms. *Annals of the New York Academy of Sciences* 1223(1), 69–81. DOI: 10.1111/j.1749-6632.2010.05925.x.

Fahrig, L. (2017) Ecological responses to habitat fragmentation per se. *Annual Review of Ecology, Evolution, and Systematics* 48(1), 1–23. DOI: 10.1146/annurev-ecolsys-110316-022612.

Fischer, J., Fazey, I., Briese, R. and Lindenmayer, D.B. (2005) Making the matrix matter: challenges in Australian grazing landscapes. *Biodiversity and Conservation* 14(3), 561–578. DOI: 10.1007/s10531-004-3916-5.

Fischer, L.K., von der Lippe, M. and Kowarik, I. (2013) Urban land use types contribute to grassland conservation: the example of Berlin. *Urban Forestry & Urban Greening* 12(3), 263–272. DOI: 10.1016/j.ufug.2013.03.009.

Fisher, D.R., Svendsen, E.S., Connolly, J., Svendsen, E.S. and Connolly, J. (2015) *Urban Environmental Stewardship and Civic Engagement: How planting trees strengthens the roots of democracy*. Routledge, Abingdon, UK.

Foderaro, L.W. (2011) *New York Moves to Stop Foraging in City's Parks*. The New York Times.

Folke, C., Hahn, T., Olsson, P. and Norberg, J. (2005) Adaptive governance of social-ecological systems. *Annual Review of Environment and Resources* 30(1), 441–473. DOI: 10.1146/annurev.energy.30.050504.144511.

Francis, R.A. and Chadwick, M.A. (2012) What makes a species synurbic? *Applied Geography* 32(2), 514–521. DOI: 10.1016/j.apgeog.2011.06.013.

Gaston, K.J., Smith, R.M., Thompson, K. and Warren, P.H. (2005) Urban domestic gardens (II): experimental tests of methods for increasing biodiversity. *Biodiversity and Conservation* 14(2), 395–413. DOI: 10.1007/s10531-004-6066-x.

Goddard, M.A., Dougill, A.J. and Benton, T.G. (2010) Scaling up from gardens: biodiversity conservation in urban environments. *Trends in Ecology & Evolution* 25(2), 90–98. DOI: 10.1016/j.tree.2009.07.016.

Godefroid, S. and Koedam, N. (2003) How important are large vs. small forest remnants for the conservation of the woodland flora in an urban context? *Global Ecology and Biogeography* 12(4), 287–298. DOI: 10.1046/j.1466-822X.2003.00035.x.

Graham, S.I., Kinnaird, M.F., O'Brien, T.G., Vågen, T.-G., Winowiecki, L.A. *et al*. (2019) Effects of land-use change on community diversity and composition are highly variable among functional groups. *Ecological Applications* 29(7), e01973. DOI: 10.1002/eap.1973.

Guldin, J.M., Smith, J.R. and Thompson, L. (1990) Stand structure of an old-growth upland hardwood forest in Overton Park, Memphis, Tennessee. In: Mitchell, R.S., Sheviak, C.J. and Leopold, D.J. (eds)

Ecosystem Management: Rare Species and Significant Habitats. New York State Museum, Albany, New York, pp. 61–66.

Güneralp, B. and Seto, K.C. (2013) Futures of global urban expansion: uncertainties and implications for biodiversity conservation. *Environmental Research Letters* 8(1), 014025. DOI: 10.1088/1748-9326/8/1/014025.

Hall, S.J., Learned, J., Ruddell, B., Larson, K.L., Cavender-Bares, J. *et al*. (2016) Convergence of microclimate in residential landscapes across diverse cities in the United States. *Landscape Ecology* 31(1), 101–117. DOI: 10.1007/s10980-015-0297-y.

Harper, J.L., Hawksworth, D.L. and Hawksworth, D.L. (1994) Biodiversity: measurement and estimation. Preface. *Philosophical Transactions of the Royal Society of London. Series B: Biological Sciences* 345, 5–12.

Hostetler, M., Allen, W. and Meurk, C. (2011) Conserving urban biodiversity? Creating green infrastructure is only the first step. *Landscape and Urban Planning* 100(4), 369–371. DOI: 10.1016/j.landurbplan.2011.01.011.

Hughes, R.M., Dunham, S., Maas-Hebner, K.G., Yeakley, J.A., Schreck, C. *et al*. (2014) A review of urban water body challenges and approaches: (1) rehabilitation and remediation. *Fisheries* 39(1), 18–29. DOI: 10.1080/03632415.2013.836500.

Hulme, P.E. (2008) Contrasting alien and native plant species-area relationships: the importance of spatial grain and extent. *Global Ecology and Biogeography* 17(5), 641–647. DOI: 10.1111/j.1466-8238.2008.00404.x.

Ignatieva, M., Stewart, G.H. and Meurk, C. (2011) Planning and design of ecological networks in urban areas. *Landscape and Ecological Engineering* 7(1), 17–25. DOI: 10.1007/s11355-010-0143-y.

Imhoff, M.L., Bounoua, L., DeFries, R., Lawrence, W.T., Stutzer, D. *et al*. (2004) The consequences of urban land transformation on net primary productivity in the United States. *Remote Sensing of Environment* 89(4), 434–443. DOI: 10.1016/j.rse.2003.10.015.

Ives, C.D., Lentini, P.E., Threlfall, C.G., Ikin, K., Shanahan, D.F. *et al*. (2016) Cities are hotspots for threatened species. *Global Ecology and Biogeography* 25(1), 117–126. DOI: 10.1111/geb.12404.

Johnson, M.T.J. and Munshi-South, J. (2017) Evolution of life in urban environments. *Science* 358(6363), eaam8327. DOI: 10.1126/science.aam8327.

Johnson, L.R., Breger, B. and Drummond, F. (2019a) Novel plant–insect interactions in an urban environment: enemies, protectors, and pollinators of invasive knotweeds. *Ecosphere* 10(11), e02885. DOI: 10.1002/ecs2.2885.

Johnson, M.L., Novem Auyeung, D.S., Sonti, N.F., Pregitzer, C.C., McMillen, H.L. *et al*. (2019b) Social-ecological research in urban natural areas: an emergent process for integration. *Urban Ecosystems* 22(1), 77–90. DOI: 10.1007/s11252-018-0763-9.

Kahn, P.H. Jr. and Weiss, T. (2017) The importance of children interacting with big nature. *Children, Youth and Environments* 27, 7–24.

Kantsa, A., Tscheulin, T., Junker, R.R., Petanidou, T. and Kokkini, S. (2013) Urban biodiversity hotspots wait to get discovered: the example of the city of Ioannina, NW Greece. *Landscape and Urban Planning* 120, 129–137. DOI: 10.1016/j.landurbplan.2013.08.013.

Kattwinkel, M., Strauss, B., Biedermann, R. and Kleyer, M. (2009) Modelling multi-species response to landscape dynamics: mosaic cycles support urban biodiversity. *Landscape Ecology* 24(7), 929–941. DOI: 10.1007/s10980-009-9371-7.

Kattwinkel, M., Biedermann, R. and Kleyer, M. (2011) Temporary conservation for urban biodiversity. *Biological Conservation* 144(9), 2335–2343. DOI: 10.1016/j.biocon.2011.06.012.

Koselka, E.P.D., Weidner, L.C., Minasov, A., Berman, M.G., Leonard, W.R. *et al*. (2019) Walking green: developing an evidence base for nature prescriptions. *International Journal of Environmental Research and Public Health* 16(22), 4338. DOI: 10.3390/ijerph16224338.

Kowarik, I. (2011) Novel urban ecosystems, biodiversity, and conservation. *Selected papers from the conference Urban Environmental Pollution: Overcoming Obstacles to Sustainability and Quality of Life (UEP2010)*, Boston, USA 159, 20-23 June 2010, pp. 1974–1983.

Kowarik, I., Pyšek, P., Prach, K., Rejmánek, M. and Wade, M. (1995) Time lags in biological invasions with regard to the success and failure of alien species. In: *Time Lags in Biological Invasions with Regard to the Success and Failure of Alien Species*. SPB Academic Pubishing, Amsterdam, The Netherlands, pp. 15–38.

Kowarik, I. and Körner, S. (eds) (2005) *Wild Urban Woodlands: New Perspectives for Urban Forestry*. Springer, Berlin, Heidelberg.

Kowarik, I., Hiller, A., Planchuelo, G., Seitz, B., von der Lippe, M. *et al.* (2019) Emerging urban forests: opportunities for promoting the wild side of the urban green infrastructure. *Sustainability* 11(22), 6318. DOI: 10.3390/su11226318.

Kühn, N. (2006) Intentions for the unintentional. *Journal of Landscape Architecture* 1(2), 46–53. DOI: 10.1080/18626033.2006.9723372.

Kühn, I. and Klotz, S. (2006) Urbanization and homogenization – comparing the floras of urban and rural areas in Germany. *Biological Conservation* 127(3), 292–300. DOI: 10.1016/j.biocon.2005.06.033.

Kühn, I., Brandl, R. and Klotz, S. (2004) The flora of German cities is naturally species rich. *Evolutionary Ecology Research* 6, 749–764.

La Sorte, F.A., Aronson, M.F.J., Williams, N.S.G., Celesti-Grapow, L., Cilliers, S. *et al.* (2014) Beta diversity of urban floras among European and non-European cities. *Global Ecology and Biogeography* 23(7), 769–779. DOI: 10.1111/geb.12159.

Lepczyk, C.A., Aronson, M.F.J., Evans, K.L., Goddard, M.A., Lerman, S.B. *et al.* (2017) Biodiversity in the city: fundamental questions for understanding the ecology of urban green spaces for biodiversity conservation. *BioScience* 67(9), 799–807. DOI: 10.1093/biosci/bix079.

Lerman, S.B., Turner, V.K. and Bang, C. (2012) Homeowner associations as a vehicle for promoting native urban biodiversity. *Ecology and Society* 17(4). DOI: 10.5751/ES-05175-170445.

Lin, B.B. and Fuller, R.A. (2013) FORUM: Sharing or sparing? How should we grow the world's cities? *The Journal of Applied Ecology* 50, 1161–1168.

Loeb, R.E. (2011) *Old Growth Urban Forests*. Springer, New York.

Lososová, Z., Chytrý, M., Tichý, L., Danihelka, J., Fajmon, K. *et al.* (2012) Biotic homogenization of central European urban floras depends on residence time of alien species and habitat types. *Biological Conservation* 145(1), 179–184. DOI: 10.1016/j.biocon.2011.11.003.

Louv, R. (2011) *The Nature Principle: Human Restoration and the End of Nature-Deficit Disorder*. Algonquin Books.

Lundberg, J. and Moberg, F. (2003) Mobile link organisms and ecosystem functioning: implications for ecosystem resilience and management. *Ecosystems* 6(1), 0087–0098. DOI: 10.1007/s10021-002-0150-4.

Lundholm, J.T. and Richardson, P.J. (2010) Mini-review: habitat analogues for reconciliation ecology in urban and industrial environments. *Journal of Applied Ecology* 47(5), 966–975. DOI: 10.1111/j.1365-2664.2010.01857.x.

Marris, E. (2013) *Rambunctious Garden: Saving Nature in a Post-Wild World.* Bloomsbury USA, New York.

McDonald, T., Gann, G.D., Jonson, J. and Dixon, K.W. (2016) *International Standards for the Practice of Ecological Restoration – Including Principles and Key Concepts*, 1st edn. Society for Ecological Restoration, Washington, DC.

McDonnell, M.J. and Hahs, A.K. (2015) Adaptation and adaptedness of organisms to urban environments. *Annual Review of Ecology, Evolution, and Systematics* 46(1), 261–280. DOI: 10.1146/annurev-ecolsys-112414-054258.

McDonnell, M.J. and Hahs, A.K. (2013) The future of urban biodiversity research: moving beyond the 'low-hanging fruit'. *Urban Ecosystems* 16(3), 397–409. DOI: 10.1007/s11252-013-0315-2.

McHale, M.R., Bunn, D.N., Pickett, S.T. and Twine, W. (2013) Urban ecology in a developing world: why advanced socioecological theory needs Africa. *Frontiers in Ecology and the Environment* 11(10), 556–564. DOI: 10.1890/120157.

McHale, M., Pickett, S., Barbosa, O., Bunn, D., Cadenasso, M. *et al.* (2015) The new global urban realm: complex, connected, diffuse, and diverse Social-Ecological systems. *Sustainability* 7(5), 5211–5240. DOI: 10.3390/su7055211.

McKinney, M.L. (2006) Urbanization as a major cause of biotic homogenization. *Biological Conservation* 127(3), 247–260. DOI: 10.1016/j.biocon.2005.09.005.

McLain, R.J., Hurley, P.T., Emery, M.R. and Poe, M.R. (2014) Gathering "wild" food in the city: rethinking the role of foraging in urban ecosystem planning and management. *Local Environment* 19(2), 220–240. DOI: 10.1080/13549839.2013.841659.

McPhearson, T., Pickett, S.T.A., Grimm, N.B., Niemelä, J., Alberti, M. *et al.* (2016) Advancing urban ecology toward a science of cities. *BioScience* 66(3), 198–212. DOI: 10.1093/biosci/biw002.

Miller, J.R. (2005) Biodiversity conservation and the extinction of experience. *Trends in Ecology & Evolution* 20(8), 430–434. DOI: 10.1016/j.tree.2005.05.013.

Miyazaki, Y. (2018) *Shinrin Yoku: The Japanese Art of Forest Bathing*. Timber Press.

Morzillo, A.T., de Beurs, K.M. and Martin-Mikle, C.J. (2014) A conceptual framework to evaluate human-wildlife interactions within coupled human and natural systems. *Ecology and Society* 19(3). DOI: 10.5751/ES-06883-190344.

Nassauer, J.I. (2011) Care and stewardship: from home to planet. *Landscape and Urban Planning* 100(4), 321–323. DOI: 10.1016/j.landurbplan.2011.02.022.

Niemelä, J. (1999) Ecology and urban planning. *Biodiversity and Conservation* 8(1), 119–131. DOI: 10.1023/A:1008817325994.

Nilon, C.H. (2011) Urban biodiversity and the importance of management and conservation. *Landscape and Ecological Engineering* 7(1), 45–52. DOI: 10.1007/s11355-010-0146-8.

Nilon, C.H., Aronson, M.F.J., Cilliers, S.S., Dobbs, C., Frazee, L.J. *et al.* (2017) Planning for the future of urban biodiversity: a global review of city-scale initiatives. *BioScience* 67(4), 332–342. DOI: 10.1093/biosci/bix012.

Ossola, A., Locke, D., Lin, B. and Minor, E. (2019) Yards increase forest connectivity in urban landscapes. *Landscape Ecology* 34(12), 2935–2948. DOI: 10.1007/s10980-019-00923-7.

Ostrom, E. (2009) A general framework for analyzing sustainability of social-ecological systems. *Science* 325(5939), 419–422. DOI: 10.1126/science.1172133.

Padullés Cubino, J., Cavender-Bares, J., Hobbie, S.E., Pataki, D.E., Avolio, M.L. *et al.* (2018) Drivers of plant species richness and phylogenetic composition in urban yards at the continental scale. *Landscape ecology* 34, 63–77.

Palmer, M.A., Kramer, J.G., Boyd, J. and Hawthorne, D. (2016) Practices for facilitating interdisciplinary synthetic research: the National Socio-Environmental Synthesis Center (SESYNC). *Current Opinion in Environmental Sustainability* 19, 111–122. DOI: 10.1016/j.cosust.2016.01.002.

Palomino, M., Taylor, T., Göker, A., Isaacs, J. and Warber, S. (2016) The online dissemination of nature–health concepts: lessons from sentiment analysis of social media relating to "nature-deficit disorder". *International Journal of Environmental Research and Public Health* 13(1), 142. DOI: 10.3390/ijerph13010142.

Pearse, W.D., Cavender-Bares, J., Hobbie, S.E., Avolio, M.L., Bettez, N. *et al.* (2018) Homogenization of plant diversity, composition, and structure in North American urban yards. *Ecosphere* 9(2), e02105. DOI: 10.1002/ecs2.2105.

Pickett, S.T.A. and Cadenasso, M.L. (2009) Altered resources, disturbance, and heterogeneity: a framework for comparing urban and non-urban soils. *Urban Ecosystems* 12(1), 23–44. DOI: 10.1007/s11252-008-0047-x.

Pickett, S.T.A., Burch, W., Dalton, S. and Foresman, T. (1997) A conceptual framework for the study of human ecosystems in urban areas. *Urban Ecosystems*.

Pickett, S.T.A., Cadenasso, M.L., Grove, J.M., Boone, C.G., Groffman, P.M. *et al.* (2011) Urban ecological systems: scientific foundations and a decade of progress. *Journal of Environmental Management* 92(3), 331–362. DOI: 10.1016/j.jenvman.2010.08.022.

Pickett, S.T.A., Cadenasso, M.L., Rosi-Marshall, E.J., Belt, K.T., Groffman, P.M. *et al.* (2017) Dynamic heterogeneity: a framework to promote ecological integration and hypothesis generation in urban systems. *Urban Ecosystems* 20(1), 1–14. DOI: 10.1007/s11252-016-0574-9.

Pimm, S.L. and Raven, P. (2000) Extinction by numbers. *Nature* 403(6772), 843–845. DOI: 10.1038/35002708.

Prange, S. and Gehrt, S.D. (2004) Changes in mesopredator-community structure in response to urbanization. *Canadian Journal of Zoology* 82(11), 1804–1817. DOI: 10.1139/z04-179.

Pregitzer, C.C., Charlop-Powers, S., Bibbo, S., Forgione, H.M., Gunther, B. *et al.* (2019) A city-scale assessment reveals that native forest types and overstory species dominate New York City forests. *Ecological Applications: a publication of the Ecological Society of America* 29(1), e01819. DOI: 10.1002/eap.1819.

Rainer, T. and West, C. (2015) *Planting in a Post-Wild World: Designing Plant Communities for Resilient Landscapes*. Timber Press, Portland, Oregon.

Rebelo, A.G., Holmes, P.M., Dorse, C. and Wood, J. (2011) Impacts of urbanization in a biodiversity hotspot: conservation challenges in metropolitan Cape Town. *South African Journal of Botany* 77(1), 20–35. DOI: 10.1016/j.sajb.2010.04.006.

Rodewald, A.D. and Kearns, L.J. (2011) Shifts in dominant nest predators along a rural-to-urban landscape gradient Cambio de Depredadores de NIDO Dominantes a lo largo de un Gradiente Rural—Urbano. *The Condor* 113, 899–906.

Rudd, H., Vala, J. and Schaefer, V. (2002) Importance of backyard habitat in a comprehensive biodiversity conservation strategy: a connectivity analysis of urban green spaces. *Restoration Ecology* 10(2), 368–375. DOI: 10.1046/j.1526-100X.2002.02041.x.

Schuttler, S.G., Sorensen, A.E., Jordan, R.C., Cooper, C. and Shwartz, A. (2018) Bridging the nature gap: can citizen science reverse the extinction of experience? *Frontiers in Ecology and the Environment* 16(7), 405–411. DOI: 10.1002/fee.1826.

Seiter, D. (2016) *Spontaneous Urban Plants*. Archer.

Seto, K.C., Güneralp, B. and Hutyra, L.R. (2012) Global forecasts of urban expansion to 2030 and direct impacts on biodiversity and carbon pools. *Proceedings of the National Academy of Sciences* 109(40), 16083–16088. DOI: 10.1073/pnas.1211658109.

Shackleton, C., Hurley, P., Dahlberg, A., Emery, M. and Nagendra, H. (2017) Urban foraging: a ubiquitous human practice overlooked by urban planners, policy, and research. *Sustainability* 9(10), 1884. DOI: 10.3390/su9101884.

Shochat, E., Warren, P.S., Faeth, S.H., McIntyre, N.E. and Hope, D. (2006) From patterns to emerging processes in mechanistic urban ecology. *Trends in Ecology & Evolution* 21(4), 186–191. DOI: 10.1016/j.tree.2005.11.019.

Shochat, E., Lerman, S.B., Anderies, J.M., Warren, P.S., Faeth, S.H. *et al*. (2010) Invasion, competition, and biodiversity loss in urban ecosystems. *BioScience* 60(3), 199–208. DOI: 10.1525/bio.2010.60.3.6.

Singh, S.D. (1969) Urban monkeys. *Scientific American* 221(1), 108–115. DOI: 10.1038/scientificamerican0769-108.

Soanes, K., Sievers, M., Chee, Y.E., Williams, N.S.G., Bhardwaj, M. *et al*. (2019) Correcting common misconceptions to inspire conservation action in urban environments. *Conservation Biology* 33(2), 300–306. DOI: 10.1111/cobi.13193.

Soga, M., Yamaura, Y., Koike, S. and Gaston, K.J. (2014) Land sharing vs. land sparing: does the compact city reconcile urban development and biodiversity conservation? *Journal of Applied Ecology* 51(5), 1378–1386. DOI: 10.1111/1365-2664.12280.

Sorensen, A.E., Jordan, R.C., Blaise, G., Brown, J., Campbell, L.K. *et al*. (2018) Drivers of public participation in urban restoration stewardship programs: linkages between environmental identity and knowledge. *Arboriculture and Urban Forestry* 44, 266–282.

Stilgoe, J.R. (1999) *Outside Lies Magic: Regaining History and Awareness in Everyday Places*. Walker Books, London.

Stillfried, M., Fickel, J., Börner, K., Wittstatt, U., Heddergott, M. *et al*. (2017) Do cities represent sources, sinks or isolated islands for urban wild boar population structure? *Journal of Applied Ecology* 54(1), 272–281. DOI: 10.1111/1365-2664.12756.

Svendsen, E. and Campbell, L. (2008) *Urban Ecological Stewardship: Understanding the Structure, Function and Network of Community-based Urban Land Management*. Cities and the Environment (CATE) 1.

Svendsen, E.S., Campbell, L.K., Fisher, D.R., Connolly, J.J.T., Johnson, M.L. *et al*. (2016) Stewardship mapping and assessment project: a framework for understanding community-based environmental stewardship. General technical report NRS-156). *US Forest Service Northern Research Station*.

Swan, C.M., Pickett, S.T.A., Szlavecz, K., Warren, P. and Willey, K.T. (2011) *Biodiversity and Community Composition in Urban Ecosystems: Coupled Human, Spatial, and Metacommunity Processes*. Oxford University Press.

Tallamy, D.W. and Darke, R. (2009) *Bringing Nature Home: How You Can Sustain Wildlife with Native Plants, Updated and Expanded*, 2nd edn. Timber Press, Portland, Oregon.

TEEB (2011) *The Economics of Ecosystems and Biodiversity Manual for Cities: Ecosystem Services in Urban Management*. United Nations Environment Programme, ICLEI Local Governments for Sustainability, Local Action for Biodiversity Programme, and International Union for the Conservation of Nature (IUCN).

Trentanovi, G., von der Lippe, M., Sitzia, T., Ziechmann, U., Kowarik, I. *et al*. (2013) Biotic homogenization at the community scale: disentangling the roles of urbanization and plant invasion. *Diversity and Distributions* 19(7), 738–748. DOI: 10.1111/ddi.12028.

Turvey, S.T., Barrett, L.A.,Hao, Y., Zhang, L., Zhang, X. *et al*. (2010) Rapidly shifting baselines in Yangtze fishing communities and local memory of extinct species. *Conservation Biology* 24(3), 778–787. DOI: 10.1111/j.1523-1739.2009.01395.x.

United Nations (2014) *World Urbanization Prospects, 2014 revision*. United Nations, Department of Economic and Social Affairs, Population Division, New York.

van den Bosch, M. and Ode Sang, Å. (2017) Urban natural environments as nature-based solutions for improved public health - A systematic review of reviews. *Environmental Research* 158, 373–384. DOI: 10.1016/j.envres.2017.05.040.

Van Rossum, F. and Triest, L. (2012) Stepping-stone populations in linear landscape elements increase pollen dispersal between urban forest fragments. *Plant Ecology and Evolution* 145(3), 332–340. DOI: 10.5091/plecevo.2012.737.

Wessolek, G. (2008) Sealing of soils. *Urban Ecology*.

Williams, N.S.G., Lundholm, J. and Scott MacIvor, J. (2014) Forum: do green roofs help urban biodiversity conservation? *Journal of Applied Ecology* 51(6), 1643–1649. DOI: 10.1111/1365-2664.12333.

Wittig, R. (2010) Biodiversity of urban-industrial areas and its evaluation – a critical review. In: *Urban Biodiversity and Design*. John Wiley & Sons, Ltd, pp. 35–55.

Wittig, R. and Becker, U. (2010) The spontaneous flora around street trees in cities—A striking example for the worldwide homogenization of the flora of urban habitats. *Flora - Morphology, Distribution, Functional Ecology of Plants* 205(10), 704–709. DOI: 10.1016/j.flora.2009.09.001.

Wittig, R., Diesing, D. and Gödde, M. (1985) Urbanophob — Urbanoneutral — Urbanophil Das Verhalten Der Arten gegenüber dem Lebensraum Stadt. *Flora* 177(5-6), 265–282. DOI: 10.1016/S0367-2530(17)30180-9.

Wrzesień, M., Denisow, B., Mamchur, Z., Chuba, M. and Resler, I. (2016) Composition and structure of the flora in intra-urban railway areas. *Acta Agrobotanica* 69(3), 1–14. DOI: 10.5586/aa.1666.

6 Protecting Pollinators in the Urban Environment

Lisa J. Kuder*, Lindsay Miller Barranco and Dennis van Engelsdorp
Department of Entomology, University of Maryland, College Park, Maryland

Introduction

With urban land area expected to triple from 2000 to 2030 (Seto *et al.*, 2012), more species will come into contact with metropolitan environments. Such areas are traditionally viewed as biological deserts with marginal conservation value, as they are associated with habitat degradation and species loss (Hall *et al.*, 2017; McKinney, 2008). However, the rapid expansion of urbanization and the global concern about biodiversity loss have led to greater awareness and interest in urban ecology. Urban planners are starting to rethink the role of cities by incorporating ecological function and conservation into design and renovation projects. Urban areas are thus increasingly recognized as having potential to protect insect diversity and their ecosystem services, particularly pollination (Baldock *et al.*, 2015; New, 2015).

Pollinators comprise a variety of organisms, ranging from bats to birds and a host of insects including beetles, butterflies, flies, moths and, of course, bees. Managed and wild bees are the most efficient pollinators thanks to their specialized pollen collecting tools and foraging behaviours (Wilson-Rich, 2014). Approximately 75% of agricultural food crops (Klein *et al.*, 2007; Aizen *et al.*, 2009) and 89% of wild flowering plants, at least to some extent, rely on bee pollination (Ollerton *et al.*, 2011). From an economic perspective, pollination of global agricultural crops is valued at an estimated $US351 billion annually (Lautenbach *et al.*, 2012). Consequently, pollination services are highly valued by humans and considered a keystone process to agriculture and natural vegetation productivity, making managed and wild bee health a priority in rural and urban areas alike (Kevan, 1991).

From a practical standpoint, unlike mammal conservation, which often requires large areas of undisturbed habitat, bees are tiny and have a relatively small ecological footprint. Also, the physical features of bee habitat (colourful flowers and warm season bunch grasses that offer seasonal interest and texture) and various attributes (minimal maintenance needed once established) can be complementary to urban aesthetics and function, thus increasing the likelihood of pollinator conservation receiving wide public acceptance and adoption (Hall *et al.*, 2017). While urban pollinators face multiple challenges including habitat loss, degradation and fragmentation, cities often support diverse bee faunas in public green spaces, and offer tremendous potential to support and expand pollinator conservation efforts (Hall *et al.*, 2017; Gardiner *et al.*, 2013; Baldock *et al.*, 2015).

Urban green space management, likewise, faces practical challenges, as it is a complex

*Corresponding author: lkuder@umd.edu

and interacting combination of social, cultural and economic factors with multiple and diverse stakeholders (Aronson *et al.*, 2017; Turo and Gardiner, 2019). Previous efforts to create pollinator habitat in metropolitan areas have been met with mixed results, eliciting both positive and negative feelings from those living nearby (Aronson *et al.*, 2017; Hoyle *et al.*, 2017; Turo and Gardiner, 2019). To achieve desired outcomes, urban ecologists must meet pollinators' needs in a way that is economically practical and respectful of societal norms and safety concerns (Kevan 1991; Gardiner *et al.*, 2013). Lessons learned from numerous case studies show that a community-guided approach is needed in design and management of public green spaces (Hunter and Hunter, 2008; Green *et al.*, 2016; Turo and Gardiner, 2019).

In this chapter, we provide guidelines on how to protect bees in the urban environment. With that goal in mind, we review the basic habitat requirements of bees, consider the factors that influence plant–pollinator interactions in urban environments, discuss urban trends that have been observed by both beekeepers and wild-bee researchers, consider the complex challenge of combining pollinator conservation with the needs and preferences of human communities, and provide a roadmap for achieving long-lasting pollinator conservation in urban environments.

Pollinator Conservation – What Do Bees Need?

Bees (Order: Hymenoptera) belong to a group in the superfamily Apoidea called *Anthophila*; a name that originates from the adjective anthophagous, which means feeding on flowers. The common honey bee (*Apis mellifera*) is only one of over 20,000 known species of bees worldwide (Wilson-Rich, 2014). Angiosperms or flowering plants, which coevolved with bees and other pollinating insects, have flourished and are now the dominant vegetation type with over 300,000 species comprising ~80% of the earth's flora (Stevenson *et al.*, 2017). Among the world's flowering plants, wide variation exists in floral morphology (i.e. in their shape and structure), phenology (bloom time) and nutritional rewards provided to pollinators (i.e. essential vitamins and minerals), so it is not surprising that this flowering plant diversity is mirrored by the wide variation observed in bee morphology and seasonality (Wilson-Rich, 2014). The basic habitat requirements of wild and managed bees are discussed below to provide guidelines for plant selection and vegetation management that supports bees and other pollinators in urban environments.

Floral diversity

Since floral and bee traits vary greatly, different groups of bees visit different types of flowers. The seven major bee families are divided into three major groups based on tongue length, an anatomical feature that determines from which flowers a bee can efficiently collect rewards (e.g. shallow or deep tubular flowers) (Wilson-Rich, 2014). Families of small-tongued bees include Andrenidae, Colletidae and Stenotritidae, medium-tongued bees Melittidae and Halictidae and long-tongued bees Apidae and Megachilidae (Wilson-Rich, 2014). Matching the pollinator species' seasonal life cycle with that of its preferred flowers is another important factor. Unlike honey bees and bumble bees (Genus: *Bombus*), which are active for the entire growing season, the majority of bee species are active as adults for only a short time (about 6 weeks), which corresponds to the bloom time of their floral hosts (Michener, 2000).

Bees also choose flowers based on the type and characteristics of the rewards offered. Flowers might offer bees nectar (a carbohydrate that fuels adult bees) and pollen (a rich source of protein necessary for larval bee development) (Wilson-Rich, 2014). Nutritional profiles of pollen and nectar vary widely. For instance, pollen grains can include ten essential amino acids with protein levels ranging from 2–60%, as well as varying amounts of carbohydrates, lipids, sterols and other micronutrients, depending on the floral species (Lee-Mäder *et al.*, 2016). Early in the phylogeny of bees, species were oligolectic (specialist) feeders, which provision their nests with pollen from one plant species, genus or family (Danforth *et al.*, 2006). While some bees such as honey bees and bumble bees have evolved to be polylectic (generalist) feeders, many species continue to have a much narrower diet breadth (Lee-Mäder *et al.*, 2016). Nectar profiles can vary widely, too, and contain primarily water and the sugars fructose, glucose and sucrose, which range in concentrations from 10% to 70% depending on the plant species and environmental conditions (temperature, precipitation, humidity and time of day).

Another factor that determines floral preference is the type of toxic defence chemical(s), such as alkaloids, terpenoids and flavonoids, which plants typically use to deter herbivores (Arnold *et al.*, 2014; Stevenson *et al.*, 2017) but is also present in floral rewards. These chemicals vary widely among plant families with equally wide-ranging effects on different groups of pollinators from beneficial to toxic (Arnold *et al.*, 2014; Stevenson *et al.*, 2017). Evolutionary adaptive responses of pollinators include avoidance, floral specificity and pollen mixing to mitigate unfavourable chemical properties (Eckhardt *et al.*, 2014; Stevenson *et al.*, 2017). Due to the complexities and spatial-temporal fluctuations of floral rewards, to thrive, bees need an abundance of flowers as well as heterogeneity. In fact, the number of floral species, density and quality of floral resources (nectar, pollen and oils) are the strongest factors structuring pollinator communities (Potts *et al.*, 2003, 2004).

Dispersal ability

Bees are naturally mobile creatures. Flight gives them the ability to travel throughout the landscape to collect needed resources for themselves and their colony members, or in the case of solitary bees, for their offspring. The maximum distance between a bees' nest and the flowers from which they collect nectar and pollen is referred to as foraging range (Zurbuchen *et al.*, 2010). Foraging distance information is key for understanding how bees will respond in an urban environment, determining the reproductive success of fragmented plant communities and designing pollinator habitat in cities (Greenleaf *et al.*, 2007).

Most foraging range research has focused on honey bees (*Apis mellifera*). For this highly social species, foraging range estimates vary widely, and seems dependent on the differences in the quality and quantity of available forage. There are reports of honey bees foraging as far as 13.5 km (Von Frisch, 1967), but most studies suggest their forage range is several hundred metres to approximately 6 km (Visscher and Seeley, 1982; Schneider, 1989; Waddington *et al.*, 1994). Bumble bees can also fly long-distance – more than 1.5 km (Osborne *et al.*, 2008b). Yet several species have shown a much smaller foraging range, ranging from 25 to 625 m (Darvill *et al.*, 2004; Wolf and Moritz, 2008; Geib *et al.*, 2015). As for the small, mostly solitary bees much less is known. Predictions based on statistical modelling suggest a strong relationship between body size and maximum travel distance, with a limited forage range of 100–200 m for small bees (i.e. sweat bees in the genus *Lasioglossum*; females are about 4.7–5 mm in length) and 1100 m for larger bees (i.e. species in the genus *Bombus*; worker bees range from 11 to 17 mm in length) (Greenleaf *et al.*, 2007).

The relatively small foraging range and dispersal ability of most bee species presents challenges for bee conservation efforts in urban areas. The inherent patchiness of habitat and physical barriers (i.e. buildings, paved surfaces, vehicles etc.) within cities can constrain a bee's mobility, limiting its ability to find mates, establish nests and/or find suitable forage. This can negatively impact genetic diversity of bee and plant populations. For instance, isolated bumble bee colonies are prone to inbreeding, which can result in sterile males and reduced immunity (Whitehorn *et al.*, 2009, 2011). Yet urban planners can limit these impacts by using geographical mapping data to plan pollinator refuges strategically within a greater landscape context. Also, existing vegetation management practices of utility and transportation rights-of-way can be modified (i.e. by reduced mowing and selective herbicide use), which reduce maintenance costs while creating pollinator-friendly corridors that increase connectivity between nesting sites and urban plant populations (Wojcik and Buchmann, 2012; Hopwood, 2013).

Suitable nesting sites

Honey bees (*Apis sp.*) are biologically unique among the 20,000 bee species that occur worldwide in that they are truly the only social organisms, with colonies consisting of as many as 50,000 individuals. They are one of the few invertebrate species that humans have domesticated, where they are managed in moveable frame hives. The majority of bee species are solitary and nest below-ground (~70%) or in small cavities of pithy stems or tunnels created by wood-boring insects (~30%) (Wilson-Rich, 2014).

While there is a paucity of data on the nesting requirements of ground-nesting bees, what is known suggests that different bee species have different nesting needs. Bees display preference for a wide range of soil types (varying ratios of sand, clay and loam), temperature and compaction (Cane, 1991). Some species also prefer nesting in the sides of slopes while others tend to nest in flat terrain. Generally, ground-nesting bees seem to prefer to nest in slopes, as this likely provides good drainage, and in less compact soil (Sardiñas and Kremen, 2014). Incorporating small patches of bare, undisturbed ground and/or areas with thin layers of mulch, free of vegetation, can also increase ground-nesting bee populations (Vaughan and Hoffman Black, 2007; Gregory and Wright, 2005; Pardee and Philpott, 2014). Unfortunately the availability of suitable ground nesting sites is scarce in urban environments,which poses a major challenge for ground-nesting bees.

Urban Drivers of Plant–Pollinator Interactions

Urbanization alters natural ecosystems in myriad numerous ways, i.e. paving over soil with impervious surface cover, introduction and persistence of exotic species, urban warming, and various forms of pollution. Heavily modified landscapes result in relatively novel ecosystems that are incubators for change and adaptation by their inhabitants, whether they are behavioural or functional. Further, each city is different, as it is shaped by its unique combination of natural resources, history and socio-economic forces. The emerging field of urban ecology is examining how plants and pollinators respond to human-modified environments and, in many cases, adapt to such changes.

Habitat loss and fragmentation

A major consequence of urban development is replacement of natural habitats with impervious surfaces. As noted earlier, the loss of exposed ground has profound effects on soil-dependent organisms, such as ground-nesting bees (Geslin *et al.*, 2016). The authors of a recent study on urban bumble bees (*Bombus spp.*) found that higher areas of impervious surfaces supported a lower density of bumble bee nests and fewer bees overall, particularly female workers which tend to forage close to their nest (Glaum *et al.*, 2017). Also, impervious cover can limit dispersal of queen bumble bees because both suitable overwintering and nesting sites are scarce, with detrimental effects on gene flow (Jha and Kremen, 2013). Impervious surface cover also reduces the abundance and richness of other wild bees (Geslin *et al.*, 2016; Hamblin *et al.*, 2018). For instance, as the impervious surface gradient ranging from 0.06% to 64.31% (corresponding to rural, semi-urban and urban) increased, the abundance and diversity of ground-nesting species significantly dropped regardless of local floral resources (Geslin *et al.*, 2016).

In addition to habitat loss, urban development creates a matrix of fragmented open spaces such as parks, vacant lots, residential yards, naturalized remnants and community gardens. Disconnected pockets of flowering plants can alter pollinator populations and foraging behaviours, and thus potentially disrupt pollinator networks (Harrison and Winfree, 2015). Plant species dependent on a particular pollinator may fail to set seed if their pollinators are absent, which, in turn, could ultimately lead to local extinctions (Pauw, 2007). In other cases, generalist species might fill the gap or serve as replacements to specialist species (Cane *et al.*, 2006; Winfree *et al.*, 2014). This emphasizes the importance of generalist pollinators, which can help buffer losses and maintain structure of urban plant-pollinator networks (Pauw, 2007; Harrison and Winfree, 2015). Fragmentation can also lead to behavioural changes that support Pyke's optimal foraging theory, which states that pollinators will either avoid isolated patches or alter their visitation behaviour to offset the energetic costs of foraging (Pyke, 1984). Highly fragmented populations of an unmanaged weed species in an urban setting exhibited low seed set and higher self-fertilization rates than weeds in continuous rural settings, demonstrating that pollinators avoided isolated patches (Cheptou and Avendaño, 2006). However, other studies have concluded that instead of avoidance, bees alter their behaviour by spending more time at each flower, thereby utilizing the flower's resources more fully, in order to compensate for energy lost searching for floral resources and travelling between plant patches (Andrieu *et al.*, 2009; Harrison and Winfree, 2015).

Non-native and managed plant species

Exotic species are often particularly abundant and diverse in highly disturbed sections of city landscapes such as roadside verges, vacant lots and railway tracks (Štajerová *et al.*, 2017). They are also common in highly managed areas such as home gardens, parks and shopping centres. The proportion of exotics in private and public green spaces can be 70% or greater, and is driven by retail availability at garden centres and customer demand for particular aesthetics (Salisbury *et al.*, 2015). The blending of plant species from multiple geographic origins frequently results in novel plant communities that drive plant–pollinator interactions via several potential mechanisms including competitive displacement, pollinator preference and altered bloom phenology (Harrison and Winfree, 2015).

Exotic plants have potential to competitively displace native plants, especially when cultivated by humans, resulting in altered plant–pollinator communities. In some regions, half of the exotic species that have become invasive were introduced intentionally through the ornamental plant trade (La Sorte *et al.*, 2014). Intentional introductions, while a long-standing practice, have intensified in modern times, as they are now done repeatedly, in large quantities and in a way that can facilitate release from enemies and competitors (Catford *et al.*, 2018). For instance, in the USA, turf grasses, which are predominantly European cool-season grasses, are introduced and managed on a massive scale (over 40 million acres), and released from enemy pressure (turf pests) and competition (other flora) through intensive chemical and maintenance inputs (Schueler, 2010). Turf grass, which is the largest cultivated crop in the USA (Schueler, 2010), arguably displaces native flora in semi-urban and urban environments, shifting plant composition and habitat quality for pollinators.

Pollinator preference can be an important mechanism for testing competitive effects between non-native plants and native plants (Harrison and Winfree, 2015). While pollinators utilize novel resources, particularly when they are in high abundance, they do not necessarily show a preference for them (Williams *et al.*, 2011). Nutrition seems to be the determining factor in plant preference versus place of origin. For instance, a native bumble bee species readily utilized pollen from exotic plants with similar nutrition profiles as the pollen collected from native plants, which suggests that bumble bees selectively choose plants (native or exotic) based on nutrient availability (Harmon-Threatt and Kremen, 2015). However, a bee's intrinsic preference for certain host plants means that some might prefer exotic species while others avoid them, particularly if their host plant is scarce or absent from the landscape (Bartomeus *et al.*, 2016). A bee's floral choices can have indirect effects not only on the individual but also other bees in the community, as the preference of each pollinator will impact other pollinators' choices (Bartomeus *et al.*, 2016). For instance, some pollinators may receive indirect benefits if exotic plants lure other pollinators from natives by decreasing competition (Bartomeus *et al.*, 2016). Conversely, pollinators may pay indirect costs if competition is increased or if exotic species displace their preferred native plants (Bartomeus *et al.*, 2016).

Another mechanism that might affect plant–pollinator interactions is a shift in bloom phenology via altered season availability of floral resources. There is some evidence that exotic species might extend or shift bloom phenology. Cultivars grown in a greenhouse tended to have faster and more abundant germination than native species, suggesting a human-mediated selection for certain phenology traits (Chrobock *et al.*, 2011). Similarly, ornamentals are often cultivated for long-season bloom (Garbuzov and Ratnieks, 2014), which can extend the forage season for pollinators. Growers might also alter plant phenology through management practices such as irrigation, tree thinning or deforestation (Harrison and Winfree, 2015). For example, tree removal can extend bloom time of spring ephemerals by allowing more sunlight to filter though the canopy (Winfree *et al.*, 2014). Changes in urban plant phenology will likely result in 'winners' and 'losers' in the pollinator community, similar to what has been observed with other groups of wildlife like birds and mammals.

Urban warming

Perhaps one of the most pronounced environmental drivers brought about by urbanization is the phenomenon known as the urban heat island effect, a process that makes cities warmer than surrounding non-urban areas (Duffy and Chown, 2016). Higher temperatures can affect both flora and fauna in multiple and synergistic ways. Bees and other insects are ectotherms, organisms that regulate body temperature via heat exchange with the surrounding environment, so are sensitive to rising temperatures (Colinet *et al.*, 2015). Changes in the degree of warming can significantly alter numerous physiological and biological processes, potentially affecting an insect's fitness, fecundity and longevity (Colinet *et al.*, 2015; Hamblin *et al.*, 2018).

Results from a study examining the effects of urban warming on wild bee communities show that temperature, not floral density, is the most important predictor of urban bee abundance and richness (Hamblin *et al.*, 2018). Bee abundance was reduced by 41% for every 1 degree Celsius rise in temperature (Hamblin *et al.*, 2018). This may be due to direct mechanisms, such as species-level variation in thermal limits or the physiological responses to warming. Given bees' sensitivity to fluctuating temperatures, warming will likely act as a filter selecting for the most heat-tolerant species (Hamblin *et al.*, 2018). More importantly, changes in pollinator composition could alter pollination services and plant reproductive success (Scaven and Rafferty, 2013). The effects of urban warming on flowering plants might also indirectly effect pollinators (Scaven and Rafferty, 2013). Plant stress due to drought or high temperatures can decrease plant fitness and the production of pollen and nectar available to pollinators (Scaven and Rafferty, 2013).

In addition to temperature, floral density can be an effective predictor of bee abundance and diversity for certain bee groups (Hamblin *et al.*, 2018). In cityscapes experiencing urban warming, only bumble bees (*Bombus spp.*) benefitted from higher floral density (Hamblin *et al.*, 2018). The implications for pollinator conservation in urban areas is that simply adding more flowers to support bees might be insufficient to fully mitigate the effects of a warmer landscape (Hamblin *et al.*, 2018). However, collectively, green infrastructure (i.e. green roofs, tree-lined streets, expanding and enhancing green space, etc.) has potential to significantly moderate rising temperatures, thereby helping urban flora and fauna cope with heat-related stress (Gill *et al.*, 2007). Urban warming has also been linked to shifts in plant phenology. In general, plants grow new leaves and bloom earlier in cities than in non-urban areas (Jochner and Menzel, 2015). There is concern that urban warming, as well as climate change, may cause a phenology mismatch between pollinators and their floral hosts (Bartomeus *et al.*, 2011). Given that many bee species are active as adults for only a short period, a mismatch could alter or disrupt plant–pollinator interactions. A comparison of phenology data for ten generalist bee species and their floral hosts, dating back to 1970, demonstrated that bee emergence seems to be keeping pace with shifts in flowering dates of host plants (Bartomeus *et al.*, 2011).

Environmental contaminants

Urban expansion is associated with a host of environmental contaminants that pollute soil, water and air, which presents a unique set of issues for city flora and fauna. Several key pollutants that may shape urban plant and insect communities, and thereby constrain potential interactions, include nitrogen, insecticides and light pollution. In cities, vehicular and industrial combustion produce atmospheric nitrogen that accumulates in soil and water as nitrate (Pickett *et al.*, 2011). Although nitrates are essential for plant growth and vigour, plants vary in their ability to assimilate and take advantage of nitrogen enrichment (Harrison and Winfree, 2015). Nitrogen deposition can act as a community filter by giving plants, such as grasses, which prefer a nutrient-rich soil, a competitive advantage over herbaceous flowering plants. The latter generally prefer a nutrient-poor soil (Harrison and Winfree, 2015). In such scenarios, as plant compositions shift from herbaceous flowering plants to grasses, there is less forage for insect pollinators (Burkle and Irwin, 2010). In the case of low-level nitrogen deposition, plant–pollinator interactions can be altered through non-filtering mechanisms, including plant investments in reproductive success, such as investments in nectar and pollen production (Burkle and Irwin, 2010; Harrison and Winfree, 2015). Increases in the quality and quantity of floral rewards could improve foraging efficiency and bee nutrition (Harrison and Winfree, 2015), contributing to overall better bee and plant fitness. Other anthropogenic inputs that can shape plant–pollinator interactions include the use of insecticides.

Millions of pounds of insecticides are used annually to maintain healthy, well manicured lawns (Schueler, 2010). Although lawn weeds such as dandelions and white clover can support diverse bee assemblages (Larson *et al.*, 2014), when treated with insecticides the pollen and nectar of flowering weeds can become toxic (Larson *et al.*, 2015). In the USA, neonicotinoids, a class of neuro-active insecticides, are frequently used to control turf pests despite their deleterious effects on non-target organisms, including bees (Rundlöf *et al.*, 2015). The National Institute of Health showed that certain neonicotinoids are a significant hazard to both managed and wild bees (Kessler *et al.*, 2015). They found that honey bees and bumble bees preferred foods containing neonicotinoid pesticides and that exposure decreased their overall food intake (Kessler *et al.*, 2015). Thus, field-relevant exposure to neonicotinoids can have detrimental effects on honey bee and wild bee health (Rundlöf *et al.*, 2015; Kessler *et al.*, 2015). To protect pollinators from non-target effects, best management practices for turf grass should include mowing flowering weeds that are inadvertently sprayed (Larson *et al.*, 2015) and adhering to all label precautions including the timing of treatment application to avoid times when bees forage (Larson *et al.*, 2017). There is also a growing movement to embrace 'spontaneous' flowering lawns through reduced mowing and a reduction in insecticide use (Larson *et al.*, 2014).

Another anthropogenic driver linked to urban pollinator declines is light pollution or skyglow. Artificial night light from street lights and other sources disrupts nocturnal pollination services performed by moths and bats (Patriarca and Debernardi, 2010; Lewanzik and Voigt, 2014; Knop *et al.*, 2017). The disruption can have detrimental effects on plant reproductive success. For example, in artificially illuminated plant– pollinator communities, nocturnal visits to plants were reduced by 62% compared to dark areas (Knop *et al.*, 2017). This resulted in a 13% overall reduction in fruit set of a focal plant even though the plant received numerous visits by diurnal pollinators (Knop *et al.*, 2017). Further, the negative effects of artificial light at night on nocturnal pollination will likely aggravate the decline of the diurnal community (Knop *et al.*, 2017). Mitigation strategies to help reduce the ecological consequences of light pollution include: maintaining and increasing natural dark areas, decreasing the 'trespass' and intensity of lighting (Gaston *et al.*, 2013) and working with engineers on developing alternatives to the current LED spectrum, which is more attractive and harmful to nocturnal pollinators than older lighting technologies (Pawson and Bader, 2014).

Bee Trends along Rural–urban Gradients

In the past two decades, we have learned that diverse wild bee fauna can persist in metropolitan areas throughout the world (Hall *et al.*, 2017), including major cities in Canada (Colla *et al.*, 2009), Europe (Banaszak-Cibicka *et al.*, 2016; Hausmann *et al.*, 2016), South America (Zanette *et al.*, 2005), South-east Asia (Threlfall *et al.*, 2015), the UK (Goulson *et al.*, 2008; Sirohi *et al.*, 2015; Baldock *et al.*, 2015) and the USA (Wojcik *et al.*, 2008; Matteson *et al.*, 2008; Fetridge *et al.*, 2008; Frankie *et al.*, 2009; Potter and LeBuhn, 2015; Molumby and Przybylowicz, 2012). Bees utilize floral resources in private and community gardens (Matteson *et al.*, 2008; Fetridge *et al.*, 2008; Kaluza *et al.*, 2016; Makinson *et al.*, 2017;

Langellotto *et al.*, 2018), city parks (McFrederick and LeBuhn, 2006), green roofs (Colla *et al.*, 2009; Tonietto *et al.*, 2011; Ksiazek *et al.*, 2014), tree-lined streets (Hausmann *et al.*, 2016; Somme *et al.*, 2016), vacant lots (Sivakoff *et al.*, 2018) and native remnants (McFrederick and LeBuhn, 2006). Interestingly, the patchy nature and continual disturbance in urban environments, whether through construction, foot traffic or heavily managed landscapes, creates a mosaic of habitats, which can potentially support diverse flora and fauna (Zerbe *et al.*, 2003), but, nevertheless, also act as a filter for organisms that can adapt quickly to dynamic and drastically modified landscapes (Sattler *et al.*, 2010; Banaszak-Cibicka and Żmihorski, 2012). A review of urban bee studies shows three major trends along rural–urban gradients: natural areas have a higher abundance and species richness of bee species than urban areas, urban environments have a higher ratio of cavity nesting bee species compared to semi-rural and rural areas, and floral bee specialists are largely scarce or absent in the built environment (Hernandez *et al.*, 2009).

Bees richness is greater in natural areas but that trend is changing

Most urban bee studies show a negative correlation between bee richness and urbanization in before-and-after urbanization studies (Dauber *et al.*, 2003; Ahrné *et al.*, 2009) and comparative studies (e.g. semi-natural, agricultural, suburban, urban) (Fetridge *et al.*, 2008; Matteson *et al.*, 2008; Geslin *et al.*, 2013; Deguines *et al.*, 2016). In addition to the effects of urban environments on individual pollinators, it is important to understand the changes that occur at the community level along rural-to-urban gradients (Banaszak-Cibicka and Żmihorski, 2012). Of the 17,857 interactions observed between pollinators and flowering plants, the number of interactions was significantly lower in urban landscapes compared to those in semi-natural and agricultural habitats (Deguines *et al.*, 2016). Further, certain types of pollinators are affected to a greater extent. For example, syrphids and solitary bees were significantly affected by urbanization, although other pollinators such as bumble bees were not (Deguines *et al.*, 2016). Overall, urbanization is associated with reduced flower visitor richness and a shift to more generalist insects, leading to simplified plant–pollinator networks and functional homogenization (Deguines *et al.*, 2016).

The opposite trend has also been noted. Recent studies show that cities, especially those that implement green infrastructure, can support greater abundance and diversity of managed and wild bees than can neighbouring agricultural landscapes (Hall *et al.*, 2017; Osborne *et al.*, 2008a; Frankie *et al.*, 2009; Gunnarsson and Federsel, 2014; Sirohi *et al.*, 2015; Baldock *et al.*, 2015). Certainly, the surrounding landscape influences how many species will be present in urban environments, either positively or negatively (Sattler *et al.*, 2010; Banaszak-Cibicka and Żmihorski, 2012). Arguably, technological advances and agricultural efficiencies are homogenizing rural environments, making them less hospitable to pollinators and the wildflowers that support them (Hall *et al.*, 2017). In such cases, urban environments are often more pollinator-friendly than rural habitats.

Urban development shifts bee communities to mostly cavity-nesting species

Cavity-nesting bee species predominate in many urban areas (Cane, 2005; Zanette *et al.*, 2005; Cane *et al.*, 2006; Matteson *et al.*, 2008). The high percentage of cavity-nesting species observed in cities is the opposite of what is typically found in semi-urban and natural areas. Generally, ground-nesting species make up 70% of the total bee fauna and cavity-nesting species the other ~30% (Vaughan and Hoffman Black, 2007). Several factors are thought to drive the higher ratio of cavity nesters in urban environments. Cavity-nesting bees might thrive in urban areas because they can use man-made structures, such as bee hotels (MacIvor and Packer, 2015) as well as various crevices and holes found in building materials. Also, cavity-nesting species' ability to find nesting sites is not negatively impacted by the high percentage of impervious surface cover associated with urban environments. As noted earlier, this puts ground-nesting species at a disadvantage. Further, it is well established that urban soils often have altered physical, chemical and structural properties compared to local non-urbanized soils, which can disrupt ecosystem functions (Pavao-Zuckerman, 2008) and, potentially, ground-nesting bee success (Xie *et al.*, 2013).

Specialist bee species decrease along the gradient of increasing urbanization

Floral specialists are nearly absent from urban areas (McFrederick and LeBuhn, 2006; Fetridge *et al.*, 2008; Frankie *et al.*, 2009), which is likely a direct result of a decrease in overall floral diversity and shifts in floral composition. Both plant and bee diversity tend to decrease in central urban core areas but, interestingly, plant diversity can increase in moderately urban environments (McKinney, 2008). However, the potential increase in plant diversity is often owing to the introduction of exotic species (McKinney, 2008). In urban areas, exotic species are likely to include both weedy species that tend to thrive in disturbed environments as well as planted ornamentals, which may not meet the nutritional requirements of pollen specialists. The lack of specialist flora can be remedied through pollinator-friendly plant selection and landscape design.

Paradigm Shift: 'Ecology in the City' to 'Ecology for the City'

Researchers increasingly acknowledge that the long-term success of pollinator conservation efforts depends on buy-in from local community members (Turo and Gardiner, 2019). Importance of the human component in ecological conservation projects is reflected in a paradigm shift within urban design and planning, whose traditional focus was 'ecology in the city' to a modern-day emphasis on 'ecology for the city' (Pickett *et al.*, 2016). 'Ecology *for* the city' embraces a holistic, multi-disciplinary approach towards sustainable ecological stewardship that brings together urban residents, city planners, decision makers and conservationists (Pickett *et al.*, 2016).

To increase long-term success of urban pollinator conservation, project managers aim to address the specific apprehensions of city residents. Residents may worry about the presence of stinging insects or have safety concerns related to unoccupied city lots (Turo and Gardiner, 2019). In addition to violent crime, vacant lots can quickly turn into a place for dumping rubbish, hiding guns, and other illegal activities such as drug sales and prostitution (Garvin *et al.*, 2013). Yet, researchers have found that urban lot greening efforts have potential to tackle safety concerns and reduce violent crime (Garvin *et al.*, 2013). As part of a Philadelphia-centred study, the Pennsylvania Horticultural Society took steps to remove debris, plant grass and trees, build a low fence with entry openings to the lot and maintain the space every two weeks. Researchers concluded that these greening efforts were associated with reductions in certain gun crimes and improvements in residents' perceptions of safety. Though this study concentrated on grass and tree planting, it is believed that wildflower plantings for pollinators have similar potential, provided these plantings are strategically placed (Hoyle *et al.*, 2017), meet the needs of local decision makers, land managers (Hoyle *et al.*, 2017) and residents (Nassauer and Raskin, 2014; Turo and Gardiner, 2019) and, through messaging, show that these spaces are being maintained by signage and other 'cues to care' (Nassauer, 1995, 2011; Nassauer and Raskin, 2014).

'Cues to care' (Nassauer, 2011), a term coined by landscape architecture professor Joan Iverson Nassauer, refers to landscape characteristics or physical objects that send messages to local communities that someone is actively caring for the land and watching over it (Nassauer, 1995, 2011; Nassauer and Raskin, 2014) and can include mown turf, colourful flowers, fences, lack of weeds, lawn ornamentation, neat edging or borders and defined planting areas (Nassauer and Raskin, 2014). Some 'cues to care,' such as mown grass or trimmed hedges, might seem counter to environmental stewardship, but indicate to the neighbourhood that someone is caring for the property, thereby encouraging others to likewise care for their land (Nassauer and Raskin, 2014). Research indicates that landscape preferences can occur on a very small scale, and can vary block-to-block, since residents are often influenced by a neighbour's choice (Brown *et al.*, 2003). There are many complexities to this picture, but community meetings, where residents can discuss their needs and desires in conversation with scientists, urban planners and decision makers, may provide key information, so that any potential stumbling-blocks to a project can be overcome in advance.

In terms of which plants best support pollinators, the question must be asked if the needs of both residents in the community and of pollinators can be reconciled and addressed simultaneously. Pollinating insects, particularly bees, benefit from native plant species and wildflowers (Williams *et al.*, 2011; Pardee and Philpott, 2014), though it is difficult to know with precision which plant species are needed for urban bees (Turo and Gardiner, 2019). Some scientists indicate that more research is needed, especially when it comes to the nutritional value of wildflower seed mixes (Garbuzov and Ratnieks, 2014; Turo and Gardiner, 2019). Bees visit commonly available, naturally occurring weedy blossoms like clover and dandelion, and are often attracted to non-native species, even though non-native species are not the optimal food source for native bees (Turo and Gardiner, 2019). Taller wildflower meadow plantings, referred to as 'prairie pockets', attract numerous pollinators (Turo and Gardiner, 2019). These floral areas often have reduced management and cutting requirements, which is important in pollinator habitats (Hall *et al.*, 2017; Wastian *et al.*, 2016) and can reduce municipality maintenance costs (Aronson *et al.*, 2017).

Drawbacks to the taller, wilder-looking native plants are that they may look messy and unkempt to local residents, who might prefer a landscape that is neat and tidy (Nassauer, 1997). There can be great variation in aesthetic preferences. Tall wildflower meadow plants are viewed positively in economically robust cities (Southon *et al.*, 2017; Turo and Gardiner, 2019). In a recent paper, researchers noted that residents' views on planting conservation projects in green spaces may vary greatly depending on the overall economic status of an individual city (Turo and Gardiner, 2019). City residents located in areas where economic expansion occurs may hold positive views of green spaces (Ives *et al.*, 2017), whereas residents of 'shrinking cities', that have too many vacant lots, may view the areas as abandoned or in decline (Nassauer and Raskin, 2014). Without community engagement, there might be resistance to pollinator planting projects that place meadows in areas that have historically experienced socio-economic injustices (Clement and Kanai, 2015).

As touched upon above, there is a need to consider the needs of both bees and people. Fig. 6.1 provides an iterative decision-making process that incorporates input from the perspectives of conservation science and community development. The decision tree encourages regular feedback via repeated site monitoring and stakeholder engagement, which is then used to improve the plan. Thus, rather than a static path forward, a community will have a dynamic design and management plan that meets the needs of humans and insect pollinators, resulting in improved long-term outcomes.

Urban Beekeeping

Interest in beekeeping has skyrocketed in recent years, particularly in urban areas. Recreational beekeepers interested in pollinating plants and producing honey are placing hives in community gardens, private lots and on rooftops (Nassauer, 1997; Southon *et al.*, 2017). Urban agriculture includes a variety of enterprises, such as beekeeping, aquaculture and small-scale vegetable and fruit crop farming in private gardens, public community gardens, school-based gardens, green roofs and vacant lots (Ives *et al.*, 2017). Urban agriculture and beekeeping go hand-in-hand in that honey bees, along with wild bees, provide pollination services for urban crops. Growing food on vacant land provides numerous social, economic and environmental benefits to cities and urban populations, such as boosting biodiversity, increasing the availability of locally grown food which limits travel time between the farm and consumer, cooling buildings and improving air quality (Clement and Kanai, 2015). For instance, Seattle, Portland, Atlanta, San Francisco, Vancouver and New York have amended local codes to approve urban beekeeping and to acknowledge the essential pollination services that bees provide (Clement and Kanai, 2015).

Urban beekeeping (Fig. 6.2), which falls under the scope of urban agriculture, can bring numerous benefits to cities and urban residents.

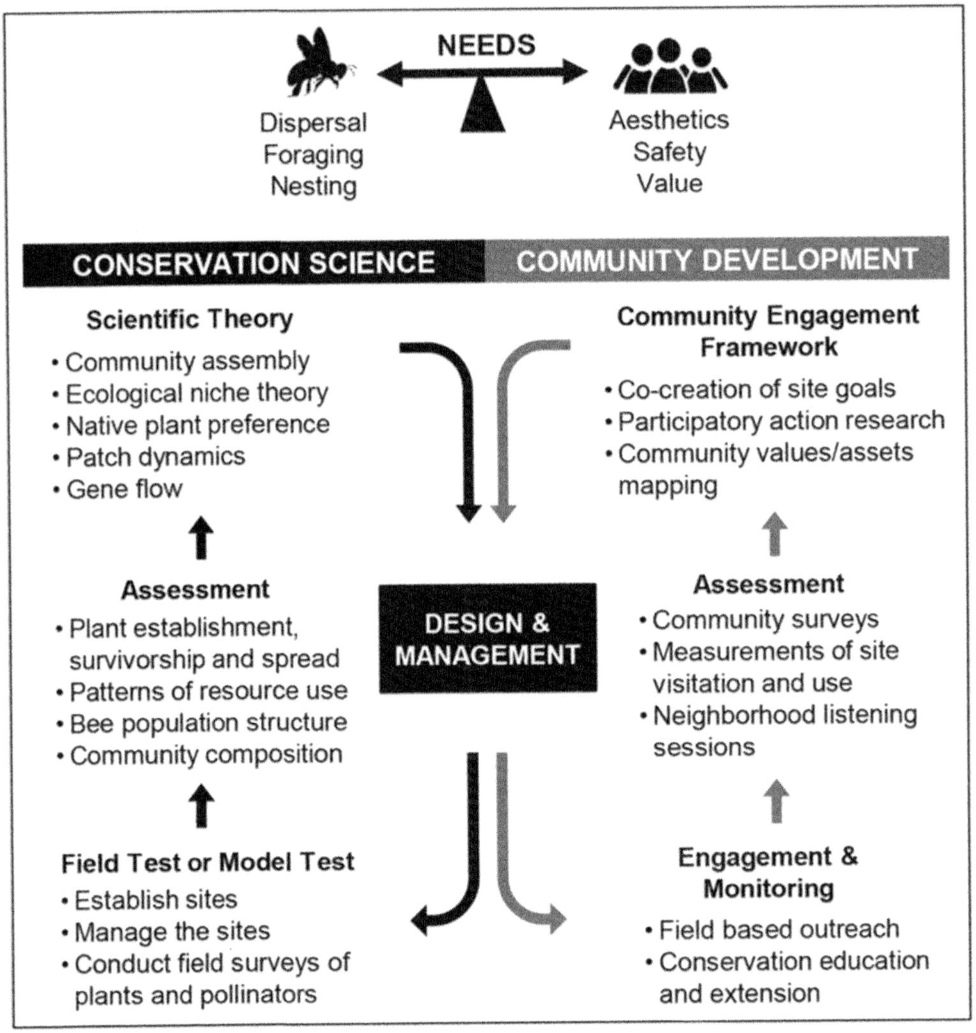

Fig. 6.1. This chart depicts one way forward in the planning and implementation of pollinator conservation projects, in that it suggests a collaborative framework that considers the needs of both people and pollinators. (From Turo and Gardiner, 2019. Reprinted by permission of John Wiley & Sons, Inc.)

Benefits include featuring beehives in city beautification projects, the conversion of vacant lots to agricultural use, thereby reducing mowing and maintenance costs to local governments, urban agricultural enterprises that can provide job training for city residents, and providing fresh food (in the form of pollination services) and honey to local residents (Peters, 2011). Further, urban gardening and agriculture provide increased food security for city residents through collective efforts, i.e. the building of strong community bonds via partnerships between local organizations, schools and businesses (Gardiner *et al.*, 2014).

In addition to the many benefits of urban beekeeping, there are also challenges. Residents may have concerns about being in close proximity to stinging insects. Cities, including Milwaukee, Baltimore, Denver, Portland and Chicago, are addressing these challenges by requiring registration of honey bee colonies with local agencies, and requiring property boundary line setbacks, proper

Fig. 6.2. Honey bee hives managed by Mandy Shaw, located on top of a building in Portland, Oregon. (Photo: Mandy Shaw, Portland Urban Beekeepers)

hive maintenance, provision of adequate water supply to hives and flyway barrier requirements, so that honey bee flight patterns do not interfere with the use of properties by adjoining neighbours (Broadway, 2009).

What will the future bring to urban beekeeping?

While mismanagement and the spread of pests and disease are very real and important issues impacting honey bee colonies and beekeepers, a secondary threat to the existence of urban beekeeping is the inability of regulations to accurately reflect the cumulative concerns of beekeepers and community members. Lack of understanding with respect to beekeeping and the needs of beekeepers coupled with the failure to consider community concerns can lead to the passage of onerous laws. The recent creation of best-management materials can assist urban communities in working toward positive outcomes that are tailored toward specific community needs. State university extension services can be instrumental in providing guidance to urban beekeepers regarding best practices for beekeeping in urban settings, thereby minimizing nuisance problems and maximizing the exchange of concerns (Melathopoulos *et al.*, 2018). These publications encourage beekeepers to speak with their neighbours about their bees, to listen to their neighbours' concerns and to be willing to provide neighbours with information about honey bees, the varying reactions to bee stings, the differences between honey bees and wasps, the benefits of urban beekeeping, and honey bee biology and behaviour (Melathopoulos *et al.*, 2018). Cities like Toronto have prepared guidelines for urban beekeeping, recognizing that local regulations often cannot keep up with increased interest in beekeeping activities. These guides are intended to encourage thoughtful decision making around beekeeping in cities (Berquist *et al.*, 2012), and make the following suggestions:

1. More decision making should be left to the municipalities allowing specific concerns to be addressed in a way that best suits the local community.
2. Regulations ought to consider how honey bee flight paths can be altered rather than passing rules that set rigid distances between the location of hives, property lines and neighbours. Bee flight paths can be, with a little ingenuity, redirected away from walkways and people traffic.
3. Voluntary best-management guidelines can be effective in reducing problems.
4. Support networks between the beekeeping industry and with governments are key.
5. The education of beekeepers and the public is essential.

(Berquist *et al.*, 2012)

Is recreational beekeeping the same as bee conservation?

Recent research shows that honey bee colonies may out-compete native wild bee species in foraging for limited nectar and pollen resources and that honey bee colonies should not be placed in conservation areas (Geldmann and González-Varo, 2018; Henry and Rodet, 2018). These authors acknowledge that honey bees play an important role in agriculture and have helped raise awareness of the problems encountered by bees overall. But conservation of wild bee species should not be confused with the agricultural management problems experienced by honey bees and that more research should be conducted to determine the impact of honey bees on wild bee species, particularly in limited forage areas (Geldmann and González-Varo, 2018). Though not used broadly by municipalities, a study conducted in Oslo, Norway, describes a mapping and zoning assessment programme used by government officials for identifying areas for optimal honey bee hive placement, so that managed honey bees are not depleting limited urban floral resources (Stange *et al.*, 2017). A global roundtable discussion between bee experts on the topic of 'Can cities save bees? How can urban habitats be made to serve pollinator conservation? How can that story be better told?' is an excellent summary of the issues facing urban beekeepers (Maddox, 2018).

A roadmap to community-based pollinator conservation

Further progress is needed on a number of fronts in order for meaningful pollinator conservation efforts to result in successful outcomes.

Past mistakes and how to avoid them

One way to improve outcomes is to allow lessons from past mistakes to inform future progress (Turo and Gardiner, 2019). Lessons learned from past projects show that if members of the community do not favour the idea of a planting conservation site, the area may be destroyed or sabotaged (Turo and Gardiner, 2019). Broad approval of a planting project by residents is key to its success and may even lead to local volunteer efforts that can help with ongoing management of a planted area (Turo and Gardiner, 2019). As noted earlier, projects tend to receive greater public acceptance when they include visual 'cues to care' such as vigilance in trash removal, regular maintenance and the presence of fencing (Nassauer and Raskin, 2014).

Past experience demonstrates that even the best-intended project, undertaken with great care to engage community members through outreach efforts, can lead to both positive and negative responses from community members (Turo and Gardiner, 2019). In one case, despite the positive experiences of some residents involved in the community outreach events for the 'pocket prairie' sites, other residents feared for their safety while passing the tall meadow plantings on the way to the bus stop in the early morning hours (Turo and Gardiner, 2019). Trying to understand how meadow planting projects may be perceived by as many community members as possible, at various stages of vegetative growth, is important (Turo and Gardiner, 2019). In meeting conservation goals, care must be taken to minimize displacement or marginalization of local residents in the name of ecological green space development, particularly when impacting communities that have experienced historical racial injustices (Turo and Gardiner, 2019).

For example, in Detroit, Michigan, a city in decline, a vast proposed green space project called 'Detroit Future City', was planned and included moving residents in low-vacancy areas to make way for urban farms and green space (Clement and Kanai, 2015). The overall plan was roundly criticized because the inclusions for the re-zoning of green space were swept up in an 'innovation landscape' marketing plan intended to attract global investment at the expense of disadvantaged local communities (Clement and Kanai, 2015). The 'Detroit Future City' project exemplifies the need for green initiatives, including pollinator meadow areas, to involve community leaders. Further, projects need to engage those with expertise in the field of sociology who can assist with community meetings, large-scale discussions about projects and surveys of residents, in order to illicit as much information and input as possible from all stakeholders (Hunter and Hunter, 2008). Adding to this discussion about overall city planning, some suggest that ecological site design may be the preferred route over purchasing tracts of land, so as not to place insect conservation at odds with urban development (Hunter and Hunter, 2008).

How to ensure pollinator conservation longevity

The longevity of urban pollinator conservation efforts tends to increase when projects are undertaken by a diverse team and when management decisions include those with hands-on management and long-term interest in the project's success. Urban conservation experts espouse the value of including entomological expertise in project site development and the management of green space projects (Hunter and Hunter, 2008). They also encourage increased active involvement in the community development process by having entomologists 'more fully assuming the identity of community developers' (Turo and Gardiner, 2019) and to forge strong collaborative bonds with urban design professionals (Hunter and Hunter, 2008; Nassauer, 2012). These diverse perspectives from pollinator habitat researchers, land managers, decision makers, community members including immediate neighbours, at-large community members and those that may walk by the areas all play a central role in the planning and long-term maintenance of a successful conservation project (Turo and Gardiner, 2019).

In the planning stages, the manner in which information is acquired and exchanged is important to the success of a project. Public input workshops and listening sessions give urban residents a crucial voice in the process (Green *et al.*, 2016). For example, aesthetic and economic concerns, such as overgrown plants, can be addressed by engaging residents in finding feasible solutions (Green *et al.*, 2016). Considering long-term maintenance plans, research indicates that when urban residents have positive encounters with nature close to home, they tend to be more engaged with wildlife and nature (Nilon, 2011; Van Velsor and Nilon, 2006). The more people are drawn into the beauty of a space, the more likely they are to become involved in the care of the place long-term (Nassauer, 1997; Hunter and Hunter, 2008). Future management plans that involve both conservationists/researchers and the local community can be the best way forward, provided funding is available.

LONG-TERM FUNDING PLANS. Cities increasingly look to reduce expenses, leaving urban land managers to implement low-maintenance sustainable planting regimes in public spaces (Aronson *et al.*, 2017). The use of annual flowers can provide season-long colour, but these need replacing each year, creating ongoing and often unsustainable costs. Perennial wildflower meadows are an option that can lead to long-term savings in lawn-maintenance care, in addition to providing ecological benefits (Hall *et al.*, 2017). In the UK, one third of urban park managers have experienced maintenance budget and staffing cuts exceeding 20% in a two-year period, at a time when park usage is rising, leading to the need for innovative and low-cost management strategies for public spaces (Heritage Lottery Fund, 2016). Thus, some UK urban park land managers are moving away from annual bedding plants in favour of low-maintenance perennial meadow plantings and are transferring management responsibilities to outside organizations.

Jurisdictions in the USA are looking to encourage biodiversity through various financial reward schemes. In the Chicago Sustainable Backyards Program, native tree plantings, rain barrels, composting and the conversion of lawns to native naturalized plantings offer financial rebates to residents. However, financial rewards alone may not be enough to encourage biodiversity in planting schemes, but may be a possible solution provided the incentives are implemented via strong engagement between government, planners and the public (Hostetler et al., 2011; Aronson et al., 2017). Advocates of urban pollinator conservation projects suggest low-cost, low-maintenance wildflower plantings for bees that also include already-present weedy vegetation (Turo and Gardiner, 2019). Partnerships with local groups that care about the neighbourhood can provide needed assistance for routine work on the ground and can assist in securing funds through endowments, tax breaks and other financial incentives (Hostetler et al., 2011; Turo and Gardiner, 2019).

The educational component can make a difference

The value of education in the conservation of pollinators cannot be over-emphasized. In the USA, the National Wildlife Federation (www.nwf.org) has created a schoolyard habitat programme that provides teachers with resources for environmental stewardship classes. Similarly, in the UK, Buglife.org (www.Buglife.org.uk) engages children in learning more about insects and other invertebrates (Hunter and Hunter, 2008). With respect to wildflowers, research indicates that once people understand the value of wildflowers, they are more interested in them (Younis *et al.*, 2010). Scientists urge greater efforts in educating the public about pollinators and bees, since many are not aware of the diversity of native bee species (Wilson *et al.*, 2017; Turo and Gardiner, 2019). Citizen science conservation projects can also help (Aronson *et al.*, 2017), but these projects require the ability of participants to correctly identify different types of bees, so more education is needed in this realm (Wilson *et al.*, 2017). A recent survey conducted in the USA found that the public has an interest in bees. Ninety-nine per cent of respondents believed that bees are important but had limited knowledge of native bee diversity and difficulty telling the difference between bees and non-bees (Wilson *et al.*, 2017). Pollinator outreach efforts, on the importance of bee diversity and the habitat requirements of bees, are already beginning to move the general public and policy makers to action. Continued efforts on this front are needed, as much remains to be done.

Conclusion

Cities offer tremendous opportunities for pollinator conservation. The recent focus on urban areas comes at a time in history when pollinators are in decline, and agricultural areas might not provide refuge for bats, butterflies, moths and bees. The body of research referenced in this chapter highlights some of the challenges ahead but also points us in an encouraging direction. And that is, pollinators can thrive in cities provided they have suitable nesting habitat, floral resources and the ability to disperse across the city matrix. Humans have great potential and the ability to assist in creation of pollinator-friendly cities. Urban planners are already beginning to rethink the built environment, and to take a multidisciplinary approach to designing cityscapes that work for both humans and pollinators. Entomologists and conservationists, who have specialized knowledge about insect ecology, can take a more active role in engaging decision makers, landscape professionals and community members. By including all stakeholders in the iterative design process, we can achieve lasting protection of pollinators in cities.

Additional Resources

Online – pollinator conservation

- Pollinator Partnership (https://pollinator.org/)
- Xerces Society for Invertebrate Conservation (https://xerces.org/pollinator-conservation)
- Bumblebee Conservation Trust (https://www.bumblebeeconservation.org/)
- Bee City USA (www.beecityUSA.org)
- The Honeybee Conservancy (https://thehoneybeeconservancy.org/)
- Buglife (https://www.buglife.org.uk/)

Online – pollinator-friendly planting guides

- Pollinator Partnership (https://pollinator.org/guides)
- Xerces (https://xerces.org/pollinator-conservation/plant-lists/)
- Lady Bird Johnson Wildflower Center (https://www.wildflower.org/collections/)
- The Royal Horticultural Society (https://www.rhs.org.uk/science/conservation-biodiversity/wildlife/plants-for-pollinators#nav-about)

Books on pollinators and bee-friendly gardens

- *The Bees in Your Backyard: A Guide to North America's Bees* by J.S. Wilson and O.J. Messinger Carril

- *Attracting Native Pollinators: The Xerces Society Guide, Protecting North America's Bees and Butterflies* – Xerces Society
- *Bee Garden: All you Need to Know in One Concise Manual* by Elke Schwarzer
- *Pollinators of Native Plants: Attract, Observe and Identify Pollinators and Beneficial Insects with Native Plants* by H.N. Holm
- *Pollinator Friendly Gardening* by Rhonda Fleming Hayes
- *Keeping Bees in Towns and Cities* by L. Dixon
- *100 Plants to Feed the Bees: Provide a Healthy Habitat to Help Pollinators Thrive* – Xerces Society
- *The Bee-Friendly Garden: Design an Abundant, Flower-Filled Yard that Nurtures Bees and Supports Biodiversity* by K. Frey and G. LeBuhn Packer
- *Garden Revolution: How Our Landscapes Can Be a Source of Environmental Change* by L. Weaner and T. Christopher

References

Ahrné, K., Bengtsson, J. and Elmqvist, T. (2009) Bumble bees (*Bombus* spp) along a gradient of increasing urbanization. *PLOS ONE* 4(5), e5574. DOI: 10.1371/journal.pone.0005574.

Aizen, M.A., Garibaldi, L.A., Cunningham, S.A. and Klein, A.M. (2009) How much does agriculture depend on pollinators? lessons from long-term trends in crop production. *Annals of Botany* 103(9), 1579–1588. DOI: 10.1093/aob/mcp076.

Andrieu, E., Dornier, A., Rouifed, S., Schatz, B. and Cheptou, P.-O. (2009) The town *Crepis* and the country *Crepis*: How does fragmentation affect a plant–pollinator interaction? *Acta Oecologica* 35(1), 1–7. DOI: 10.1016/j.actao.2008.07.002.

Arnold, S.E.J., Idrovo, M.E.P., Arias, L.J.L., Belmain, S.R. and Stevenson, P.C. (2014) Herbivore defence compounds occur in pollen and reduce bumblebee colony fitness. *Journal of Chemical Ecology* 40(8), 878–881. DOI: 10.1007/s10886-014-0467-4.

Aronson, M.F.J., Lepczyk, C.A., Evans, K.L., Goddard, M.A., Lerman, S.B. *et al.* (2017) Biodiversity in the city: key challenges for urban green space management. *Frontiers in Ecology and the Environment* 15(4), 189–196. DOI: 10.1002/fee.1480.

Baldock, K.C.R., Goddard, M.A., Hicks, D.M., Kunin, W.E., Mitschunas, N. *et al.* (2015) Where is the UK's pollinator biodiversity? the importance of urban areas for flower-visiting insects. *Proceedings of the Royal Society B: Biological Sciences* 282(1803), 20142849. DOI: 10.1098/rspb.2014.2849.

Banaszak-Cibicka, W. and Żmihorski, M. (2012) Wild bees along an urban gradient: winners and losers. *Journal of Insect Conservation* 16(3), 331–343. DOI: 10.1007/s10841-011-9419-2.

Banaszak-Cibicka, W., Ratyńska, H. and Dylewski, Ł. (2016) Features of urban green space favourable for large and diverse bee populations (Hymenoptera: Apoidea: Apiformes). *Urban Forestry & Urban Greening* 20, 448–452. DOI: 10.1016/j.ufug.2016.10.015.

Bartomeus, I., Ascher, J.S., Wagner, D., Danforth, B.N., Colla, S. *et al.* (2011) Climate-associated phenological advances in bee pollinators and bee-pollinated plants. *Proceedings of the National Academy of Sciences* 108(51), 20645–20649. DOI: 10.1073/pnas.1115559108.

Bartomeus, I., Fründ, J. and Williams, N.M. (2016) Invasive plants as novel food resources, the pollinators' perspective: biological invasions and animal behaviour. In: Sol, D. and Weis, J. (eds) *Biological Invasions and Behaviour*. Cambridge University Press, Cambridge, UK, pp. 119–132.

Berquist, M., Bird, A., Dean, M.G., Law, R.B., Lee, S. *et al.* (2012) *Towards a new approach to beekeeping policy in urban Ontario*. University of Toronto, pp. 1–31.

Broadway, M. (2009) Growing urban agriculture in North American cities: the example of Milwaukee. *Focus on Geography* 52(3-4), 23–30 DOI: 10.1111/j.1949-8535.2009.tb00251.x.

Brown, B., Perkins, D.D. and Brown, G. (2003) Place attachment in a revitalizing neighborhood: individual and block levels of analysis. *Journal of Environmental Psychology* 23(3), 259–271. DOI: 10.1016/S0272-4944(02)00117-2.

Burkle, L.A. and Irwin, R.E. (2010) Beyond biomass: measuring the effects of community-level nitrogen enrichment on floral traits, pollinator visitation and plant reproduction. *Journal of Ecology* 98(3), 705–717. DOI: 10.1111/j.1365-2745.2010.01648.x.

Cane, J.H. (1991) Soils of ground-nesting bees (Hymenoptera: Apoidea): texture, moisture, cell depth and climate. *Journal of the Kansas Entomology Society* 64, 406–413.
Cane, J.H. (2005) Bees, pollination, and the challenges of sprawl. In: *Nature in Fragments: the Legacy of Sprawl*. Columbia University Press, New York, pp. 109–124.
Cane, J.H., Minckley, R.L., Kervin, L.J., Roulston, T'ai H.. and Williams, N.M. (2006) Complex responses within a desert bee guild (Hymenoptera: Apiformes) to urban habitat fragmentation. *Ecological Applications* 16(2), 632–644. DOI: 10.1890/1051-0761(2006)016[0632:CRWADB]2.0.CO;2.
Catford, J.A., Bode, M. and Tilman, D. (2018) Introduced species that overcome life history tradeoffs can cause native extinctions. *Nature Communications* 9(1), 1–7. DOI: 10.1038/s41467-018-04491-3.
Cheptou, P.-O. and Avendaño V, L.G. (2006) Pollination processes and the Allee effect in highly fragmented populations: consequences for the mating system in urban environments. *The New phytologist* 172(4), 774–783. DOI: 10.1111/j.1469-8137.2006.01880.x.
Chrobock, T., Kempel, A., Fischer, M. and van Kleunen, M. (2011) Introduction bias: cultivated alien plant species germinate faster and more abundantly than native species in Switzerland. *Basic and Applied Ecology* 12(3), 244–250. DOI: 10.1016/j.baae.2011.03.001.
Clement, D. and Kanai, M. (2015) The Detroit Future City: how pervasive neoliberal urbanism exacerbates racialized spatial injustice. *American Behavioral Scientist* 59, 369–385.
Colinet, H., Sinclair, B.J., Vernon, P. and Renault, D. (2015) Insects in fluctuating thermal environments. *Annual Review of Entomology* 60(1), 123–140. DOI: 10.1146/annurev-ento-010814-021017.
Colla, S.R., Willis, E. and Packer, L. (2009) Can green roofs provide habitat for urban beeds (Hymenoptera: Apidae)? *Cities and the Environment* 2(1), 1–12. DOI: 10.15365/cate.2142009.
Danforth, B.N., Sipes, S., Fang, J. and Brady, S.G. (2006) The history of early bee diversification based on five genes plus morphology. *Proceedings of the National Academy of Sciences* 103(41), 15118–15123. DOI: 10.1073/pnas.0604033103.
Darvill, B., Knight, M.E. and Goulson, D. (2004) Use of genetic markers to quantify bumblebee foraging range and nest density. *Oikos* 107(3), 471–478. DOI: 10.1111/j.0030-1299.2004.13510.x.
Dauber, J., Hirsch, M., Simmering, D., Waldhardt, R., Otte, A. *et al.* (2003) Landscape structure as an indicator of biodiversity: matrix effects on species richness. *Agriculture, Ecosystems & Environment* 98(1-3), 321–329. DOI: 10.1016/S0167-8809(03)00092-6.
Deguines, N., Julliard, R., de Flores, M. and Fontaine, C. (2016) Functional homogenization of flower visitor communities with urbanization. *Ecology and Evolution* 6(7), 1967–1976. DOI: 10.1002/ece3.2009.
Duffy, G.A. and Chown, S.L. (2016) Urban warming favours C4 plants in temperate European cities. *Journal of Ecology* 104(6), 1618–1626. DOI: 10.1111/1365-2745.12652.
Eckhardt, M., Haider, M., Dorn, S. and Müller, A. (2014) Pollen mixing in pollen generalist solitary bees: a possible strategy to complement or mitigate unfavourable pollen properties? *Journal of Animal Ecology* 83(3), 588–597. DOI: 10.1111/1365-2656.12168.
Fetridge, E.D., Ascher, J.S. and Langellotto, G.A. (2008) The bee fauna of residential gardens in a suburb of New York City (Hymenoptera: Apoidea). *Annals of the Entomological Society of America* 101(6), 1067–1077. DOI: 10.1603/0013-8746-101.6.1067.
Frankie, G.W., Thorp, R.W., Hernandez, J., Rizzardi, M., Ertter, B. *et al.* (2009) Native bees are a rich natural resource in urban California gardens. *California Agriculture* 63(3), 113–120. DOI: 10.3733/ca.v063n03p113.
Garbuzov, M. and Ratnieks, F.L.W. (2014) Quantifying variation among garden plants in attractiveness to bees and other flower-visiting insects. *Functional Ecology* 28(2), 364–374. DOI: 10.1111/1365-2435.12178.
Gardiner, M.M., Burkman, C.E. and Prajzner, S.P. (2013) The value of urban vacant land to support arthropod biodiversity and ecosystem services. *Environmental Entomology* 42(6), 1123–1136. DOI: 10.1603/EN12275.
Gardiner, M.M., Prajzner, S.P., Burkman, C.E., Albro, S. and Grewal, P.S. (2014) Vacant land conversion to community gardens: influences on generalist arthropod predators and biocontrol services in urban greenspaces. *Urban Ecosystems* 17(1), 101–122. DOI: 10.1007/s11252-013-0303-6.
Garvin, E.C., Cannuscio, C.C. and Branas, C.C. (2013) Greening vacant lots to reduce violent crime: a randomised controlled trial. *Injury Prevention* 19(3), 198–203. DOI: 10.1136/injuryprev-2012-040439.
Gaston, K.J., Bennie, J., Davies, T.W. and Hopkins, J. (2013) The ecological impacts of nighttime light pollution: a mechanistic appraisal. *Biological Reviews* 88(4), 912–927. DOI: 10.1111/brv.12036.

Geib, J.C., Strange, J.P. and Galen, C. (2015) Bumble bee nest abundance, foraging distance, and host-plant reproduction: implications for management and conservation. *Ecological Applications* 25(3), 768–778. DOI: 10.1890/14-0151.1.

Geldmann, J. and González-Varo, J.P. (2018) Conserving honey bees does not help wildlife. *Science* 359(6374), 392–393. DOI: 10.1126/science.aar2269.

Geslin, B., Gauzens, B., Thébault, E. and Dajoz, I. (2013) Plant pollinator networks along a gradient of urbanisation. *PLOS ONE* 8(5), e63421. DOI: 10.1371/journal.pone.0063421.

Geslin, B., Le Féon, V., Folschweiller, M., Flacher, F., Carmignac, D. *et al.* (2016) The proportion of impervious surfaces at the landscape scale structures wild bee assemblages in a densely populated region. *Ecology and Evolution* 6(18), 6599–6615. DOI: 10.1002/ece3.2374.

Gill, S.E., Handley, J.F., Ennos, A.R. and Pauleit, S. (2007) Adapting cities for climate change: the role of the green infrastructure. *Built Environment* 33(1), 115–133. DOI: 10.2148/benv.33.1.115.

Glaum, P., Simao, M.-C., Vaidya, C., Fitch, G. and Iulinao, B. (2017) Big city *Bombus*: using natural history and land-use history to find significant environmental drivers in bumble-bee declines in urban development. *Royal Society Open Science* 4(5), 170156–170166. DOI: 10.1098/rsos.170156.

Goulson, D., Lye, G.C. and Darvill, B. (2008) Decline and conservation of bumble bees. *Annual Review of Entomology* 53(1), 191–208. DOI: 10.1146/annurev.ento.53.103106.093454.

Green, O.O., Garmestani, A.S., Albro, S., Ban, N.C., Berland, A. *et al.* (2016) Adaptive governance to promote ecosystem services in urban green spaces. *Urban Ecosystems* 19(1), 77–93. DOI: 10.1007/s11252-015-0476-2.

Greenleaf, S.S., Williams, N.M., Winfree, R. and Kremen, C. (2007) Bee foraging ranges and their relationship to body size. *Oecologia* 153(3), 589–596. DOI: 10.1007/s00442-007-0752-9.

Gregory, S. and Wright, I. (2005) Creation of patches of bare ground to enhance the habitat of ground-nesting bees and wasps at Shotover Hill, Oxfordshire, England. *Conservation Evidence* 2, 139–141.

Gunnarsson, B. and Federsel, L.M. (2014) Bumblebees in the city: abundance, species richness and diversity in two urban habitats. *Journal of Insect Conservation* 18(6), 1185–1191. DOI: 10.1007/s10841-014-9729-2.

Hall, D.M., Camilo, G.R., Tonietto, R.K., Ollerton, J., Ahrné, K. *et al.* (2017) The city as a refuge for insect pollinators. *Conservation Biology* 31(1), 24–29. DOI: 10.1111/cobi.12840.

Hamblin, A.L., Youngsteadt, E. and Frank, S.D. (2018) Wild bee abundance declines with urban warming, regardless of floral density. *Urban Ecosystems* 21(3), 419–428. DOI: 10.1007/s11252-018-0731-4.

Harmon-Threatt, A.N. and Kremen, C. (2015) Bumble bees selectively use native and exotic species to maintain nutritional intake across highly variable and invaded local floral resource pools. *Ecological Entomology* 40(4), 471–478. DOI: 10.1111/een.12211.

Harrison, T. and Winfree, R. (2015) Urban drivers of plant-pollinator interactions. *Functional Ecology* 29(7), 879–888. DOI: 10.1111/1365-2435.12486.

Hausmann, S.L., Petermann, J.S. and Rolff, J. (2016) Wild bees as pollinators of city trees. *Insect Conservation and Diversity* 9(2), 97–107. DOI: 10.1111/icad.12145.

Henry, M. and Rodet, G. (2018) Controlling the impact of the managed honeybee on wild bees in protected areas. *Scientific Reports* 8(1), 9308–9317. DOI: 10.1038/s41598-018-27591-y.

Heritage Lottery Fund (2016) *State of the UK Public Parks*. Heritage Lottery Fund, London.

Hernandez, J.L., Frankie, G.W. and Thorp, R.W. (2009) Ecology of urban bees: a review of current knowledge and directions for future study. *Cities and the Environment* 2(1), 1–15. DOI: 10.15365/cate.2132009.

Hopwood, J.L. (2013) Roadsides as habitat for pollinators: management to support bees and butterflies.In: *Proceedings 7th International Conference on Ecology and Transport (IOECT)*.

Hostetler, M., Allen, W. and Meurk, C. (2011) Conserving urban biodiversity? Creating green infrastructure is only the first step. *Landscape and Urban Planning* 100(4), 369–371. DOI: 10.1016/j.landurbplan.2011.01.011.

Hoyle, H., Jorgensen, A., Warren, P., Dunnett, N. and Evans, K. (2017) "Not in their front yard": the opportunities and challenges of introducing perennial urban meadows: a local authority stakeholder perspective. *Urban Forestry & Urban Greening* 25, 139–149. DOI: 10.1016/j.ufug.2017.05.009.

Hunter, M.R. and Hunter, M.D. (2008) Designing for conservation of insects in the built environment. *Insect Conservation and Diversity* 1, 189–196.

Ives, C.D., Giusti, M., Fischer, J., Abson, D.J., Klaniecki, K. *et al.* (2017) Human–nature connection: a multidisciplinary review. *Current Opinion in Enviromental Sustainability* 26, 106–113.

Jha, S. and Kremen, C. (2013) Urban land use limits regional bumble bee gene flow. *Molecular Ecology* 22(9), 2483–2495. DOI: 10.1111/mec.12275.

Jochner, S. and Menzel, A. (2015) Urban phenological studies – past, present, future. *Environmental Pollution* 203, 250–261. DOI: 10.1016/j.envpol.2015.01.003.

Kaluza, B.F., Wallace, H., Heard, T.A., Klein, A.-M. and Leonhardt, S.D. (2016) Urban gardens promote bee foraging over natural habitats and plantations. *Ecology and Evolution* 6(5), 1304–1316. DOI: 10.1002/ece3.1941.

Kessler, S.C., Tiedeken, E.J., Simcock, K.L., Derveau, S., Mitchell, J. *et al.* (2015) Bees prefer foods containing neonicotinoid pesticides. *Nature* 521(7550), 74–76.See also: Kessler, S.C., Tiedeken, E.J., Simcock, K.L., Derveau, S., Mitchell, J., Softley, S., Radcliffe, A., Stout, J.C. and Wright, G.A. Correction: Corrigendum: Bees prefer foods containing neonicotinoid pesticides. *Nature* 533: 278. DOI: 10.1038/nature14414.

Kevan, P.G. (1991) Pollination: keystone process in sustainable global productivity. *Acta Horticulturae* 288, 103–110. DOI: 10.17660/ActaHortic.1991.288.11.

Klein, A.-M., Vaissière, B.E., Cane, J.H., Steffan-Dewenter, I., Cunningham, S.A. *et al.* (2007) Importance of pollinators in changing landscapes for world crops. *Proceedings of the Royal Society B: Biological Sciences* 274(1608), 303–313. DOI: 10.1098/rspb.2006.3721.

Knop, E., Zoller, L., Ryser, R., Gerpe, C., Hörler, M. *et al.* (2017) Artificial light at night as a new threat to pollination. *Nature* 548(7666), 206–209. DOI: 10.1038/nature23288.

Ksiazek, K., Tonietto, R. and Ascher, J.S. (2014) Ten bee species new to green roofs in the Chicago area. *The Great Lakes Entomologist* 47, 87–92.

La Sorte, F.A., Aronson, M.F.J., Williams, N.S.G., Celesti-Grapow, L., Cilliers, S. *et al.* (2014) Beta diversity of urban floras among European and non-European cities. *Global Ecology and Biogeography* 23(7), 769–779. DOI: 10.1111/geb.12159.

Langellotto, G., Melathopoulos, A., Messer, I., Anderson, A., McClintock, N. *et al.* (2018) Garden pollinators and the potential for ecosystem service flow to urban and peri-urban agriculture. *Sustainability* 10(6), 2047–2062. DOI: 10.3390/su10062047.

Larson, J.L., Kesheimer, A.J. and Potter, D.A. (2014) Pollinator assemblages on dandelions and white clover in urban and suburban lawns. *Journal of Insect Conservation* 18(5), 863–873. DOI: 10.1007/s10841-014-9694-9.

Larson, J.L., Redmond, C.T. and Potter, D.A. (2015) Mowing mitigates bioactivity of neonicotinoid insecticides in nectar of flowering lawn weeds and turfgrass guttation. *Environmental Toxicology and Chemistry* 34(1), 127–132. DOI: 10.1002/etc.2768.

Larson, J.L., Dale, A., Held, D., McGraw, B., Richmond, D.S. *et al.* (2017) Optimizing pest management practices to conserve pollinators in turf landscapes: current practices and future research needs. *Journal of Integrated Pest Management* 8(1), 1–10. DOI: 10.1093/jipm/pmx012.

Lautenbach, S., Seppelt, R., Liebscher, J. and Dormann, C.F. (2012) Spatial and temporal trends of global pollination benefit. *PLOS ONE* 7(4), e35954. DOI: 10.1371/journal.pone.0035954.

Lee-Mäder, E., Fowler, J., Vento, J. and Hopwood, J. (2016) *100 Plants to Feed the Bees: Provide a Healthy Habitat to Help Pollinators Thrive*. The Xerces Society for Invertebrate Conservation, Washington, DC.

Lewanzik, D. and Voigt, C.C. (2014) Artificial light puts ecosystem services of frugivorous bats at risk. *Journal of Applied Ecology* 51(2), 388–394. DOI: 10.1111/1365-2664.12206.

MacIvor, J.S. and Packer, L. (2015) 'Bee hotels' as tools for native pollinator conservation: a premature verdict? *PLOS ONE* 10(3), e0122126. DOI: 10.1371/journal.pone.0122126.

Maddox, D. (2018) Can cities save bees? How can urban habitats be made to serve pollinator conservation? How can that story be better told? 03/26/19]. Available at: https://www.thenatureofcities.com/2016/01/20/can-cities-save-bees-how-can-urban-habitats-be-made-to-serve-pollinator-conservation-how-can-that-story-be-better-told/

Makinson, J.C., Threlfall, C.G. and Latty, T. (2017) Bee-friendly community gardens: impact of environmental variables on the richness and abundance of exotic and native bees. *Urban Ecosystems* 20(2), 463–476. DOI: 10.1007/s11252-016-0607-4.

Matteson, K.C., Ascher, J.S. and Langellotto, G.A. (2008) Bee richness and abundance in New York City urban gardens. *Annals of the Entomological Society of America* 101(1), 140–150. DOI: 10.1603/0013-8746(2008)101[140:BRAAIN]2.0.CO;2.

McFrederick, Q.S. and LeBuhn, G. (2006) Are urban parks refuges for bumble bees *Bombus* spp. (Hymenoptera: Apidae)? *Biological Conservation* 129(3), 372–382. DOI: 10.1016/j.biocon.2005.11.004.

McKinney, M.L. (2008) Effects of urbanization on species richness: a review of plants and animals. *Urban Ecosystems* 11(2), 161–176. DOI: 10.1007/s11252-007-0045-4.

Melathopoulos, A., Rodia, R. (M.)., Holt, J. and Sagili, R.R. (2018) *Residential Beekeeping: Best-practice Guidelines for Nuisance-free Beekeeping in Oregon*. Oregon State University Extension Service.

Michener, C.D. (2000) *The Bees of the World*. Vol. 1. Johns Hopkins University Press.

Molumby, A. and Przybylowicz, T. (2012) Bees (Hymenoptera: Apoidea) of the Chicago area: diversity and habitat use in an urbanized landscape. *The Great Lakes Entomologist* 45, 79–98.

Nassauer, J.I. (1995) Messy ecosystems, orderly frames. *Landscape Journal* 14(2), 161–170. DOI: 10.3368/lj.14.2.161.

Nassauer, J.I. (2012) Landscape as medium and method for synthesis in urban ecological design. *Landscape and Urban Planning* 106(3), 221–229. DOI: 10.1016/j.landurbplan.2012.03.014.

Nassauer, J.I. (1997) Cultural sustainability: aligning aesthetics and ecology. In: Nassauer, J.I. (ed.) *Placing Nature. Culture and Landscape Ecology*. Island Press, Washington, D.C.

Nassauer, J.I. (2011) Care and stewardship: from home to planet. *Landscape and Urban Planning* 100(4), 321–323. DOI: 10.1016/j.landurbplan.2011.02.022.

Nassauer, J.I. and Raskin, J. (2014) Urban vacancy and land use legacies: a frontier for urban ecological research, design, and planning. *Landscape and Urban Planning* 125, 245–253. DOI: 10.1016/j.landurbplan.2013.10.008.

New, T.R. (2015) *Insect Conservation and Urban Environments*. Springer Publishing Company.

Nilon, C.H. (2011) Urban biodiversity and the importance of management and conservation. *Landscape and Ecological Engineering* 7(1), 45–52. DOI: 10.1007/s11355-010-0146-8.

Ollerton, J., Winfree, R. and Tarrant, S. (2011) How many flowering plants are pollinated by animals? *Oikos* 120(3), 321–326. DOI: 10.1111/j.1600-0706.2010.18644.x.

Osborne, J.L., Martin, A.P., Carreck, N.L., Swain, J.L., Knight, M.E. *et al.* (2008a) Bumblebee flight distances in relation to the forage landscape. *Journal of Animal Ecology* 77(2), 406–415. DOI: 10.1111/j.1365-2656.2007.01333.x.

Osborne, J.L., Shortall, C.R., Shortall, C.R., Todd, A.D., Goulson, D. *et al.* (2008b) Quantifying and comparing bumblebee nest densities in gardens and countryside habitats. *Journal of Applied Ecology* 45(3), 784–792. DOI: 10.1111/j.1365-2664.2007.01359.x.

Pardee, G.L. and Philpott, S.M. (2014) Native plants are the bee's knees: local and landscape predictors of bee richness and abundance in backyard gardens. *Urban Ecosystems* 17(3), 641–659. DOI: 10.1007/s11252-014-0349-0.

Patriarca, E. and Debernardi, P. (2010) Bats and light pollution. Centro Regionale Chirotteri, Turin. Available at: http://www.centroregionalechirotteri.org/download/eurobats/Bats%20and%20light%20pollution.pdf

Pauw, A. (2007) Collapse of a pollination web in small conservation areas. *Ecology* 88(7), 1759–1769. DOI: 10.1890/06-1383.1.

Pavao-Zuckerman, M.A. (2008) The nature of urban soils and their role in ecological restoration in cities. *Restoration Ecology* 16(4), 642–649. DOI: 10.1111/j.1526-100X.2008.00486.x.

Pawson, S.M. and Bader, M.K.-F. (2014) Led lighting increases the ecological impact of light pollution irrespective of color temperature. *Ecological Applications* 24(7), 1561–1568. DOI: 10.1890/14-0468.1.

Peters, K.A. (2011) Current and emerging issues in the new urban agriculture: a case study. *Journal of Food Law & Policy* 7, 297–356.

Pickett, S.T.A., Cadenasso, M.L., Grove, J.M., Boone, C.G., Groffman, P.M. *et al.* (2011) Urban ecological systems: scientific foundations and a decade of progress. *Journal of Environmental Management* 92(3), 331–362. DOI: 10.1016/j.jenvman.2010.08.022.

Pickett, S.T.A., Cadenasso, M.L., Childers, D.L., Mcdonnell, M.J. and Zhou, W. (2016) Evolution and future of urban ecological science: ecology in, of, and for the city. *Ecosystem Health and Sustainability* 2(7), e01229. DOI: 10.1002/ehs2.1229.

Potter, A. and LeBuhn, G. (2015) Pollination service to urban agriculture in San Francisco, CA. *Urban Ecosystems* 18(3), 885–893. DOI: 10.1007/s11252-015-0435-y.

Potts, S.G., Vulliamy, B., Dafni, A., Ne'eman, G. and Willmer, P. (2003) Linking bees and flowers: how do floral communities structure pollinator communities? *Ecology* 84(10), 2628–2642. DOI: 10.1890/02-0136.

Potts, S.G., Vulliamy, B., Roberts, S., O'Toole, C., Dafni, A. *et al.* (2004) Nectar resource diversity organises flower-visitor community structure. *Entomologia Experimentalis et Applicata* 113(2), 103–107. DOI: 10.1111/j.0013-8703.2004.00212.x.

Pyke, G.H. (1984) Optimal foraging theory: a critical review. *Annual Review of Ecology and Systematics* 15(1), 523–575. DOI: 10.1146/annurev.es.15.110184.002515.

Rundlöf, M., Andersson, G.K.S., Bommarco, R., Fries, I., Hederström, V. *et al*. (2015) Seed coating with a neonicotinoid insecticide negatively affects wild bees. *Nature* 521(7550), 77–80. DOI: 10.1038/nature14420.

Salisbury, A., Armitage, J., Bostock, H., Perry, J., Tatchell, M. and Thompson, K. *et al*. (2015) Editor's choice: enhancing gardens as habitats for flower-visiting aerial insects (pollinators): should we plant native or exotic species? *Journal of Applied Ecology* 52(5), 1156–1164. DOI: 10.1111/1365-2664.12499.

Sardiñas, H.S. and Kremen, C. (2014) Evaluating nesting microhabitat for ground-nesting bees using emergence traps. *Basic and Applied Ecology* 15(2), 161–168. DOI: 10.1016/j.baae.2014.02.004.

Sattler, T., Borcard, D., Arlettaz, R., Bontadina, F., Legendre, P. *et al*. (2010) Spider, bee, and bird communities in cities are shaped by environmental control and high stochasticity. *Ecology* 91(11), 3343–3353. DOI: 10.1890/09-1810.1.

Scaven, V.L. and Rafferty, N.E. (2013) Physiological effects of climate warming on flowering plants and insect pollinators and potential consequences for their interactions. *Current Zoology* 59(3), 418–426. DOI: 10.1093/czoolo/59.3.418.

Schneider, S.S. (1989) Spatial foraging patterns of the African honey bee, *Apis mellifera scutellata*. *Journal of Insect Behavior* 2(4), 505–521. DOI: 10.1007/BF01053351.

Schueler, T.R. (2010) The clipping point: turf cover estimates for the Chesapeake Bay watershed and management implications. Chesapeake Stormwater Network Technical Bulletin. Available at: http://chesapeakestormwater. net/2012/03/technical-bulletin-no-8-the-clipping-point/

Seto, K.C., Güneralp, B. and Hutyra, L.R. (2012) Global forecasts of urban expansion to 2030 and direct impacts on biodiversity and carbon pools. *Proceedings of the National Academy of Sciences* 109(40), 16083–16088. DOI: 10.1073/pnas.1211658109.

Sirohi, M.H., Jackson, J., Edwards, M. and Ollerton, J. (2015) Diversity and abundance of solitary and primitively eusocial bees in an urban centre: a case study from Northampton (England). *Journal of Insect Conservation* 19(3), 487–500. DOI: 10.1007/s10841-015-9769-2.

Sivakoff, F., Prajzner, S. and Gardiner, M. (2018) Unique bee communities within vacant lots and urban farms result from variation in surrounding urbanization intensity. *Sustainability* 10(6), 1926–1942. DOI: 10.3390/su10061926.

Somme, L., Moquet, L., Quinet, M., Vanderplanck, M., Michez, D. *et al*. (2016) Food in a row: urban trees offer valuable floral resources to pollinating insects. *Urban Ecosystems* 19(3), 1149–1161. DOI: 10.1007/s11252-016-0555-z.

Southon, G.E., Jorgensen, A., Dunnett, N., Hoyle, H. and Evans, K.L. (2017) Biodiverse perennial meadows have aesthetic value and increase residents' perceptions of site quality in urban green-space. *Landscape and Urban Planning* 158, 105–118. DOI: 10.1016/j.landurbplan.2016.08.003.

Štajerová, K., Šmilauer, P., Brůna, J. and Pyšek, P. (2017) Distribution of invasive plants in urban environment is strongly spatially structured. *Landscape Ecology* 32(3), 681–692. DOI: 10.1007/s10980-016-0480-9.

Stange, E., Zulian, G., Rusch, G., Barton, D. and Nowell, M. (2017) Ecosystem services mapping for municipal policy: ESTIMAP and zoning for urban beekeeping. *One Ecosystem* 2(4), e14014, 1–28. DOI: 10.3897/oneeco.2.e14014.

Stevenson, P.C., Nicolson, S.W. and Wright, G.A. (2017) Plant secondary metabolites in nectar: impacts on pollinators and ecological functions. *Functional Ecology* 31(1), 65–75. DOI: 10.1111/1365-2435.12761.

Threlfall, C.G., Walker, K., Williams, N.S.G., Hahs, A.K., Mata, L. *et al*. (2015) The conservation value of urban green space habitats for Australian native bee communities. *Biological Conservation* 187, 240–248. DOI: 10.1016/j.biocon.2015.05.003.

Tonietto, R., Fant, J., Ascher, J., Ellis, K. and Larkin, D. (2011) A comparison of bee communities of Chicago green roofs, parks and prairies. *Landscape and Urban Planning* 103(1), 102–108. DOI: 10.1016/j.landurbplan.2011.07.004.

Turo, K.J. and Gardiner, M.M. (2019) From potential to practical: conserving bees in urban public green spaces. *Frontiers in Ecology and the Environment* 17(3), 167–175. DOI: 10.1002/fee.2015.

Van Velsor, S.W. and Nilon, C.H. (2006) A qualitative investigation of the urban African-American and Latino adolescent experience with Wildlife. *Human Dimensions of Wildlife* 11(5), 359–370. DOI: 10.1080/10871200600894944.

Vaughan, M. and Hoffman Black, S. (2007) *Enhancing Nest Sites for Native Bee Crop Pollinators. Agroforestry note 34*. USDA National Agroforestry Center, Lincoln.

Visscher, P.K. and Seeley, T.D. (1982) Foraging strategy of honeybee colonies in a temperate deciduous forest. *Ecology* 63(6), 1790–1801. DOI: 10.2307/1940121.

Von Frisch, K. (1967) *The Dance Language and Orientation of Bees*. Harvard University Press.

Waddington, K.D., Herbert, T.J., Visscher, P.K. and Richter, M.R. (1994) Comparisons of forager distributions from matched honey bee colonies in suburban environments. *Behavioral Ecology and Sociobiology* 35(6), 423–429. DOI: 10.1007/BF00165845.

Wastian, L., Unterweger, P.A. and Betz, O. (2016) Influence of the reduction of urban lawn mowing on wild bee diversity (Hymenoptera, Apoidea). *Journal of Hymenoptera* 49, 51–63.

Whitehorn, P.R., Tinsley, M.C., Brown, M.J.F., Darvill, B. and Goulson, D. (2009) Impacts of inbreeding on bumblebee colony fitness under field conditions. *BMC Evolutionary Biology* 9, 152–160. DOI: 10.1186/1471-2148-9-152.

Whitehorn, P.R., Tinsley, M.C., Brown, M.J.F., Darvill, B. and Goulson, D. (2011) Genetic diversity, parasite prevalence and immunity in wild bumblebees. *Proceedings of the Royal Society B: Biological Sciences* 278(1709), 1195–1202. DOI: 10.1098/rspb.2010.1550.

Williams, N.M., Cariveau, D., Winfree, R. and Kremen, C. (2011) Bees in disturbed habitats use, but do not prefer, alien plants. *Basic and Applied Ecology* 12(4), 332–341. DOI: 10.1016/j.baae.2010.11.008.

Wilson-Rich, N. (2014) *The Bee: A Natural History*. Princeton University Press.

Wilson, J.S., Forister, M.L. and Carril, O.M. (2017) Interest exceeds understanding in public support of bee conservation. *Frontiers in Ecology and the Environment* 15(8), 460–466. DOI: 10.1002/fee.1531.

Winfree, R., Williams, N.M., Dushoff, J. and Kremen, C. (2014) Species abundance, not diet breadth, drives the persistence of the most linked pollinators as plant-pollinator networks disassemble. *The American Naturalist* 183(5), 600–611. DOI: 10.1086/675716.

Wojcik, V.A. and Buchmann, S. (2012) Pollinator conservation and management on electrical transmission and roadside rights-of-way: a review. *Journal of Pollination Ecology* 7, 16–26.

Wojcik, V.A., Frankie, G.W., Thorp, R.W. and Hernandez, J.L. (2008) Seasonality in bees and their floral resource plants at a constructed urban bee habitat in Berkeley, California. *Journal of the Kansas Entomological Society* 81(1), 15–28. DOI: 10.2317/JKES-701.17.1.

Wolf, S. and Moritz, R.F.A. (2008) Foraging distance in *Bombus terrestris* L. (Hymenoptera: Apidae). *Apidologie* 39(4), 419–427. DOI: 10.1051/apido:2008020.

Xie, Z., Qiu, J. and Chen, X. (2013) Decline of nest site availability and nest density of underground bees along a distance gradient from human settlements. *Entomological Science* 16(2), 170–178. DOI: 10.1111/ens.12009.

Younis, A., Riaz, A., Saleem, S. and Hameed, M. (2010) Potential use of wild flowers in urban landscape. In: Gianquinto, P.G. and Orsini, F. (eds) *Proceedings of 2nd International Conference on Landscape and Urban Horticulture*, pp. 229–233. DOI: 10.17660/ActaHortic.2010.881.29.

Zanette, L.R.S., Martins, R.P. and Ribeiro, S.P. (2005) Effects of urbanization on Neotropical wasp and bee assemblages in a Brazilian Metropolis. *Landscape and Urban Planning* 71(2-4), 105–121. DOI: 10.1016/j.landurbplan.2004.02.003.

Zerbe, S., Maurer, U., Schmitz, S. and Sukopp, H. (2003) Biodiversity in Berlin and its potential for nature conservation. *Landscape and Urban Planning* 62(3), 139–148. DOI: 10.1016/S0169-2046(02)00145-7.

Zurbuchen, A., Landert, L., Klaiber, J., Müller, A., Hein, S. *et al.* (2010) Maximum foraging ranges in solitary bees: only few individuals have the capability to cover long foraging distances. *Biological Conservation* 143(3), 669–676. DOI: 10.1016/j.biocon.2009.12.003.

7 Urban Ecology as an Integrative Science and Practice

Steward T.A. Pickett[1]*, J. Morgan Grove[2], Shannon L. LaDeau[1], Emma J. Rosi[1] and Mary L. Cadenasso[3]

[1]Cary Institute of Ecosystem Studies, Millbrook, New York, USA; [2]USDA Forest Service, Baltimore, Maryland, USA; [3]University of California Davis, Department of Plant Sciences, Davis, California, USA

Abstract:

Urban ecology began as a biological science practised within cities. As such, it sought out forests, abandoned or undeveloped land, meadows and surface waters as analogues of the places outside of cities where ecologists had worked for decades. In 1997, the call for proposals by the National Science Foundation's Long-Term Ecological Research (LTER) programme for urban research sites required integration of biological and physical structures and processes with social structures and processes. The Baltimore Ecosystem Study, referred to as BES, was one of the two successful responses to that call, and thus worked along with the Central Arizona Phoenix LTER to craft a new, integrated approach to the understanding of urban ecological systems. The conceptual and visionary framework to support this integration is termed 'ecology of the city', which for North America is a new kind of social-ecological-technological systems (SETS) research. The ecology of the city has required BES to establish interactions across disciplinary boundaries, linking with sociology, geography, anthropology, economics, engineering, climatology, public health and urban design. Furthermore, to meet the challenges of the region in which the project is embedded, BES has had to become transdisciplinary, that is to move beyond the scope of individual or even linked scientific disciplines. Transdisciplinary research engages with communities, non-governmental organizations and people from federal, state, county and city governments to answer questions that are of interest to science, while at the same time addressing needs of society. As a result, BES exemplifies growth beyond the traditional disciplinary focus of ecology on biological habitats within urban areas. That growth first linked various social, biological and physical science disciplines and, ultimately, was integrated with policy, management and design concerns. We trace the development of BES to illustrate integrative science and practice.

Roots of Our Social-ecological Science

Interdisciplinary fundamentals of urban ecology, laid out in this chapter, begin in the 1980s with Dr Mark McDonnell and Dr Richard Pouyat (McDonnell *et al.*, 1997). Like all histories of

*Corresponding author: picketts@caryinstitute.org

major developments in any academic field of study, this is a story about people and the ideas they brought together. These colleagues were engaged in a project to understand the structure and function of New York City's last remaining primeval forest on the grounds of the New York Botanical Garden in the Bronx. This was new territory for American ecology. Mark, Rich and their colleague Dr Carl White had discovered puzzling dynamics of soil nitrogen in that ancient but urban forest. Comparison between the urban forest with those outside the city emerged as a way to investigate the unexpected results. Mark and Rich devised a sampling transect from the Bronx northward on the same bedrock to examine and compare closed-canopy, oak-dominated forests isolated within landscapes of decreasing urbanization. The sampling design was a literal belt transect across the metropolitan landscape of New York. The sampling transect was laid out to discover whether there was a gradient of urban effects on forest structure and dynamics (McDonnell *et al.*, 2012). The hypothesis was that urban effects seen in the forest in the Bronx would decline along the 120 km, or thereabouts, transect. The project was called Urban-Rural Gradient Ecology (URGE; McDonnell *et al.*, 1997). Many of the specific scientific results have been summarized elsewhere (e.g. Pouyat *et al.*, 1995, 2009; Baxter *et al.*, 2002; McDonnell and Hahs, 2008; Carreiro *et al.*, 2009). When Steward Pickett joined McDonnell and Pouyat in 1987 they sought to employ gradients of stress and disturbance as a system of ideas intended to explain the impact of environmental and ecological extremes on the organization and function of ecosystems, to urban–rural comparisons. The ecological theories of stress and disturbance gradients (Fox *et al.*, 2011) had not previously been applied to urban systems. Connecting these theories to urban ecology has proven to be a lasting legacy of the URGE project, in that an urban–rural gradient perspective has been used widely in urban ecology (e.g. Adler and Tanner, 2013; Douglas and James, 2014).

After a few years of concentrating on the novel biology and geochemistry of forests along the New York metropolitan transect, the URGE group began to wonder about the social drivers of the emerging ecological patterns, by accessing and using data sources such as the decennial US census for basic human statistics. That is where the adventure of integration began. This chapter lays out how the URGE project established a way of thinking that was later linked to an ecologically informed, socially connected community forestry project in Baltimore. This linkage would become the core of a new integrative approach to urban ecology (Grove *et al.*, 2013). The adventure is described here in terms of how and why ecology was linked with other natural sciences, with social sciences, and with the applied professions of engineering and urban design.

First Steps with Social Science

The New York transect stimulated the URGE team to reach out to social scientists who might contribute by helping to explain the ecological patterns it was discovering. A national jaunt by McDonnell and colleagues to visit likely social scientists at various institutions exposed the team to brilliant and productive researchers, but perhaps not surprisingly, they were all too busy to join. The interactions convinced McDonnell and team even more strongly of the need to link with social sciences. Fortunately, Cary Institute was able to provided funds to hire a postdoc. Kimberly Medley, a geographer, joined the team to analyse human population, road density and traffic flows in blocks of territory surrounding the forests along the URGE transect. Although all the geographic variables were correlated with the biogeochemical processes in the forests, the functional variable of traffic flow exhibited the greatest explanatory power (Medley *et al.*, 1995). Given the role of automobiles as sources of the atmospheric nitrogen and heavy metals that were expected to alter forest nutrient cycling (Pouyat *et al.*, 2009), the higher correlation with traffic stimulated further thinking about social drivers of ecosystem processes. The possibilities were many, including property ownership, public and private management practices, transportation policy and regulations, as well as economic resources and their allocation. What might be the constellation of social factors? How might the social factors interact with the ecology? How might ecological structure and process feed back to the social factors? Although these became important questions for the URGE team;

they were beyond the expertise of a primarily biologically trained research group.

Our affiliation with Dr Morgan Grove was particularly important to meeting our project goals because his graduate training had been under the mentorship of Dr William 'Bill' Burch, Jr, who was a pioneer in social ecology, i.e. the study of relationships between people and their environment (Burch, 1971). Members of this expanded group worked well together. This fledgling community was clearly a rare intellectual resource that had great potential. Since 1989, Grove and Burch had been practising the science of social ecology and community forestry in Baltimore, Maryland (Grove and Carrera, 2019). Upon visiting Baltimore with Grove and Burch, the biologically trained ecologists discovered the 'power of place' in social ecology. Place is important for two reasons. First is its conceptual role in co-locating social processes, ecological processes and the interactions among them. Place, in this conceptual sense, is key to understanding complex systems. Second, place as a shared concern by researchers, managers and policy makers, helps create the social and institutional networks based on trust, sometimes referred to as social capital (Vanderbilt and Gaiser, 2017), that are required to conduct and apply research in urban areas. Trust is something that is achieved over time from common concerns, mutual respect and common experiences. Grove and Burch had built that trust over their decade of work in Baltimore (Grove and Carrera, 2019). Now they were sharing that hard-earned capital with ecological colleagues originally trained as biologists.

The expanded URGE team struggled to win the approval of the National Science Foundation for the establishment of regular ecological research proposals for urban ecology in Baltimore and New York. The funding community at the time was somewhat hostile toward urban ecology research. Urban ecology focuses on complex, non-linear systems, having feedbacks involving social attitudes, institutional motivations, political constraints, human perceptions, and even fashion. Thus, reviewers' expectations about linear hypotheses, idealized theory and ready falsification were a poor match to the work that had to be done. More problematically, reviewers often opined that ecology just wasn't done in cities. However, the LTER programme had been stimulated by its first decadal programme review, conducted in 1993, to consider bringing urban ecology into the fold of its roughly 20 projects focusing on 'natural' ecosystems throughout the USA and its territories (Risser and Lubchenco, 2010; Collins, 2019). The LTER programme, led at the time by programme officer Dr Scott Collins, decided to consider 'up to two' urban LTER sites. Integration of ecological and social sciences was a key part of the 1997 call for proposals. That call is quoted in its entirety by Pickett *et al.* (2019a). The Baltimore Ecosystem Study is the result of the authors' response to that call for integrated social-ecological urban proposals.

Patch Hierarchies as Integrative Frameworks for Urban Ecology

The authors of the first BES proposal used two fundamental concepts from modern ecology to integrate social-ecological systems research. One was patch dynamics and the other was watershed theory (Cadenasso *et al.*, 2006). Both of these concepts invoke hierarchies. Patches are parts of nested hierarchies. That is, patches, at one scale, may be aggregated into more inclusive patches at larger scales, or may be disaggregated into smaller or constituent patches. Also, watersheds can be aggregated into larger catchments, or can be dissected into smaller tributary catchments (Fig. 7.1). Such hierarchies are an important integrative tool in ecology (Allen and Starr, 1982) and are critical for integrating the disciplinary traditions contributing to BES.

Patch dynamics

Patch dynamics is a theory from community ecology, ecosystem ecology and landscape ecology that addresses the form, causes, consequences and changes of spatial heterogeneity (Pickett and White, 1985; Wu and Loucks, 1995). The recently recognized discipline of land change science (e.g. Meyfroidt *et al.*, 2018) fundamentally represents a patch dynamic approach. A patch dynamic approach is based on a pervasive concern of the science of ecology with spatial heterogeneity (Scheiner and Willig,

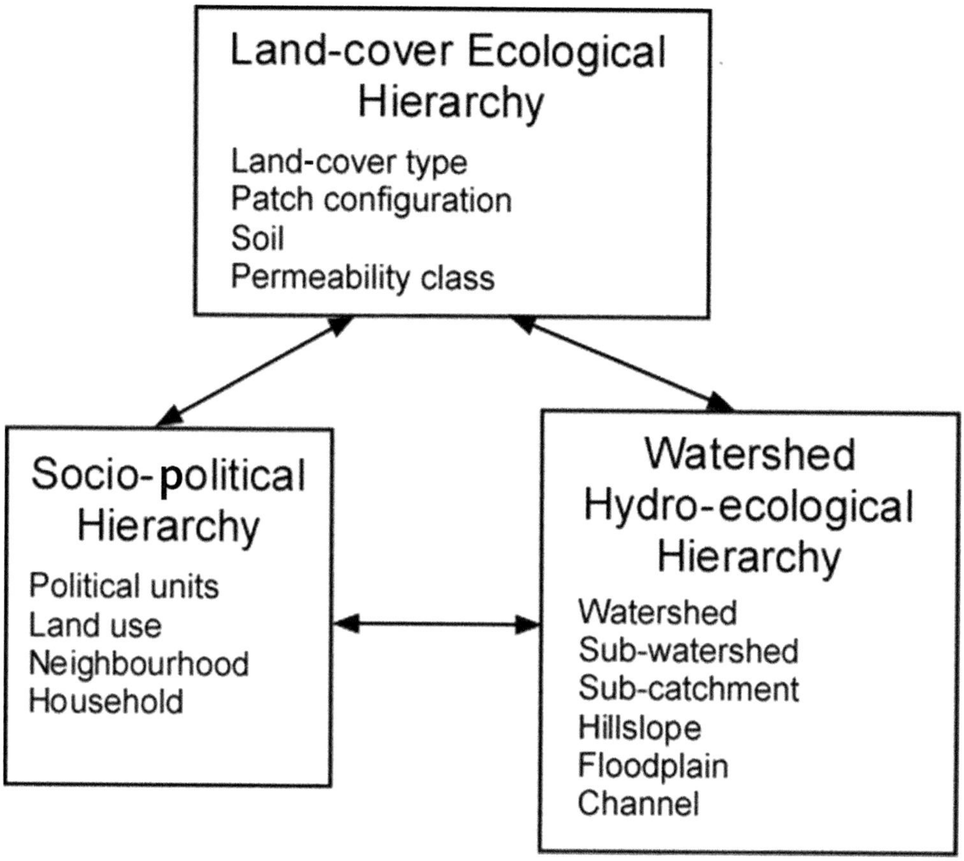

Fig. 7.1. The initial spatial, interdisciplinary and nested hierarchical frameworks used by the Baltimore Ecosystem Study in 1997. The conceptual frameworks assume that urban areas integrate social, landscape and watershed hierarchies, and that the urban area constitutes a social-ecological system composed of heterogeneous patches incorporating elements of all three hierarchies. Patch dynamics is a conceptual tool that provides a complementary expression of the spatial heterogeneity, functional connections among the hierarchies, and dynamics of the patchworks involved. The human ecosystem framework is an additional framework (figure 7.2) that indicates the social structures and social processes, and their relationships to the biophysical components of the urban social-ecological system.

2011). Evolution similarly relies on heterogeneity in many of its key tenets, such as the spatial distribution of genetic differentiation or the role of geographic isolation in speciation (e.g. Rivkin *et al.*, 2019).

Social sciences also use spatial heterogeneity as an explanatory tool (Shevky and Bell, 1955; Grove and Burch Jr., 1997; Gottdiener and Hutchison, 2011). Because social groups, neighbourhoods, institutions, policies and social legacies are all spatially distributed, patch dynamics is a major tool for integrating social and ecological sciences in cities. The terminologies of the different disciplines concerned with spatial heterogeneity can be translated across disciplinary boundaries during integration. Patch dynamics can be expressed in urban areas as nested social units, such as neighbourhoods, census blocks and administrative units. Likewise, patch dynamics can be conveyed in nested typologies of soils, vegetation covers and the built environment.

Some critics seem to interpret patch dynamics as being a descriptive and static approach. However, patch dynamics is concerned not only with structure but, equally, with the degree and configuration of transfers among patches, as well as how the patches themselves, and hence their functioning, change over time, i.e. its dynamics. Furthermore, depending on the criteria that are used to differentiate patches, there can be several kinds of patch arrays that can exist within a given area (Cadenasso *et al.*, 2003). A patch array is a spatial assemblage resolvable as patches that can exist on any spatial scale. Different arrays can be generated depending on the questions research seeks to resolve or the applied motivations of the research project (Cadenasso *et al.*, 2013). For example, administrative patches such as census tracts provide one kind of patch array in an urban place. However, that same spatial scope can be mapped as small watersheds defined by drainage infrastructure, or alternatively as patches with dense tree cover, versus patches of isolated trees, herbaceous cover or lacking vegetation. Finally, that area can be divided by neighbourhood associations or territories of concern to different environmental NGOs. Each of these criteria will entail a specific patch array or mosaic. How these mosaics change and how they relate to each other are important research questions.

Watershed approach

Watersheds are spatial units of land tied together by the flow of surface and sub-surface water. In biological and physical sciences, watersheds have helped us and can help us, going forward, in understanding processes of ecosystem ecology (Likens, 1985). Both biological and constructed features are parts of watersheds. Consequently, watersheds can be used to measure the impact of both constructed and biological features upstream of their confluence with receiving bodies of water, such as larger streams, lakes or coasts (Groffman *et al.*, 2003, Groffman *et al.*, 2019). There were three reasons to employ the watershed concept in BES. First, watersheds have definable physical boundaries so they serve as a unifying spatial extent for data collection. Such boundaries were familiar to residents and decision makers in Baltimore because of the strength of the civic watershed associations and the powerful connection of the city to the Chesapeake Bay watershed. Second, the watershed combines terrestrial and aquatic perspectives on spatial patchiness in a functional way (Likens, 1992). Finally, it is a concept that reviewers of an interdisciplinary proposal might recognize from their own perspective.

Part of the strong foundation for using watersheds emerged from the hydrological laboratories across the USA, established to assess forest management strategies. For example, within the LTER network, watersheds have been central for the sites at Andrews Experimental Forest, Coweeta, Harvard Forest and Luquillo. An important inspiration for BES was the use of the watershed by the Hubbard Brook LTER site located in the White Mountains of New Hampshire. Hubbard Brook was founded to use the watershed approach to test the impact of different timber-harvesting practices on the quality and quantity of water exiting manipulated watersheds. In Baltimore, the authors needed to add to such biophysical concerns by assessing social attitudes, perceptions, decisions and investments. We made these assessments in mid-sized watersheds across the Baltimore region that contrasted in the degree and form of urban cover, land use and social characteristics. In this way, we expanded the integrative power of watersheds to urbanized regions. Mid-sized watersheds that differ in land uses and kinds of stormwater and sanitary infrastructure provided contrasts that are expected to affect ecosystem function in different ways.

Integrating frameworks and guiding questions

The concepts of patch dynamics and watersheds, based as they are on spatial differentiation, easily link to the concerns of urban scholars, designers and planners with the fine-grained and changing heterogeneity of cities and suburbs (Shane, 2005). Patch dynamics and watersheds are also linked to the concerns about water quality of environmental activists and policy makers in Baltimore (Pickett *et al.*, 2007b). Consequently, the patch dynamic framework would have power in basic science as well as efficacy in connecting with residents and decision makers in Baltimore (Cadenasso *et al.*, 2008).

These concepts suggested the three guiding original BES questions about urban areas as integrated social-ecological systems:

1. How do the spatial structures of socio-economic, ecological and physical features of an urban area relate to one another, and how do they change through time?
2. What are the fluxes of energy, matter, human capital, built capital and social capital (cf. Hassler and Kohler, 2014) in an urban system; how do they relate to one another, and how do they change over the long term?
3. How can people develop and use an understanding of the metropolis as an ecological system to improve the quality of their environment, and reduce pollution elsewhere?

The first two questions are straightforward applications of the idea that patch dynamics and watersheds are functional units in urban systems. The third question emerges from the knowledge that people, civic organizations and decision makers in Baltimore are concerned about the quality of their local environments, and the health of downstream systems, such as the Chesapeake Bay. These questions address biological, physical and social features and processes of the urban system. Thus, they are in themselves integrative. However, they also invoke the 'human ecosystem' as an inclusive framework (Cadenasso *et al.*, 2006). This framework was introduced to the BES team by Burch and colleagues (Machlis *et al.*, 1997). The BES team added several important ecological details to the framework in order to better link with the interests of biological and physical scientists (Pickett *et al.*, 1997; Fig. 7.2). This framework is described most fully in the 2017 book by Burch *et al.* (2017). The human ecosystem identifies the kinds of social, economic, behavioural and biophysical structures that make up any urban area. Furthermore, the abstract connections among the various general components of the framework can be fleshed out by specific models of flows, influences and detailed processes of interest. So the human ecosystem serves as a framework for the specific models needed for an integrated understanding of urban social-ecological systems.

A Culture of Integration and Synthesis

The culture of BES has facilitated integration. By culture we refer to both specialized training and 'the totality of socially transmitted behavior patterns, arts, beliefs, institutions, and all other products of human work and thought' (Pickett, 1999). First, several of the founders of BES articulated a philosophy of exploration across disciplinary boundaries (Pickett *et al.*, 1999). This culture emphasized the creative and open-minded impulse in contrast to the more conservative and defensive perspectives that are sometimes observed in disciplinary science. Furthermore, the project identified 'habits of mind' focused on synthesis, on overcoming narrow disciplinary differentiation, and on avoiding strictly reductionist stances in the origination and testing of hypotheses (Pickett *et al.*, 2007a). In particular, the integrative 'habits of mind' include (i) a long-term commitment to synthesis; (ii) the use of analogy and radical juxtaposition of seemingly disparate ideas; (3) vanquishing the 'eureka myth' of instantaneous discovery (i.e. moments or circumstances during which there is an epiphany leading to total clarity); and (iv) improved use of a diversity of people, perspectives and knowledge domains in the research community (Longino, 1990; Pickett, 1999). This final attribute has been particularly evident in BES. Talent and skill sets were sought broadly, and participants from many academic, civic and governmental institutions were involved in BES from the beginning. This allowed the project not to be limited by the members of particular departments or universities, or particular government agencies. The project was a 'big tent', embracing members from many states in the USA and indeed several foreign countries. Budgetary constraints dictated that no contributors could be fully supported by project funds. Consequently, a certain level of commitment had to be intellectual rather than financial. The work of BES was leveraged by other sources of support beyond the base grant.

The 2018 20th annual meeting of BES participants, institutional partners and interested citizens provided an opportunity to ask those involved in BES what they thought the most important finding or contribution of BES had been over those 20 years. Although the informal

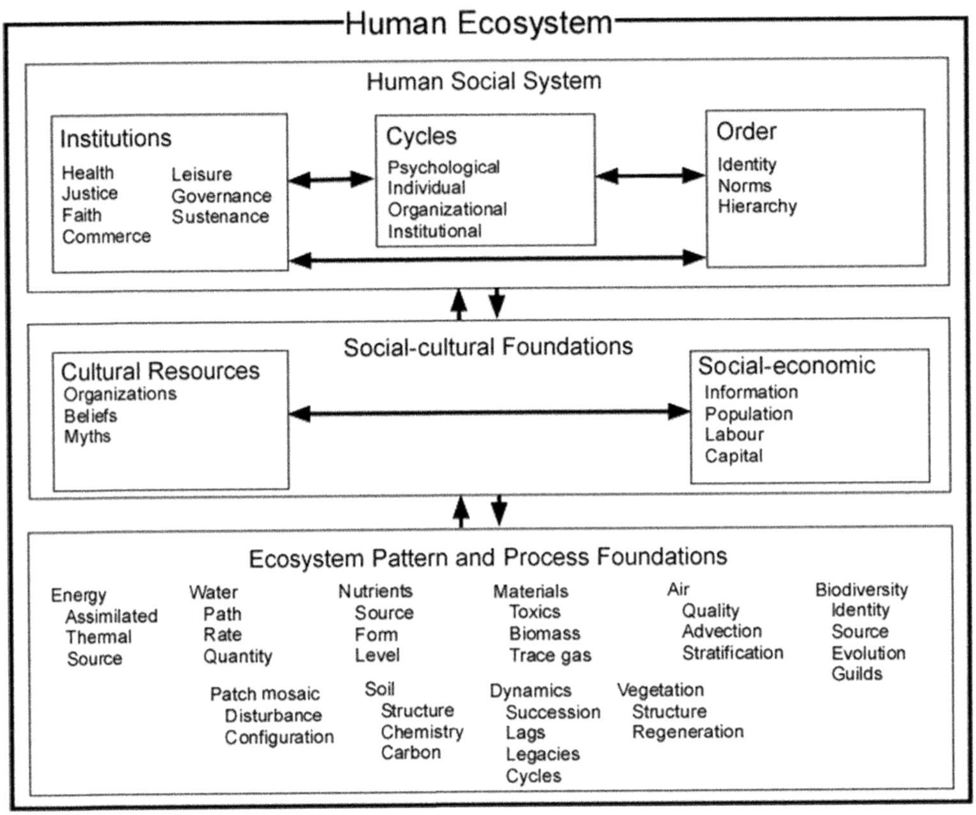

Fig. 7.2. A representation of the human ecosystem framework, which suggests the structural content and general interactive links that constitute entire urban ecosystems and their component neighbourhoods, watersheds or other spatially differentiated units. A fuller discussion of the human ecosystem framework is provided by Burch *et al.* (2017).

survey was not statistically rigorous, the majority said that the most valuable contribution of BES was BES itself, as an intellectual network and community of practice (Lave and Wenger, 1991). A community of practice is a group of people who are committed to shared learning and action about a place or problem. The BES community of practice has aimed to advance urban ecological science and its application. Thus, attention to a positive culture of integration has been both a tool and an outcome of BES.

Examples of the Reach beyond Disciplines

This section gives examples of research in BES that reached across disciplines. In some cases, the research emerged primarily from one discipline but had implications and connections with other disciplines. In other cases, the research was interdisciplinary from the start. In addition to various kinds of ecological science, the examples draw upon sociology, economics, geography, engineering, climatology, public health and urban design. We do not have space to present all disciplinary perspectives and links used in BES in this chapter. Many other examples are given in the book by Pickett *et al.* (2019b).

Structure and function of urban lands and waters

Ecology is one of the pillars of BES, reflecting broad, interdisciplinary concerns such as systems approaches, distribution and adaptation of organisms, and the control of ecosystem material fluxes. We present two examples that emerged from, but went beyond, ecology. One is a new integrated classification of urban land covers, and the other is a multi-disciplinary understanding of riparian function in urban watersheds. Traditionally, classification of urban areas combines land use and land cover to map the area into patches of residential, commercial, industrial, parks etc. This industry standard for urban classification is based on the methodology of Anderson *et al.* (1976), which was intended to describe land use at continental scales. In addition, the Anderson scheme assumes that urban, agricultural and biologically dominated wild or rural lands are distinct and mutually exclusive categories. For both of these reasons, we judged this commonly used approach to be inadequate for urban systems such as Baltimore because we needed to describe the spatial heterogeneity *within* the metropolis. The merging of structure and function within the standard approach was also problematic because it did not allow us to use land cover as an independent variable to test the link between system structure and function in social-ecological comparisons. We also wished to examine the implications of the assumption that urban areas were actually hybrids of biological, physical and social processes and products. Consequently, Cadenasso *et al.* (2007) developed a new classification, called HERCULES. This classification assumes that urban areas comprise land covers that are hybrids of vegetation, various kinds of surfaces, and buildings. Such thinking had been introduced by Ridd (1995) and seemed promising for BES, given the desire to integrate biological and social phenomena. Using the HERCULES classification of land cover generated better statistical prediction of nitrate dynamics in streams than those generated by using the usual land-use classifications. HERCULES is more finely *conceptually* resolved than land-use/land-cover classifications in the Anderson *et al.* (1976) mould (see above) and it assumes hybridity of social, physical and biological components of urban cover (Cadenasso, 2013). In addition, the land-use/land-cover maps available at the start of BES were based on 30 m pixels, which was too coarse-grained to support many of the desired mechanistic ecological comparisons. Current aerial and satellite imagery support classification at spatial scales of less than a metre resolution, and when combined with the fine conceptual resolution of HERCULES provide a new lens on urban system structure (Zhou *et al.*, 2014a).

Functional integration of watersheds is exemplified in two ways. First, BES demonstrated that riparian zones, which were expected to remove nitrate pollution from streams, did not function that way in the city and suburbs. The amount and pathway of urban stormwater flow bypassed the floodplains of urban streams. This starved those streamside areas of the environmental conditions required for the denitrifying bacteria that could otherwise remove the polluting nitrate (Groffman *et al.*, 2003; Groffman, 2012).

Functional integration between human actions and ecosystem processes in urban watersheds is illustrated by pharmaceuticals and personal care products (Rosi-Marshall *et al.*, 2013, 2015). These compounds include prescription medications purposefully discarded in the waste-water stream, metabolites or unmetabolized legal and illicit drugs, and residues of products for personal hygiene and care. Many aspects of the human ecosystems of urban areas are involved in the interactions of pharmaceuticals and personal care products with stream metabolism and the ecology of aquatic organisms. Intentional disposal of drugs, accidental contamination, faulty infrastructure and the fact that neither storm drains nor sanitary sewage treatment are designed to clear these compounds from the environment are factors that contribute to complex patterns of contamination. Such contaminants affect the evolution and development of aquatic organisms and the productivity of aquatic ecosystems. Therefore, this emerging issue for urban watersheds requires policy attention (Rosi-Marshall and Kelly, 2015).

Social drivers and outcomes of environmental decisions

Sociology has a deep history in urban areas in the USA, dating at least to the Chicago School of the early 20th century (Hawley, 1986). The social sciences deal with the aggregations into which people sort and organize, the nature and motivations of choices made by people, the role of power and status, and the distribution of resources based on social structures, among other concerns. An example of the marriage of social and ecological concerns is our new theory of 'the ecology of prestige' (Grove *et al.*, 2014; Locke *et al.*, 2018). This concept argues that key environmental decisions, by tenants and homeowners, can be driven by their desire to adopt and reinforce neighbourhood aesthetics, and, as a result, garner prestige in their neighbourhood. This rationale influences the management of front gardens more than back gardens, which are regarded as private space (Locke *et al.*, 2018). Management of the front garden affects the amount and timing of fertilizer application, the use of pesticides and, indeed, the species of plants used in landscaping. Such seemingly aesthetic choices affect run-off of potentially polluting nutrients and the capacity of parcels to support native species biodiversity, for example. The ecology of prestige thus links social sciences and ecological sciences.

An emerging concern is the relationship between social status and the condition of neighbourhoods and their environmental hazards, risks and opportunities. For example, social variables including infrastructure condition and maintenance, related to neighbourhood wealth, have been used to explain differences in where urban mosquito communities breed (LaDeau *et al.*, 2013; Little *et al.*, 2017) and the resulting pattern of human exposure to mosquitoes in different neighbourhoods (Goodman *et al.*, 2018). In neighbourhoods of lower household incomes, greater occurrence of building abandonment and associated dumping of household or construction debris provide habitat or containers that capture rainwater and so support early-season mosquito development. In contrast, mosquito populations in higher-income neighbourhoods with low infrastructure abandonment experience higher numbers later in the summer, as a consequence of watering and irrigating gardens and planters (Becker *et al.*, 2014). Although wealth was a good correlate of the differences in mosquito populations, it was actually people's lifestyles, as reflected in maintenance of outdoor plantings, that were the proximal cause of higher summer mosquito populations in the wealthier neighbourhoods. Also related to wealth and to the perceived lack of value of houses in some low-income neighbourhoods were the large numbers of buildings that stand as ruins in such areas. When roofs decay and collapse, the ruins collect standing water, which supports mosquito populations in poorer neighbourhoods. While these lower-income neighbourhoods had up to three times more biting mosquitoes over a season, human residents were more likely to spend time outside on the street-side front stoops that were often too hot for mosquito activity. Analysis of blood in mosquito guts demonstrated that mosquitoes in wealthier neighbourhoods were more likely to bite human residents than those inhabiting lower-income neighbourhoods, in which they were most likely to feed on rats (Goodman *et al.*, 2018).

Urban sociology has been linked to mainstream economic theories as well. For instance, the 'usual suspects' among social data are population density, income and level of education. This triumvirate of factors reflects a conception of cities and their residents as drivers of industrial production. Understanding the human and social capital of a city via these three measures was crucial for assessing and predicting the productive capacity of the population gathered within it. Such a view of cities still has some relevance, but contemporary social science recognizes that cities now have different or additional purposes (Haberl *et al.*, 2006; Simone, 2010; Shane, 2011). Since World War II, consumption has become a major social motivation. In many ways, the suburbanization that characterized that timespan is itself a manifestation of consumption, of housing and property, of appliances, of automobiles, and perhaps also of leisure. This new theory of cities has required new kinds of data that account for social organization arranged around consumption. Lifestyle and life stage classifications serve this need (Boone *et al.*, 2014; Moore, 2015). BES researchers have led the way in applying such social classifications to an integrated understanding of urban areas as social-ecological systems (Grove, 1999; Grove *et al.*, 2015). These measures have succeeded in explaining environmental variables such as the amount of tree canopy in neighbourhoods, or how residents perceive stormwater utility fees.

Ecological economics within a metropolitan region

Economics has been a useful explanatory variable since at least the 1800s, when theories of declining bid rents along gradients of decreasing urban density from city centres were promulgated. Research in BES has contributed to two new ways of thinking in this context (Irwin, 2010; Irwin and Wrenn, 2014; Irwin *et al.*, 2019). First, it has shown how fine-scale amenities or disamenities can influence the sale price of houses in Baltimore. According to transacted prices of houses close to, as opposed to distant from, stormwater detention ponds, these infrastructures appear to be viewed as disamenities by home buyers. Houses close to detention ponds netted lower sale prices than those farther away (Irwin *et al.*, 2019). In contrast, houses close to restored stream reaches garnered higher sale prices than houses close to stream reaches that had not been restored. Other BES research has examined how housing prices and the presence of parks and crime combine as factors of urban ecological economics (Troy and Grove, 2008).

The examples above have relied on newly amassed data from decades of actual sale prices of houses. However, not all environmentally impactful decisions can be explained using transacted sale values. In the core of Baltimore there is an immense number of abandoned houses and vacant lots that have no effective value on the open market. Therefore, BES economists are developing new models for dealing with transfers of property and decisions concerning demolition (Chen *et al.*, 2011). Rather than transacted prices, economic decisions based on city programmes for restoration of properties, or the use of demolition to help revitalize neighbourhoods are being tested (Pickett *et al.*, 2019b). Demolition of vacant and uninhabitable rowhouses by the city is part of the strategy for improving neighbourhood liveability. An environmentally significant finding is that multiple demolitions are required to stimulate revitalization of the remaining properties in a neighbourhood. Environmental benefits of such revitalization are various and potentially quite large, ranging from tree planting, generation of large and contiguous local green-space amenities, and reduction of illegal dumping of refuse and debris from outside the neighbourhood. The social cohesion in such revitalized neighbourhoods improves as well.

Living histories of environmental and social differentiation

Geography has provided important spatial and historical perspectives on environmental drivers of development and outcomes of urban form in Baltimore. Two of Baltimore's major watersheds, the Gwynns Falls and the Jones Falls, differ in the amount of parkland and disposition of different contemporary land uses such as residential, commercial and transportation infrastructure. These differences are explained by the early history of navigation and water-powered industry (Bain and Buckley, 2019). Jones Falls, nearer the centre of what would become the city of Baltimore and navigable deep into the city, became a primarily commercial corridor, with mills along that navigable length. Its flood plain later became a site for substantial railway infrastructure. In contrast, Gwynns Falls was less important for navigation, instead becoming a major source of water power for early industry. This included a large iron forge that required high rates of forest clearance to fuel the furnaces. Two estates above the fall line in the Gwynns Falls watershed would later become the largest contiguous park in the city. Not only did the steep slopes discourage development in these areas, but also the ownership was in the form of large parcels, mirroring early land grant boundaries. These linkages of social and biological and physical features through time help explain the current distribution of green space and other land covers in the city (Bain and Buckley, 2019).

Historical geography has revealed that the usual expectation of the association of Toxics Release Inventory (TRI) sites with poor communities of colour does not hold in contemporary Baltimore (Boone, 2002, 2008). Rather, old contaminated industrial and commercial sites are statistically associated with traditionally white working-class neighbourhoods, primarily in south Baltimore. However, this unexpected finding still reflects a history of segregation and racial exclusion. In the era of the industrial city in which people largely walked to work,

African-Americans were segregated to neighbourhoods more distant from work in factories and in the port than were white workers. In all cases, however, the contemporary association of neighbourhoods with TRI sites is related to low educational attainment regardless of race (Grove *et al.*, 2018). Historical analysis shows that the patterns of segregation in Baltimore not only have contemporary environmental ramifications, but that they also have deep, systematic roots (Lord and Norquist, 2010; Buckley *et al.*, 2019). Although Baltimore had fine-scale racial heterogeneity in the late 1800s, the great migration of African-Americans, from farther south, was accompanied by official moves to isolate black residents in certain neighbourhoods. This led, in 1911, to the passage of the first municipal ordinance in the USA forbidding mixing of white and African-American residents in the same neighbourhoods. Although the Supreme Court invalidated this law in 1917, a drumbeat of formal and informal measures continued to enforce racial segregation. These include the 1930s redlining by the federal Home Owners' Loan Corporation rating African-American and new immigrant neighbourhoods as unworthy of mortgage investment. Secondly, there was the formation and use of so-called neighbourhood improvement associations to advocate for racial and commercial exclusions from their communities (Buckley and Boone, 2011). Finally, there was the use of blockbusting (Orser, 1994). Blockbusting is the practice of the real estate industry of persuading owners to sell property cheaply by playing on the fear that people of another race or class were moving into the neighbourhood, and then profiting by reselling the property at a higher price. Yet another practice was the routing of limited access highways through disempowered neighbourhoods (Giguere, 2009). The neighbourhoods that were subject to these restraints and discrimination were, in many cases, those that currently have few environmental amenities (Grove *et al.*, 2018).

Heat archipelagoes and social risk

The tendency of urban areas to have different thermal regimes than their surroundings is one of the most secure generalizations in the science of cities (Peng *et al.*, 2012). This phenomenon is termed the urban heat island (UHI). It is based on the workings of the thermal energy budget; the theory that energy is absorbed, converted from one form to another, stored and emitted by material things. Interactions among incoming and outgoing radiation, materials and land adjacent to cities generally result in cities being warmer than their surroundings, where those surroundings support dense vegetation. In contrast, something that was initially surprising to researchers was that if a city is itself well watered but embedded in an arid matrix, that city may be warmer than the surroundings at different times of day, compared to moist-climate cities (Heisler *et al.*, 2019). These contrasting results are all understood using the principles of the thermal energy budget.

How does the energy budget play out as an integration between biophysical sciences and social sciences? First, within a city, the arrangement, sizes and types of buildings, the nature and configuration of paved areas, tree cover and water bodies are patchy. Such patchiness means that the energy budget of specific places will be distinct based on their material composition and their effect on winds. So although the city may, on the whole, be warmer than its surroundings, some areas within the city will be still warmer than others. This heterogeneity within the heat island can be viewed as 'heat archipelagoes'. Thus, some places and some people will experience more heat loading than others, whereas other places will have greater capacity for cooling off in the evening or night than others will have (Wong *et al.*, 2013; Zhou *et al.*, 2014b).

A now classic social study of mortality and morbidity during a heat wave is the work of Kleinenberg (2015), on the 1995 heat wave in Chicago. Social networks are an important form of social capital (Osborne *et al.*, 2016; Romolini *et al.*, 2016). Kleinenberg found that neighbourhoods with reduced social networks due to disinvestment, poverty and racism were those that had the greatest number of deaths and hospitalizations over and above the normal rates and which thus were attributable to the heat wave. Interestingly, impoverished neighbourhoods that retained high social capital due to cultural traditions, did not experience excess deaths and heat-related illness in Chicago.

Neighbours checked on the isolated and elderly, and those otherwise vulnerable persons were also not afraid to seek relief outside of their homes. Extending this thinking to Baltimore showed the exposure of different neighbourhoods to high levels of heat stress. In Baltimore, neighbourhoods that were poor, predominantly minority, housed high numbers of elderly or experienced low educational attainment were exposed to the highest land surface temperatures (Huang *et al.*, 2011). Research in Phoenix, under the auspices of the LTER there and reflecting different racial demographics, also found that minority neighbourhoods with lower incomes were more exposed to the effects of urban heat islands (Jenerette *et al.*, 2011). Thus, there were literal hotspots across these cities that were associated not only with biophysical characteristics but also with social features, both of which contribute to environmental inequity (Boone and Buckley, 2016).

The interdisciplinary connections relating to heat might better be summarized by acknowledging that an urban area does not have a single heat island when examined at the scale of different neighbourhoods. Rather, a city presents a complicated 'heat archipelago', with great differences in the dynamics of the energy budget from place to place. When coupled with spatial differences in race, economics and social cohesion, the heat archipelago generates a patchwork of vulnerabilities that communities, health providers, social activists and city agencies need to be aware of and address (Schwarz *et al.*, 2018).

Ecological urban design

Ecological linkages to urban design have had a long history in Europe. In the USA, two milestones in integration of ecological knowledge with design are the work of Ian McHarg in *Design with Nature* (McHarg, 1969; Steiner, 2011) and of Anne Whiston Spirn (Spirn, 1984). These examples demonstrate that this linkage is relevant to new development and to revitalization of older urban neighbourhoods. McHarg engaged biological and physical scientists in his teaching and practice and focused on ecologically informed establishment of new ex-urban developments. In contrast, Spirn delved deeply into the literature of ecology and environmental science, relevant to cities (Spirn, 2012), and speaks of design in cities as having to be cognizant of the 'deep enduring context' of the place. Some of her work focused on risks associated with buried streams in existing city neighbourhoods; and she also worked with engineers on problems of wind in the city.

The authors' own attempts to further integrate ecology and urban design were founded on training jointly conducted across these two disciplines (McGrath *et al.*, 2007; McGrath, 2013), and later on an interdisciplinary conference jointly led by urban designers and ecologists (Pickett *et al.*, 2013). Among the authors of this chapter, Grove has formal training in architecture and in social ecology. Pickett helped train Alex Felson, who already had a master's degree in landscape architecture, and earned a PhD in ecology with Pickett. Grove, Pickett and Cadenasso interacted with Brian McGrath, an architect, and Victoria Marshall, a landscape architect, in urban design studios where they taught, first at Columbia University's Graduate School of Architecture, Planning and Preservation, and later at Parsons, the New School for Design. A course on visual ecology by Marshall and Pickett at Parsons was an additional case of cross-disciplinary teaching.

These interdisciplinary interactions employed several integrative strategies. First, they focused on students, who are often acknowledged to be key actors in interdisciplinary projects because of their openness, their focus on learning new things and the fact that the questions they ask often challenge disciplinary assumptions. Thus, students can draw disciplinary specialists into evaluations of fundamentals in a way that promotes interaction. Second, students, especially in a studio setting, are focused on particular real-world problems. The focus on solving real problems often softens disciplinary defensiveness. Third, design studios usually focus on a specific place or region. Field trips that are focused on discussions with clients, residents, policy makers and designers committed to those places provide a 'neutral ground' for the disciplinary instructors. The combination of neutral ground with the openness that student learning enforces on the group dynamics is important.

The idea that urban designs, as interventions in the urban fabric, constitute experiments that can expose direct and indirect environmental effects is an important motivation for pursuing ecological urban design and follow-up (Felson *et al.*, 2013). The follow-up strategy emphasizes the need to perform rigorous and ongoing monitoring of the ecological and social effects of designs once instituted. This integration of the work of ecologists and the work of designers throughout the life of projects, from inception to occupation and use (Felson, 2013), is needed to achieve urban sustainability.

Like many interdisciplinary projects, our interactions aimed at advancing ecological urban design have taken place on decadal timeframes. It has taken a long time to understand the contrasting visual, spatial, experimental and verbal approaches to identifying and solving problems. Seeing these approaches at work on the same issue has shown our diverse team the value of each. The long time that we have been interacting over ecological urban design has given us ample opportunity to read and discuss each other's literatures and to go on field trips together. One kind of activity was to visit exhibits of design history and projects, which revealed and helped connect the different ways in which designers and scientists use models (McGrath *et al.*, 2019). An important tool for integration is the *Patch Atlas* (Marshall *et al.*, 2020), a book exploring how the HERCULES classification system can be used to interpret both design and environmental themes in Baltimore. It is significant that many of these interactions took advantage of informal, relaxed settings, often involving 'fine food and drink' (Grove *et al.*, 2015).

Application as Integration

The case of integration via ecological urban design leads us to consider application as an integrative opportunity more generally. The best and most effective application involves dialogue with those who are responsible for planning, engineering and carrying out applied projects (Grove and Carrera, 2019). Use of deep dialogue contrasts with application as delivery of scientific insights, often without any substantial interaction with those who might either benefit or be adversely affected (Childers *et al.*, 2014). In BES, dialogue focused on application has been important for several reasons (Grove *et al.*, 2016). Scientists have benefitted from working on solutions to real problems. Ongoing dialogue allows scientists to learn about projects before the fact, permitting before-and-after measurements that can discover mechanisms or document outcomes. In other words, scientists can access public or private projects that can be studied like experiments by interacting with practitioners. An example is the partnership with the Baltimore City Department of Public Works (DPW), which allowed us to take advantage of improvements to the sanitary sewer system mandated by the US Environmental Protection Agency (EPA). Measuring key locations before and after the sewer retrofits showed what the ecological effects of the contamination by sewage had been. Another benefit of partnering with DPW resulted in joint research projects. Because these projects were conducted with municipal leaders, the insights gained as a result of a transdisciplinary process known as co-production (London *et al.*, 2017) could be readily incorporated into best stormwater management practices. Examples include the establishment of green infrastructure projects to improve stormwater management via street maintenance and the identification of hotspots of pollution (Groffman *et al.*, 2010). Scientists also may gain access to sources of data that they would not normally be able to use or even know about.

Not only scientists but also practitioners benefit from dialogue focused on application (Sayre *et al.*, 2013). Discussing a project before it is installed or finished may help design a monitoring scheme that has greater power to transform legal or public discourse. Scientists may have the resources to do *post facto* assessment, which is important because such after-the-fact monitoring is rarely included in municipal budgets. Practitioners gain access to a community of researchers that may be enthusiastic to conduct analyses that municipal partners are not mandated or supported to do. An example is the analysis of long-term records of salinity in Baltimore's drinking water reservoirs that had originated from road de-icing (Kaushal, 2016). Although the city had been collecting such data for decades, departmental staff had not had the time to analyse them. Lessons learned included the steady increase in salt contamination in urban streams over decades, even during the summer season, suggesting that there is a mechanism of storage yet to be discovered. In addition, the levels of chloride concentration in streams are at the threshold of toxicity to many aquatic plants and animals.

A further example of integration emerging from application is work with the Baltimore City DPW and the Parks & People Foundation (PPF) on green stormwater infrastructure in underserved neighbourhoods in west Baltimore (Cadenasso *et al.*, 2008; Hager *et al.*, 2013). PPF supported and organized projects to improve local environments and to boost civic and social revitalization of these neighbourhoods. Conversations between PPF, DPW and BES led to a joint project to monitor the contribution of the green infrastructure installations to stormwater quality and quantity in smaller watersheds in the area. The example combines concern with engineered infrastructure and green infrastructure. The watersheds were delimited by the storm drain network. BES scientists installed instruments in the storm drain system to monitor flow and collected water samples for quality analysis. The DPW provided access and personnel to help install and operate the monitoring stations. PPF provided connections to the community and managed the participatory process of selecting and designing locations for the green infrastructure. The variability of stormwater flow, the confounding effects of other management actions and the relatively low number of installations thwarted a clear biophysical signal from the study. However, the social benefits were clear. Community engagement in the process improved understanding and sensitivity to environmental quality in the neighbourhoods (Hager *et al.*, 2013).

Application can take ecological knowledge into entirely new arenas. An example is the research by BES and USDA Forest Service scientists on the presence and change of urban tree canopy and its relationship to social and structural patchiness in Baltimore (Schwarz *et al.*, 2015; Locke *et al.*, 2013; Raciti *et al.*, 2015). Policy makers decided to double tree canopy over two decades because city agencies came to recognize that tree canopy had many benefits, including amelioration of heat extremes, mitigation of stormwater flow and quality, and associations with social cohesion. The first step for BES involved helping the city to understand where and under what sorts of property regimes and ownerships land was available to achieve its goal of doubling tree canopy. In particular, BES helped the city to realize that there was not sufficient open land available for tree planting on public lands and rights of way to meet the canopy goal. Consequently, much of the effort would have to engage diverse private landholders as well. Furthermore, BES had added new dimensions to the standard demographic and economic data available through the census, so that policy makers and managers could understand how lifestyle and life stage contributed to the social heterogeneity of the lands where trees might be planted (Holtan *et al.*, 2014). This allowed planners and managers to address the specific concerns and needs of the full diversity of tenants and owners who held contrasting values, had differential access to information, and had different capacities for dealing with newly expanded tree populations (Locke *et al.*, 2013).

A creative outcome of this knowledge about the social ecology of urban tree planting was the recognition that it was a business and training opportunity. BES researchers were familiar with the interest of not-for-profits and city agencies to help neighbourhoods burdened with high abandonment and vacancy of residential parcels. BES participants in the urban tree canopy analysis also brought knowledge and networks of practice about timber as a profitable commodity along with the knowledge, mentioned above, that demolition of a minimum number of abandoned units was required for neighbourhood revitalization for local residents. The integrative step was the realization that the abandoned rowhouses could be considered sources of commercially viable timber. This combination of insights stimulated an entrepreneurial approach to neighbourhood needs (Baltimore Wood Project, n.d.). The result was a strategy of carefully deconstructing vacant rowhouses and salvaging the lumber and masonry for productive reuse. Local residents, many of whom had not had the opportunity to be gainfully employed before, or who were disadvantaged in job searches because of modest educational attainment or criminal records, are trained in the skills for careful deconstruction and processing the recovered timber. Some of the wood comprising the interior frames of the brick rowhouses had been milled in the 19th century from large, old yellow pine and other old-growth timber species, the likes of which have not existed in the wild for more than a century. The final step in the chain was engagement of a high-end furniture maker and retailer to design and market the finished products. This entrepreneurial chain is based on the combined social-ecological knowledge generated by BES, but also serves a community requiring skills, empowerment and opportunities for productive engagement with the larger economy and civic world (Hines *et al.*, 2018).

Synthesis

The history of social-ecological research in BES suggests several insights about the motivations and mechanisms of interdisciplinary research in urban systems. We do not expect this trajectory to be the same in all urban research and engagement projects, but we do believe that at least some of the insights may be helpful to others as they design or pursue interdisciplinary urban projects (Fig. 7.3).

The first step toward integration of the various disciplines and perspectives emerged from the realization that the traditional approach of ecology to seek out familiar, green spaces as research targets in cities, suburbs and exurbs, while valuable, might yield limited understanding of urban areas as comprehensive, dynamic systems. Consequently, we articulated a theoretical stance labelled 'ecology of the city', which demanded integration across disciplines for its greatest success. This included, but went well beyond, understanding the multifaceted social context of ecology in the city.

Early explorations of the ecology of the city as an integrated social-ecological pursuit convinced us of the power of place. This power rested on two pillars. First, the particular places had distinctive histories of environmental and social dynamics, and these not only interacted with each other as a specific chronology but also generated legacies that persist and continue to influence the environment and the social dynamics in contemporary time. Second, to conduct ecological research in an urban system, we had to gain the trust of communities, neighbourhoods, organizations and institutions. This required enough time to have repeated mutually respectful conversations, and to share the insights and interests of multiple parties.

An additional tool for integration is the existence of theories and conceptual frameworks that are capable of being linked at the various scales of interest across city-suburban-exurban

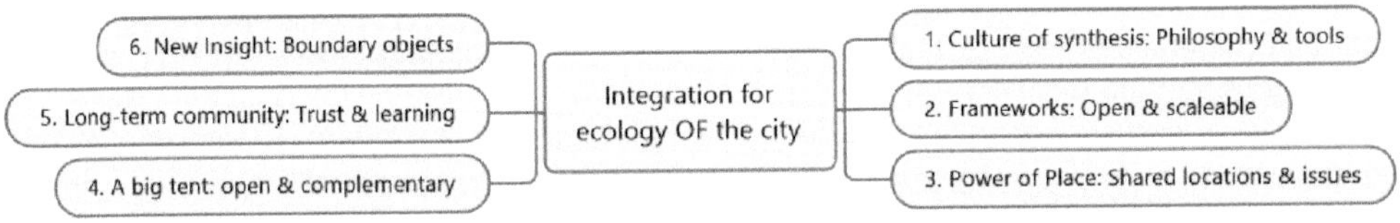

Fig. 7.3. A concept map of key attributes of social-ecological integration as illustrated by the Baltimore Ecosystem Study. The largest integrative goal appears in the centre, with strategies, mechanisms or philosophies arrayed around it and numbered clockwise from one o'clock. The first mechanism is understanding and establishing a culture of synthesis and exploring the tools that can put the cultural philosophy to work. Second is linking open-ended and scaleable frameworks that represent the main disciplines in the project. Third is exploiting the power of place, which uses shared locations and shared issues about those locations to focus research, education and application. Intellectual and activist networks also focus on specific places, and these must be engaged. Fourth is the idea that an urban ecology research project is a 'big tent', welcoming of new participation and inviting of complementary perspectives and backgrounds. Fifth is recognizing that long times are needed to develop trust among the participants and to learn from each other. Sixth is an emerging insight, not explored in the paper, that many of the tools for integration exploit 'boundary objects', which are places, concerns, models and metaphors that disparate groups can share to begin more substantive dialogue and action. Boundary objects invite exploring the deep theory and assumptions of a project without requiring that all participants share a firm, well-developed theory or paradigm at the outset.

systems and which support the contributions of individuals and partnered disciplines. Of course, it takes time, patience and forbearance in the intellectual network to understand and link the frameworks. But, starting with open, scalable frameworks was a key step for BES. We used watersheds, patch dynamics and nested institutional structures in this way (Fig. 7.1). We took pains to study what it meant to aspire to a culture of integration and what specific tactics might emerge from such a culture. That culture was expressed materially in regular meetings in Baltimore that assembled members of our distributed research and educational team along with metropolitan practitioners. An additional and powerful tool was periodic field trips in which participants from different academic disciplines, community groups, NGOs and government agencies participated. Shared data, shared research sites and an encouragement of interdisciplinary publications were also important aspects of our culture of synthesis.

Although many of our research projects involved foundational disciplines as starting points, we encouraged early and consistent involvement of participants representing complementary and sometimes seemingly unrelated disciplines. As a result, for example, we have developed a novel classification of urban land covers appropriate to an ecology of the city. Similarly, land–water interactions have been key foci for interdisciplinary work melding biological, hydrological, engineering, social and economic contributions.

Many research projects have sought to link environmental and social-economic factors as both drivers and responses, often showing how the cascades of interaction play out over the long term. There are several examples. One is the ecology of prestige as a way to understand decisions about lawn maintenance and its contribution to ecological processes in the city. A second example is how the interactions among social differentiation and land management result in spatial and temporal patterns of the abundance of human-biting mosquitoes that can transmit disease. Finally, neighbourhoods with contrasting wealth and degrees of political empowerment, different types of lifestyle and characterized by different distributions of life stage have proved to be important to understanding the reciprocal relationships between people and their environment in different locations.

Issues of social equity have emerged as important factors in Baltimore ecology. Social and legal histories have proved to be important in understanding contemporary patterns of Toxics Release Inventory sites and race- and class-related environmental equity. Patterns of racial segregation, access to environmental amenities and control of local environmental quality are historically linked through policy and neighbourhood-level decisions (Grove *et al.*, 2018). Biodiversity, stormwater quality and heat hazard are examples of environmental outcomes of segregation. Increasing attention to the ecological implications of the social processes and patterns of segregation in American cities is a new frontier for ecological research and application. Although this is related to the well-established field of environmental justice, it identifies a specific research need integrating the social and political phenomenon of racial segregation and ecological concerns with ecosystem and landscape structure and function.

Urban ecological design has proved to be a useful driver of integration. For example, interactions between ecologists and designers in teaching architectural studios and theory classes have helped understand the HERCULES land-cover classification in new ways. Urban design has provided many tools for integrating environmental understanding with the rich variety of thinking about cities that are a part of urban design, architecture and planning. In other words, ecological urban design has opened a door to the richness of urbanism and related philosophies for ecology.

Finally, use-inspired basic research, or the engagement of an integrative research culture and strategy with the concerns of communities, agencies, NGOs and even businesses has provided a rewarding and important practical platform for integration. It has been extremely rewarding to see our ecological knowledge help motivate a policy concern with urban tree canopy, the mechanisms of achieving urban tree-planning goals, and the revitalization of neighbourhoods in which there is extraordinary vacancy. The reach of this desire for revitalization into local job training and new economies that help residents in underserved neighbourhoods is perhaps the greatest reward for integrated social-ecological approaches to ecology of the city.

Acknowledgements

We acknowledge support of the National Science Foundation's LTER programme for support from 1997 to 2021, most recently through grant number DEB-1637661 and DEB-1855277. We thank Dan Childers, Director of the CAP LTER for helpful comments on an earlier draft of the chapter.

References

Adler, F.R. and Tanner, C.J. (2013) *Urban Ecosystems: Ecological Principles for the Built Environment*. Cambridge University Press.

Allen, T.F.H. and Starr, T.B. (1982) *Hierarchy: Perspectives for Ecological Complexity*. University of Chicago Press, Chicago, Illinois.

Anderson, J.R., Hardy, E.E., Roach, J.T. and Witmer, R.E. (1976) *A Land Use and Land Cover Classification System for Use with Remote Sensor Data*. US Government Printing Office, Washington DC.

Bain, D.J. and Buckley, G.L. (2019) The history and legacies of mercantile, industrial, and sanitary transformations. In: Pickett, S.T.A., Cadenasso, M.L., Grove, J.M., Irwin, E.G., Rosi, E.J., *et al.* (eds) *Science for the Sustainable City: Empirical Insights from the Baltimore School of Urban Ecology*. Yale University Press, New Haven, Connecticut, pp. 47–60.

Baltimore Wood Project (n.d.). Available at: http://www.baltimorewoodproject.org/ (accessed 25 February 2019).

Baxter, J.W., Pickett, S.T.A., Dighton, J. and Carreiro, M.M. (2002) Nitrogen and phosphorus availability in oak forest stands exposed to contrasting anthropogenic impacts. *Soil Biology and Biochemistry* 34(5), 623–633. DOI: 10.1016/S0038-0717(01)00224-3.

Becker, B., Leisnham, P.T. and LaDeau, S.L. (2014) A tale of two city blocks: differences in immature and adult mosquito abundances between socioeconomically different urban blocks in Baltimore (Maryland, USA). *International Journal of Environmental Research and Public Health* 11(3), 3256–3270. DOI: 10.3390/ijerph110303256.

Boone, C.G. (2002) An assessment and explanation of environmental inequity in Baltimore. *Urban Geography* 23(6), 581–595. DOI: 10.2747/0272-3638.23.6.581.

Boone, C.G. (2008) Environmental justice as process and new avenues for research. *Environmental Justice* 1(3), 149–154. DOI: 10.1089/env.2008.0530.

Boone, C.G. and Buckley, G.L. (2016) Historical approaches to environmental justice. In: Holyfield, R. (ed.) *Routledge Handbook of Environmental Justice*. Routledge, London.

Boone, C.G., Redman, C.L., Blanco, H., Haase, D., Koch, J. *et al.* (2014) Reconceptualizing land for sustainable urbanity. In: Seto, K.C. and Reenberg, A. (eds) *Rethinking Urban Land Use in a Global Era*. MIT Press, Cambridge, pp. 313–330.

Buckley, G.L. and Boone, C.G. (2011) "To promote the material and moral welfare of the community": neighborhood improvement associations in Baltimore, Maryland 1900–1945. In: Rodger, R. and Massard-Guilbaud, G. (eds) *Environmental and Social Justice in the City: Historical Perspectives*. White Horse Press, Winwick UK, pp. 43–65.

Buckley, G.L., Boone, C.G. and Lord, C. (2019) Environmental justice and environmental history. In: Pickett, S.T.A., Cadenasso, M.L., Grove, J.M., Irwin, E.G., Rosi, E.J., *et al.* (eds) *Science for the Sustainable City: Empirical Insights from the Baltimore School of Urban Ecology*. Yale University Press, New Haven, Connecticut, pp. 61–70.

Burch, W.R., Jr. (1971) *Daydreams and Nightmares; a Sociological Essay on the American Environment*. Harper & Row, New York.

Burch, W.R., Jr., Machlis, G.E. and Force, J.E. (2017) *The Structure and Dynamics of Human Ecosystems: Toward a Model for Understanding and Action*. Yale University Press, New Haven, Connecticut.

Cadenasso, M.L. (2013) Designing ecological heterogeneity. In: *Urban Design Ecologies: AD reader*. John Wiley & Sons, Hoboken, New Jersey, pp. 271–281.

Cadenasso, M.L., Pickett, S.T.A., Weathers, K.C. and Jones, C.G. (2003) A framework for a theory of ecological boundaries. *BioScience* 53(8), 750–758. DOI: 10.1641/0006-3568(2003)053[0750:AFFATO]2.0.CO;2.

Cadenasso, M.L., Pickett, S.T.A. and Grove, M.J. (2006) Integrative approaches to investigating human-natural systems: the Baltimore ecosystem study. *Natures Sciences Sociétés* 14(1), 4–14. DOI: 10.1051/nss:2006002.

Cadenasso, M.L., Pickett, S.T.A. and Schwarz, K. (2007) Spatial heterogeneity in urban ecosystems: reconceptualizing land cover and a framework for classification. *Frontiers in Ecology and the Environment* 5(2), 80–88. DOI: 10.1890/1540-9295(2007)5[80:SHIUER]2.0.CO;2.

Cadenasso, M.L., Pickett, S.T.A., Groffman, P.M., Band, L.E., Brush, G.S. *et al.* (2008) Exchanges across land-water-scape boundaries in urban systems: strategies for reducing nitrate pollution. *Annals of the New York Academy of Sciences* 1134, 213–232. DOI: 10.1196/annals.1439.012.

Cadenasso, M.L., Pickett, S.T.A., McGrath, B. and Marshall, V. (2013) Ecological heterogeneity in urban ecosystems: reconceptualized land cover models as a bridge to urban design. In: Pickett, S.T.A., Cadenasso, M.L. and McGrath, B. (eds) *Resilience in Ecology and Urban Design: Linking Theory and Practice for Sustainable Cities*. Springer, New York, pp. 107–129.

Carreiro, M.M., Pouyat, R.V., Tripler, C.E. and Zhu, W. (2009) Carbon and nitrogen cycling in soils of remnant forests along urban-rural gradients: case studies in the New York metropolitan area and Louisville, Kentucky. In: McDonnell, M.J., Hahs, A. and Breuste, J. (eds) *Ecology of Cities and Towns: A Comparative Approach*. Cambridge University Press, New York, pp. 308–328.

Chen, Y., Irwin, E.G. and Jayaprakash, C. (2011) Incorporating spatial complexity into economic models of land markets and land use change. *Agricultural and Resource Economics Review* 40(3), 321–340. DOI: 10.1017/S1068280500002811.

Childers, D.L., Pickett, S.T.A., Grove, J.M., Ogden, L. and Whitmer, A. (2014) Advancing urban sustainability theory and action: challenges and opportunities. *Landscape and Urban Planning* 125, 320–328. DOI: 10.1016/j.landurbplan.2014.01.022.

Collins, S.L. (2019) Foreword. In: Pickett, S.T.A., Cadenasso, M.L., Grove, J.M., Irwin, E.G. and Rosi, E.J., *et al.* (eds) *Science for the Sustainable City: Empirical Insights from the Baltimore School of Urban Ecology*. Yale University Press, New Haven, Connecticut, pp. ix–0.

Douglas, I. and James, P. (2014) *Urban Ecology*. Routledge, New York.

Felson, A.J. (2013) The design process as a framework for collaboration between ecologists and designers. In: Pickett, S.T.A., Cadenasso, M.L. and McGrath, B. (eds) *Resilience in Ecology and Urban Design: Linking Theory and Practice for Sustainable Cities*. Springer, New York, pp. 365–382.

Felson, A.J., Bradford, M.A. and Terway, T.M. (2013) Promoting earth stewardship through urban design experiments. *Frontiers in Ecology and the Environment* 11(7), 362–367. DOI: 10.1890/130061.

Fox, G.A., Scheiner, S.M. and Willig, M.R. (2011) Ecological gradient theory: a framework for aligning data and models. In: Scheiner, S.M. and Willig, M.R. (eds). *The Theory of Ecology*. University of Chicago Press, Chicago, Illinois, pp. 283–307

Giguere, A.M. (2009) *"...and never the twain shall meet": Baltimore's east-west expressway and the construction of the "Highway to Nowhere"*. Ohio University, Department of Geography, College of Arts and Sciences, Athens, Ohio.

Goodman, H., Egizi, A., Fonseca, D.M., Leisnham, P.T. and LaDeau, S.L. (2018) Primary blood-hosts of mosquitoes are influenced by social and ecological conditions in a complex urban landscape. *Parasites & Vectors* 11(1). DOI: 10.1186/s13071-018-2779-7.

Gottdiener, M. and Hutchison, R. (2011) *The New Urban Sociology*, 4th edn. Westview Press, Philadelphia, Pennsylvania.

Groffman, P.M. (2012) *Nitrogen cycling in urban wetlands. 97th ESA Annual Convention*.

Groffman, P.M., Bain, D.J., Band, L.E., Belt, K.T., Brush, G.S. *et al.* (2003) Down by the riverside: urban riparian ecology. *Frontiers in Ecology and the Environment* 1(6), 315–321. DOI: 10.1890/1540-9295(2003)001[0315:DBTRUR]2.0.CO;2.

Groffman, P.M., Stylinski, C., Nisbet, M.C., Duarte, C.M., Jordan, R. *et al.* (2010) Restarting the conversation: challenges at the interface between ecology and society. *Frontiers in Ecology and the Environment* 8(6), 284–291. DOI: 10.1890/090160.

Groffman, P.M., Band, L.E., Belt, K.T., Bettez, N.D., Bhaskar, A. *et al.* (2019) Applying the watershed approach to urban ecosystems. In: Pickett, S.T.A., Cadenasso, M.L., Grove, J.M., Irwin, E.G., Rosi, E.J., *et al.* (eds) *Science for the Sustainable City: Empirical Insights from the Baltimore School of Urban Ecology*. Yale University Press, New Haven, Connecticut, pp. 155–173.

Grove, J.M. (1999) New tools for exploring theory and methods in human ecosystem and landscape analyses: computer modeling, remote sensing and geographic information systems. In: Cordell, H.K. and Bergstrom, J.C. (eds) *Integrating Social Sciences with Ecosystem Management:*

Human Dimensions in Assessment, Policy, and Management. Sagamore, Champaign, Illinois, pp. 219–236.

Grove, J.M. and Burch, W.R. (1997) A social ecology approach and applications of urban ecosystem and landscape analyses: a case study of Baltimore, Maryland. *Urban Ecosystems* 1(4), 259–275. DOI: 10.1023/A:1018591931544.

Grove, J.M. and Carrera, J. (2019) Lessons learned from the origin and design of long-term social-ecological research. In: Pickett, S.T.A., Cadenasso, M.L., Grove, J.M., Irwin, E.G., Rosi, E.J., *et al.* (eds) *Science for the Sustainable City: Empirical Insights from the Baltimore School of Urban Ecology*. Yale University Press, New Haven, Connecticut, pp. 289–306.

Grove, J.M., Pickett, S.T.A., Whitmer, A. and Cadenasso, M.L. (2013) Building an urban LTSER: the case of the Baltimore ecosystem study and the D.C./B.C. ULTRA-Ex project. In: Singh, J.S., Haberl, H., Chertow, M., Mirtl, M. and Schmid, M. (eds) *Long Term Socio-ecological Research: Studies in Society: Nature Interactions across Spatial and Temporal Scales*. Springer, New York, pp. 369–408.

Grove, J.M., Locke, D.H. and O'Neil-Dunne, J.P.M. (2014) An ecology of prestige in New York City: examining the relationships among population density, socio-economic status, group identity, and residential canopy cover. *Environmental Management* 54(3), 402–419. DOI: 10.1007/s00267-014-0310-2.

Grove, M., Cadenasso, M.L., Pickett, S.T.A., Machlis, G. and Burch, W.R., Jr. (2015) *The Baltimore School of Urban Ecology: Space, Scale, and Time for the Study of Cities*. Yale University Press, New Haven, Connecticut.

Grove, J.M., Childers, D.L., Galvin, M., Hines, S., Muñoz-Erickson, T. *et al.* (2016) Linking science and decision making to promote an ecology for the city: practices and opportunities. *Ecosystem Health and Sustainability* 2(9), e01239. DOI: 10.1002/ehs2.1239.

Grove, M., Ogden, L., Pickett, S., Boone, C., Buckley, G. *et al.* (2018) The legacy effect: understanding how segregation and environmental injustice unfold over time in Baltimore. *Annals of the American Association of Geographers* 108(2), 524–537. DOI: 10.1080/24694452.2017.1365585.

Haberl, H., Winiwarter, V., Andersson, K., Ayres, R.U., Boone, C. *et al.* (2006) From LTER to LTSER: conceptualizing the socioeconomic dimension of long-term socioecological research. *Ecology and Society* 11(2). DOI: 10.5751/ES-01786-110213.

Hager, G.W., Belt, K.T., Stack, W., Burgess, K., Grove, J.M. *et al.* (2013) Socioecological revitalization of an urban watershed. *Frontiers in Ecology and the Environment* 11(1), 28–36. DOI: 10.1890/120069.

Hassler, U. and Kohler, N. (2014) Resilience in the built environment. *Building Research & Information* 42(2), 119–129. DOI: 10.1080/09613218.2014.873593.

Hawley, A. (1986) *Human Ecology*. University of Chicago Press, Chicago.

Heisler, G.M., Brazel, A. and Cadenasso, M.L. (2019) Urban influences on the atmospheres of Baltimore, Maryland, and Phoenix, Arizona. In: Pickett, S.T.A., Cadenasso, M.L., Grove, J.M., Irwin, E.G., Rosi, E.J., *et al.* (eds) *Science for the Sustainable City: Empirical Insights from the Baltimore School of Urban Ecology*. Yale University Press, New Haven, Connecticut, pp. 174–204.

Hines, S., Srinivasan, N., Marshall, L. and Grove, M. (2018) Reclaiming wood, lives, and communities: how do we turn a waste stream into an asset that revitalizes cities? Available at: https://www.learngala.com/cases/urbanwood

Holtan, M.T., Dieterlen, S.L. and Sullivan, W.C. (2014) Social life under cover: tree canopy and social capital in Baltimore, Maryland. *Environment and Behavior* 47, 502–525.

Huang, G., Zhou, W. and Cadenasso, M.L. (2011) Is everyone hot in the city? Spatial pattern of land surface temperatures, land cover and neighborhood socioeconomic characteristics in Baltimore, MD. *Journal of Environmental Management* 92(7), 1753–1759. DOI: 10.1016/j.jenvman.2011.02.006.

Irwin, E.G. (2010) New directions for urban economic models of land use change: incorporating spatial dynamics and heterogeneity. *Journal of Regional Science* 50(1), 65–91. DOI: 10.1111/j.1467-9787.2009.00655.x.

Irwin, E.G. and Wrenn, D.H. (2014) An assessment of empirical modeling of land use. In: Wu, J. and Duke, C. (eds) *The Oxford Handbook of Land Economics*. Oxford University Press, pp. 327–351.

Irwin, E.G., Grove, J.M., Irwin, N., Klaiber, H.A., Towe, C. *et al.* (2019) Effects of disamenities and amenities on housing markets and locational choices. In: Pickett, S.T.A., Cadenasso, M.L., Grove, J.M., Irwin, E.G., Rosi, E.J., *et al.* (eds) *Science for the Sustainable City: Empirical Insights from the Baltimore School of Urban Ecology*. Yale University Press, New Haven, Connecticut, pp. 92–111.

Jenerette, G.D., Harlan, S.L., Stefanov, W.L. and Martin, C.A. (2011) Ecosystem services and urban heat riskscape moderation: water, green spaces, and social inequality in Phoenix, USA. *Ecological Applications* 21(7), 2637–2651. DOI: 10.1890/10-1493.1.

Kaushal, S.S. (2016) Increased salinization decreases safe drinking water. *Environmental Science & Technology* 50(6), 2765–2766. DOI: 10.1021/acs.est.6b00679.

Kleinenberg, E. (2015) *Heat Wave: A Social Autopsy of Disaster in Chicago*, 2nd edn. University of Chicago Press, Chicago.

LaDeau, S.L., Leisnham, P.T., Biehler, D. and Bodner, D. (2013) Higher mosquito production in low-income neighborhoods of Baltimore and Washington, DC: understanding ecological drivers and mosquito-borne disease risk in temperate cities. *International Journal of Environmental Research and Public Health* 10(4), 1505–1526. DOI: 10.3390/ijerph10041505.

Lave, J. and Wenger, E. (1991) *Situated Learning: Legitimate Peripheral Participation*. Cambridge University Press, New York.

Likens, G.E. (1985) An experimental approach for the study of ecosystems: the fifth Tansley lecture. *The Journal of Ecology* 73(2), 381–396. DOI: 10.2307/2260481.

Likens, G.E. (1992) *The ecosystem approach: its use and abuse*. Ecology Institute, Oldendorf/Luhe.

Little, E., Biehler, D., Leisnham, P.T., Jordan, R., Wilson, S. *et al.* (2017) Socio-ecological mechanisms supporting high densities of *Aedes albopictus* (Diptera: Culicidae) in Baltimore, MD. *Journal of Medical Entomology* 54(5), 1183–1192. DOI: 10.1093/jme/tjx103.

Locke, D.H., Grove, J.M., Galvin, M., O'Neil-Dunne, J.P.M. and Murphy, C. (2013) Applications of urban tree canopy assessment and prioritization tools: supporting collaborative decision making to achieve urban sustainability goals. *Cities and the Environment* 6(Article 7).

Locke, D.H., Roy Chowdhury, R., Grove, J.M., Martin, D.G., Goldman, E. *et al.* (2018) Social norms, yard care, and the difference between front and back yard management: examining the landscape mullets concept on urban residential lands. *Society & Natural Resources* 31(10), 1169–1188. DOI: 10.1080/08941920.2018.1481549.

London, J.K., Schwarz, K., Cadenasso, M.L., Cutts, B.B., Mason, C. *et al.* (2017) Weaving community: university research and action partnerships for environmental justice. *Action Research* 1476750316678915.

Longino, H.E. (1990) *Science as Social Knowledge: Values and Objectivity in Scientific Inquiry*. Princeton University Press, Princeton.

Lord, C. and Norquist, K. (2010) Cities as emergent systems: race as a rule in organized complexity. *Environmental Law* 40, 551–596.

Machlis, G.E., Force, J.E. and Burch, W.R. (1997) The human ecosystem Part I: the human ecosystem as an organizing concept in ecosystem management. *Society & Natural Resources* 10(4), 347–367.

Marshall, V., Cadenasso, M.L., McGrath, B. and Pickett, S.T.A. (2020) *Patch Atlas: Integrating Design Principles and Ecological Knowledge for Cities as Complex Systems*. Yale University Press, New Haven, Connecticut.

McDonnell, M.J. and Hahs, A.K. (2008) The use of gradient analysis studies in advancing our understanding of the ecology of urbanizing landscapes: current status and future directions. *Landscape Ecology* 23(10), 1143–1155. DOI: 10.1007/s10980-008-9253-4.

McDonnell, M.J., Pickett, S.T.A., Groffman, P., Bohlen, P., Pouyat, R.V. *et al.* (1997) Ecosystem processes along an urban-to-rural gradient. *Urban Ecosystems* 1, 21–36. DOI: 10.1023/A:1014359024275.

McDonnell, M.J., Hahs, A.K. and Pickett, S.T.A. (2012) Exposing an urban ecology straw man: critique of Ramalho and Hobbs. *Trends in Ecology & Evolution* 27(5), 255–256. DOI: 10.1016/j.tree.2012.01.009.

McGrath, B.P., Marshall, V., Cadenasso, M.L., Grove, J.M. and Pickett, S.T.A., *et al.*. (eds) (2007) *Designing Patch Dynamics*. Columbia University School of Architecture, Planning and Preservation, New York.

McGrath, B. (2013) *Urban Design Ecologies: AD Reader*. John Wiley & Sons Ltd, Chichester, UK.

McGrath, B., Marshall, V., Pickett, S.T.A., Cadenasso, M.L. and Grove, J.M. (2019) Ecological urban design: theory, research, and praxis. In: Pickett, S.T.A., Cadenasso, M.L., Grove, J.M., Irwin, E.G., Rosi, E.J., *et al.* (eds) *Science for the Sustainable City: Empirical Insights from the Baltimore School of Urban Ecology*. Yale University Press, New Haven, Connecticut, pp. 307–338.

McHarg, I.L. (1969) *Design with Nature*. Natural History Press, Garden City, New York.

Medley, K.E., McDonnell, M.J. and Pickett, S.T.A. (1995) Forest-landscape structure along an urban-to-rural gradient. *The Professional Geographer* 47(2), 159–168. DOI: 10.1111/j.0033-0124.1995.00159.x.

Meyfroidt, P., Roy Chowdhury, R., de Bremond, A., Ellis, E.C., Erb, K.-H. *et al.* (2018) Middle-range theories of land system change. *Global Environmental Change* 53, 52–67. DOI: 10.1016/j.gloenvcha.2018.08.006.

Moore, J. (2015) Ecological footprints and lifestyle archetypes: exploring dimensions of consumption and the transformation needed to achieve urban sustainability. *Sustainability* 7(4), 4747–4763. DOI: 10.3390/su7044747.

Orser, W.E. (1994) *Blockbusting in Baltimore: the Edmonson Village Story*. University Press of Kentucky, Lexington.

Osborne, C., Baldwin, C. and Thomsen, D. (2016) Contributions of social capital to best practice urban planning outcomes. *Urban Policy and Research* 34(3), 212–224. DOI: 10.1080/08111146.2015.1062361.

Peng, S., Piao, S., Ciais, P., Friedlingstein, P., Ottle, C. *et al.* (2012) Surface urban heat island across 419 global big cities. *Environmental Science & Technology* 46(2), 696–703. DOI: 10.1021/es2030438.

Pickett, S.T.A. (1999) The culture of synthesis: habits of mind in novel ecological integration. *Oikos* 87(3), 479–487. DOI: 10.2307/3546812.

Pickett, S.T.A. and White, P.S. (eds) (1985) *The Ecology of Natural Disturbance and Patch Dynamics*. Academic Press, Orlando, Florida.

Pickett, S.T.A., Burch Jr., W.R., Dalton, S.E., Foresman, T.W., Grove, J.M. *et al.* (1997) A conceptual framework for the study of human ecosystems in urban areas. *Urban Ecosystems* 1(4), 185–199. DOI: 10.1023/A:1018531712889.

Pickett, S.T.A., Burch, W.R. and Grove, J.M. (1999) Interdisciplinary research: maintaining the constructive impulse in a culture of criticism. *Ecosystems* 2(4), 302–307. DOI: 10.1007/s100219900081.

Pickett, S.T.A., Belt, K.T., Galvin, M.F., Groffman, P.M., Grove, J.M. *et al.* (2007a) Watersheds in Baltimore, Maryland: understanding and application of integrated ecological and social processes. *Journal of Contemporary Water Research & Education* 136(1), 44–55. DOI: 10.1111/j.1936-704X.2007.mp136001006.x.

Pickett, S.T.A., Kolasa, J. and Jones, C.G. (2007b) *Ecological Understanding: The Nature of Theory and the Theory of Nature*, 2nd edn. Academic Press, San Diego, California.

Pickett, S.T.A., Cadenasso, M.L. and McGrath, B. (eds) (2013) *Resilience in Ecology and Urban Design: Linking Theory and Practice for Sustainable Cities*. Springer, New York.

Pickett, S.T.A., Cadenasso, M.L., Band, L.E., Berkowitz, A.R., Brush, G.S. *et al.* (2019a) Goals of the Baltimore school of urban ecological science. In: Pickett, S.T.A., Cadenasso, M.L., Grove, J.M., Irwin, E.G., Rosi, E.J., *et al.* (eds) *Science for the Sustainable City: Empirical Insights from the Baltimore School of Urban Ecology*. Yale University Press, New Haven, Connecticut, pp. 3–23.

Pickett, S.T.A., Cadenasso, M.L., Grove, J.M., Irwin, E.G., Rosi, E.J., *et al.* (eds) (2019b) *Science for the Sustainable City. Empirical Insights from the Baltimore School of Urban Ecology*. Yale University Press, New Haven, Connecticut.

Pouyat, R.V., McDonnell, M.J. and Pickett, S.T.A. (1995) Soil characteristics of oak stands along an urban-rural land-use gradient. *Journal of Environmental Quality* 24(3), 516–526. DOI: 10.2134/jeq1995.00472425002400030019x.

Pouyat, R. V., Carreiro, M.M., Groffman, P.M. and Pavao-Zuckerman, M.A. (2009) Investigative approaches to urban biogeochemical cycles: New York metropolitan area and Baltimore as case studies. In: McDonnell, M.J., Hahs, A. and Breuste, J. (eds) *Ecology of Cities and Towns: A Comparative Approach*. Cambridge University Press, New York, pp. 329–351.

Raciti, S., Galvin, M.F., Grove, J.M., O'Neil-Dunne, J.P.M., Todd, A. *et al.* (2015) *Urban Tree Canopy Goal Setting: A Guide for Chesapeake Bay Communities*. USDA Forest Service, Annapolis, Maryland.

Ridd, M.K. (1995) Exploring a V-I-S (vegetation-impervious surface-soil) model for urban ecosystem analysis through remote sensing: comparative anatomy for cities. *International Journal of Remote Sensing* 16(12), 2165–2185. DOI: 10.1080/01431169508954549.

Risser, P.G. and Lubchenco, J. (2010) Ten-Year Review of the National Science Foundation's Long-Term Ecological Research Program. Available at: https://lternet.edu/wp-content/uploads/2010/12/ten-year-review-of-LTER.pdf

Rivkin, L.R., Santangelo, J.S., Alberti, M., Aronson, M.F.J., de Keyzer, C.W. *et al.* (2019) A roadmap for urban evolutionary ecology. *Evolutionary Applications* 12(3), 384–398. DOI: 10.1111/eva.12734.

Romolini, M., Bixler, R. and Grove, J. (2016) A social-ecological framework for urban stewardship network research to promote sustainable and resilient cities. *Sustainability* 8(9), 956. DOI: 10.3390/su8090956.

Rosi-Marshall, E.J. and Kelly, J.J. (2015) Antibiotic stewardship should consider environmental fate of antibiotics. *Environmental Science & Technology* 49(9), 5257–5258. DOI: 10.1021/acs.est.5b01519.

Rosi-Marshall, E.J., Kincaid, D.W., Bechtold, H.A., Royer, T.V., Rojas, M. *et al.* (2013) Pharmaceuticals suppress algal growth and microbial respiration and alter bacterial communities in stream biofilms. *Ecological Applications* 23(3), 583–593. DOI: 10.1890/12-0491.1.

Rosi-Marshall, E.J., Snow, D., Bartelt-Hunt, S.L., Paspalof, A. and Tank, J.L. (2015) A review of ecological effects and environmental fate of illicit drugs in aquatic ecosystems. *Journal of Hazardous Materials* 282, 18–25. DOI: 10.1016/j.jhazmat.2014.06.062.

Sayre, N.F., Kelty, R., Simmons, M., Clayton, S., Kassam, K.-A. *et al.* (2013) Invitation to earth stewardship. *Frontiers in Ecology and the Environment* 11(7), 339–339. DOI: 10.1890/1540-9295-11.7.339.

Scheiner, S.M. and Willig, M.R. (2011) A general theory of ecology. In: Scheiner, S.M. and Willig, M.R. (eds) *The Theory of Ecology*. University of Chicago Press, Chicago, Illinois, pp. 3–18.

Schwarz, K., Fragkias, M., Boone, C.G., Zhou, W., McHale, M. *et al.* (2015) Trees grow on money: urban tree canopy cover and environmental justice. *Plos One* 10(4), e0122051. DOI: 10.1371/journal.pone.0122051.

Schwarz, K., Berland, A. and Herrmann, D.L. (2018) Green, but not just? rethinking environmental justice indicators in shrinking cities. *Sustainable Cities and Society* 41, 816–821. DOI: 10.1016/j.scs.2018.06.026.

Shane, D.G. (2005) *Recombinant Urbanism: Conceptual Modeling in Architecture*. John Wiley & Sons, Hoboken, New Jersey.

Shane, D.G. (2011) *Urban Design since 1945: A Global Perspective*. John Wiley & Sons, Chichester, UK.

Shevky, E. and Bell, W. (1955) *Social Area Analysis: Theory, Illustrative Application and Computational Procedure*. Stanford University Press, Palo Alto.

Simone, A. (2010) *City Life from Jakarta to Dakar: Movements at the Crossroads*. Routledge, New York.

Spirn, A.W. (1984) *The Granite Garden: Urban Nature and Human Design*. Basic Books, New York.

Spirn, A.W. (2012) Ecological urbanism: a framework for the design of resilient cities. Available at: http://www.annewhistonspirn.com/pdf/Spirn-EcoUrbanism-2012.pdf

Steiner, F. (2011) *Design for a Vulnerable Planet*. University of Texas Press, Austin, Texas.

Troy, A. and Grove, J.M. (2008) Property values, parks, and crime: a hedonic analysis in Baltimore, MD. *Landscape and Urban Planning* 87(3), 233–245. DOI: 10.1016/j.landurbplan.2008.06.005.

Vanderbilt, K. and Gaiser, E. (2017) The international long term ecological research network: a platform for collaboration. *Ecosphere* 8(2), e01697. DOI: 10.1002/ecs2.1697.

Wong, K.V., Paddon, A. and Jimenez, A. (2013) Review of world urban heat islands: many linked to increased mortality. *Journal of Energy Resources Technology* 135(2), 022101. DOI: 10.1115/1.4023176.

Wu, J. and Loucks, O.L. (1995) From balance of nature to hierarchical patch dynamics: a paradigm shift in ecology. *The Quarterly Review of Biology* 70(4), 439–466. DOI: 10.1086/419172.

Zhou, W., Cadenasso, M., Schwarz, K. and Pickett, S. (2014a) Quantifying spatial heterogeneity in urban landscapes: integrating visual interpretation and object-based classification. *Remote Sensing* 6(4), 3369–3386. DOI: 10.3390/rs6043369.

Zhou, W., Qian, Y., Li, X., Li, W. and Han, L. (2014b) Relationships between land cover and the surface urban heat island: seasonal variability and effects of spatial and thematic resolution of land cover data on predicting land surface temperatures. *Landscape Ecology* 29(1), 153–167. DOI: 10.1007/s10980-013-9950-5.

8 Ecology of Urban Climates: The Need for Landscape Biophysics in Cities

Elsa Youngsteadt[1]* and Adam J. Terando[2,1]
[1]*Department of Applied Ecology, North Carolina State University, Raleigh, North Carolina, United States;* [2]*US Geological Survey, Southeast Climate Adaptation Science Center, Raleigh, North Carolina, United States*

Introduction

Climate is a key driver of global species distributions, biodiversity and ecosystem productivity (Qian and Ricklefs, 2008; Chu *et al.*, 2016). Urbanization and climate change are altering local and global climate patterns, respectively, at unprecedented rates. There is a pressing need to understand and predict how species distributions, diversity and ecosystem services will change as a result. Certain parallels between urbanization and climate change have led to the widespread suggestion that the study of each can provide insights into the other (Ziska *et al.*, 2003; Carreiro and Tripler, 2005; Grimm *et al.*, 2008; Farrell *et al.*, 2015; Youngsteadt *et al.*, 2015; Pincebourde *et al.*, 2016; Lahr *et al.*, 2018), potentially accelerating progress toward understanding and predicting the biotic changes they are causing.

In this chapter we will review and evaluate the climatic and ecological parallels between urbanization and climate change, with an emphasis on temperature and its effects on terrestrial ectotherms, a group of organisms thought to be particularly sensitive to climate change (Deutsch *et al.*, 2008). We will summarize patterns of urban warming and how they relate to global climate change projections, then discuss the potential strengths and limitations of urban climates as a source of insight into biotic responses to climate change and highlight future research directions. Future work will benefit from explicit consideration of the mechanisms by which urbanization and urban warming drive ectotherms' body temperatures and activity periods. Urban ecosystems may provide a valuable first approximation of the dramatic effects of warming on certain terrestrial systems, and continued advances in urban thermal biology will promote a mechanistic understanding of how organisms use, respond to and adapt to the unique thermal landscapes of cities.

Patterns of Urban Warming

The most familiar climatological effect of cities, and the focus of this chapter, is the urban heat island (UHI), a phenomenon in which temperatures within cities are warmer than those in surrounding non-urban areas. The UHI effect has been documented since the early 1800s, and is attributed primarily to increased absorption and trapping of solar radiation by urban surfaces, reduced evaporative cooling, and anthropogenic

*Corresponding author: ekyoungs@ncsu.edu

 Urban Ecology: Its Nature and Challenges (ed. P. Barbosa)

heat inputs from combustion and human metabolism (Howard, 1833; Mills, 2008). As a prior review noted, although the UHI 'is beguilingly easy to conceptualize and measure, it is a complicated phenomenon that depends on the size, density, building practices, location, season, air flows, and many other factors of the built environment' (that is, human-made surroundings and infrastructure) (Phelan *et al.*, 2015). The complex field of urban climatology has been recently reviewed in a textbook, which we recommend to those seeking a thorough introduction to the subject (Oke *et al.*, 2017).

The UHI comprises four UHI types, each of which is defined by the urban-rural temperature difference in a particular physical substrate: subsurface (soils and underground structures); surface (built environment and vegetation); canopy (air below roof or tree height); and boundary layer (air above the canopy to the top of the boundary layer) (Kuttler, 2008; Oke *et al.*, 2017). The surface UHI is the core phenomenon that largely drives the other UHI types. We briefly review the drivers and patterns of each UHI type, excluding the boundary layer UHI, which is less directly relevant to terrestrial urban organisms.

Surface UHI

During the day, pavement and other manufactured materials that make up the built environment heat up because they absorb more radiative energy and can store more heat than vegetated surfaces or soils (Oke *et al.*, 2017). In addition, the height and configuration of buildings impact the exposure of these surfaces to solar radiation and wind. Finally, moisture availability is generally lower from urban surfaces, including construction materials and compacted soils, causing more stored heat to be released to the atmosphere through sensible, rather than latent, heat fluxes. (In other words, in the absence of moisture, surface energy fluxes drive temperature change rather than evaporation.) At night, surfaces cool, transferring heat to the air and to the subsurface, thereby generating the canopy and subterranean UHI. One key feature of the surface UHI is the much higher variability of surface temperatures compared to surrounding rural areas due to greater variability in the geometry, evaporative demand and thermal, radiative and aerodynamic properties of city environments (Oke *et al.*, 2017).

To inform urban planning and identify drivers of urban warming, correlates of surface UHI intensity have been extensively studied at local to global scales via satellite imagery. Satellite-based infrared sensors are now able to detect land surface temperatures on kilometre and sub-kilometre scales worldwide, allowing for comparative analysis of multiple cities using consistent methods. Studies often quantify surface UHI as the difference between the average surface temperature of urban pixels and surrounding non-urban pixels. Thus, the magnitude of the UHI depends not only on the properties of the city itself, but also on the choice of the surrounding reference area and the conditions therein, a complication that affects measurement of all types of UHI. Satellite-based studies have detected several consistent correlates of surface UHI intensity, which tends to be greatest during the day, in the summer and in forested biomes. We summarize each of these correlates below, for later comparison to other UHI types and to global warming.

Diurnal cycle

During the day, the surface UHI is generally more intense, and more variable among cities, than it is at night. For example, a study of 38 cities in the USA found average summer daytime UHIs of 5–9°C (excluding deserts, see below), compared to 2–3°C at night (Imhoff *et al.*, 2010). Daytime and night-time surface UHI intensities are uncorrelated with each other across cities, and are driven by different processes (Peng *et al.*, 2012; Clinton and Gong, 2013; Zhao *et al.*, 2014; Zhou *et al.*, 2017). Daytime surface temperatures primarily result from direct solar heating; night-time temperature depends on the rate at which surfaces release stored heat to the atmosphere and the subsurface.

Biome

On a global scale, the magnitude of daytime surface UHI is often best predicted by characteristics of the background ecosystem: precipitation (Zhao *et al.*, 2014), relative amount of vegetation in urban versus non-urban areas

(Peng *et al.*, 2012), or biome (Zhang *et al.*, 2010; Clinton and Gong, 2013; Imhoff *et al.*, 2010). A study of more than 3200 cities around the globe detected an annual average surface UHI intensity around 3.8°C for cities that were surrounded by forested biomes, about 1.9°C compared to surrounding grass and scrub biomes, and 0°C in arid and semi-arid biomes (where urban surfaces are often cooler than their surroundings during the day) (Zhang *et al.*, 2010; Clinton and Gong, 2013; Imhoff *et al.*, 2010).

Season

At mid and high latitudes, daytime surface UHI intensity is almost always greater in summer than in winter, likely because of the solar angle and because urban-rural vegetation differences are minimized in the winter (e.g. Imhoff *et al.*, 2010). This difference fades at low latitudes, where solar angles and vegetation cover are relatively constant (Zhang *et al.*, 2010; Peng *et al.*, 2012). Nevertheless, studies of individual tropical cities have detected seasonal variation in surface UHI intensity, with maximum intensity in dry (Suazo *et al.*, 2019) or rainy (Pereira Filho and Karam, 2016) seasons; other studies have simply been unable to measure rainy-season surface UHI at low latitudes due to high cloud cover (Tran *et al.*, 2006). At night, surface UHI intensity shows very little seasonal variation, regardless of latitude (Peng *et al.*, 2012).

Latitude

Although latitude may modulate the seasonality of the surface UHI, it does not predict its overall magnitude on a global scale (Peng *et al.*, 2012; Clinton and Gong, 2013). Zhang *et al.* (2010) binned global cities by latitude and detected differences among bins (with minimum intensities at 10–30° north), but without a clear trend from pole to equator.

City size

Several studies have found surface UHI intensity to increase with city size. For example, a study of 5000 European settlements found that urban land area was the best predictor of surface UHI, with a 0.4°C increase in daytime UHI for each doubling of the urban land area (Zhou *et al.*, 2017). Some studies agree (Zhang *et al.*, 2010; Zhou *et al.*, 2013), whereas others find this effect only at night – and, indeed, show city size to be the most important predictor of night-time surface UHI (Clinton and Gong, 2013; Zhao *et al.*, 2014). In larger cities, cooler upwind air has further to travel across the urban area; as it accumulates heat, it is less able to cool warm surfaces in the centre of large cities (Zhao *et al.*, 2014).

Within-city patterns

While global studies are valuable for describing broad patterns in the distribution of urban surface warming, their use of annual or seasonal averages compresses tremendous within-city variation in surface temperature (Pincebourde *et al.*, 2016). For example, a snapshot of surface-temperature variation in Baltimore County, Maryland, on a single summer morning revealed differences of nearly 17°C among census blocks (Huang *et al.*, 2011). Within-city surface-temperature variation is driven by local variation in amounts of vegetation and surface thermal admittance (that is, surfaces' ability to absorb and release heat), as well as the timing of solar exposure in urban canyons (Shiflett *et al.*, 2017; Oke *et al.*, 2017).

Surface UHI: caveats and summary

There are several caveats to deriving ecological (and climatological) interpretations of the surface UHI. First, the top-down view from satellites over-represents the temperatures of horizontal surfaces such as rooftops, which may not represent the conditions experienced by organisms below roof height (Arnfield, 2003). Second, satellite measurements generally perform best in cloud-free conditions. However, seasonal and regional patterns of cloud cover and precipitation are themselves important components of climate. At some times, and in some places, measurements taken on cloud-free days represent atypical conditions (Santamouris, 2015). Finally, large-scale studies that characterize individual cities with annual or seasonal averages of UHI intensity typically miss the fine-scale thermal heterogeneity that organisms actually experience on scales as small as a few metres (Pincebourde *et al.*, 2016). Satellite-based assessments of the surface UHI have nevertheless provided important perspective on patterns of thermal variation through space and time, and provide both global and local perspectives on when and where organisms can be exposed to the hottest urban surfaces.

Canopy UHI

The air below roof height in a city is known as the urban canopy layer, and its UHI is typically characterized by air temperature at 'screen height' or 'shelter height' (the standard height of weather station thermometers, typically at 1.5–2 m above the ground, also known as surface air temperature) (Jin and Dickinson, 2010). Like the surface UHI, canopy UHI is measured as a relative phenomenon; that is, the surface air temperature difference between an urban site and a non-urban reference site. Because air temperature measurements are taken at point locations, the choice of instrument location has a profound effect on the observed magnitude of the canopy UHI (Stewart, 2011; Stewart and Oke, 2012).

The spatial and temporal patterns of surface and canopy UHI are distinct. Unlike surfaces, air is highly mobile, and its temperature is driven in part by movement of air and its interaction with the many surfaces it encounters (Jin *et al.*, 2005). The air temperature at a given thermometer, then, can be influenced by surfaces several hundred metres upwind (Stewart and Oke, 2012). This mixing effect results in air temperatures being less spatially structured and less variable than surface temperatures, especially during the day when turbulent mixing is greatest (Oke *et al.*, 2017). In addition to receiving heat from urban surfaces, the urban canopy layer may be warmed by anthropogenic heat (such as that released from automobiles and HVAC systems) and by a minor 'urban greenhouse effect', wherein air pollution traps heat within the urban atmosphere (Oke *et al.*, 1991). Known correlates of canopy UHI are described in the following sections.

Diurnal cycle

Whereas the magnitude of the surface UHI is maximized during the day, the canopy UHI most often peaks at night (Arnfield, 2003). Compared to most non-urban surfaces, urban materials absorb more total heat during the day and release it more gradually to the air after sunset, resulting in maximum urban-rural air temperature differences at night. For example, a decade-long study of the UHI in Berlin found an average annual night-time canopy UHI of about 3°C, compared to about 0.5°C during the day, with occasional differences of up to 10°C (Fenner *et al.*, 2014). These results are typical of widely reported urban canopy warming patterns in temperate, forested biomes (Arnfield, 2003; Oke *et al.*, 2017), but many individual cities depart from this typical cycle and report canopy UHI maxima during the daytime; for example, due to rush-hour traffic (Santamouris, 2015).

Biome

Like the surface UHI, the canopy UHI tends to be best developed in cities located within temperate forested biomes, and less so in scrub and desert (Oke *et al.*, 2017).

Season

Like the surface UHI, the canopy UHI of cities within temperate forest biomes tends to be most intense in the summer (Arnfield, 2003; Santamouris, 2015). This effect is attributed to seasonal differences in wind and cloud cover, wherein the long, sunny days and calm nights of summer promote canopy UHI formation, but the short, cloudy days and windy nights of winter do not (Fortuniak *et al.*, 2006). For example, Fenner *et al.* (2014) found a summer night-time average canopy UHI of >6°C in Berlin, compared to the overall annual average of 1.5°C.

In tropical regions, the canopy UHI tends to be more intense in dry seasons than wet seasons (Chow and Roth, 2006; Santamouris, 2015). This pattern is attributed to the thermal behaviour of rural soils, which cool more slowly at night when wet, and thus behave more like urban materials (Roth, 2007). This pattern contrasts with the inconsistent seasonal results reported for the surface UHI in the tropics, and additional research may be needed to resolve the seasonal drivers of urban surface energy exchanges at low latitudes.

Latitude

Unlike recent global studies of surface UHI, global reviews of canopy UHI have detected latitudinal patterns. A review of maximum canopy UHI intensity in 150 cities found that latitude accounted for 6% of variation in intensity, which increased from an average of 3.5°C at the equator to nearly 7°C at 65° latitude (Wienert and Kuttler, 2005). Much of this variation was attributed to latitudinal variation in anthropogenic heat production and the size of the studied cities (Roth, 2007). Others have suggested that use of thinner and more reflective building materials in developing tropical countries also reduces UHI intensity (Chow and Roth, 2006).

City size

A positive relationship between urban population and canopy UHI intensity is essentially universal (although its slope varies regionally) (Oke, 1973; Roth, 2007; Santamouris, 2015). Population size and density are not the driving factors, but are thought to be a proxy for other features such as the area of a city and the geometry of its core, where geometry refers to the three-dimensional spatial properties of a city. More populous cities have deeper urban canyons that trap outgoing long-wave radiation (Oke, 1973; Roth, 2007).

Within-city patterns

Within-city variation in air temperatures can be profound, although it is dampened compared to surface temperatures. For example, a comparison of residential areas to native habitat remnants within cities found that, across seasons and years, residential areas averaged 2.2°C cooler than native habitats in arid US cities, whereas in mesic cities, residential areas were warmer than remnant native habitats (Hall *et al.*, 2016). Similar patterns are widely observed in city parks and surrounding urbanized areas. In mesic cities, cool temperatures from parks can spill over and mitigate urban warming up to 1 km away in the surrounding matrix (Bowler *et al.*, 2010).

Canopy UHI: caveats and summary

The history of canopy UHI studies is marred by uneven quality of documentation. A recent review of 190 studies of nocturnal canopy UHI, published from 1950 to 2007, found that nearly half of the reported warming intensities were scientifically indefensible due to poor experimental design or incomplete descriptions of methods and thermometer siting (Stewart, 2011). Moreover, weather station instruments are often sparsely distributed, such that temperatures are not measured on a spatial scale relevant to organisms. Some projects have partly overcome spatial limitations by installing dense networks of inexpensive sensors throughout urban areas, sometimes at the expense of severely biased measurements due to poor solar shielding (Terando *et al.*, 2017). An additional caveat is that air temperature at screen height, although useful to gauge UHI at heights most relevant to human experience, is not likely to represent conditions experienced by many terrestrial ectotherms. Despite the difficulties in estimating the whole-canopy UHI, certain summary points are clear. Like the surface UHI, the canopy UHI tends to be most intense in mesic biomes. But, in contrast with the surface UHI, its magnitude tends to be smaller, time-lagged, and most intense at high latitudes.

Subsurface UHI

Heat accumulated in urban surfaces also moves into soil and subterranean materials. Relative to the surface UHI, the subsurface UHI develops with time lags that increase with depth: diurnal variation is evident down to about 50 cm, and annual and seasonal variation down to about 10 m (Zhan *et al.*, 2014). At depths of 10–100+ m, substrate temperatures reflect decadal and longer temperature patterns derived from historical climates and land-use changes (Pollack *et al.*, 1998).

Subsurface UHIs are typically sensed with thermometers in bore holes or wells; they have recently also been modelled from remotely sensed surface temperatures (Zhan *et al.*, 2014). We are not aware of any broad geographic comparisons among cities that relate subsurface UHI magnitudes to city size, background climate, latitude or geology. Diurnal and seasonal cycles of the subsurface UHI have, however, been characterized in individual cities, and are closely linked to surface patterns but with time lags, such that warming peaks in different months at different depths (Shi *et al.*, 2012; Zhan *et al.*, 2014).

Within-city spatial variation in the subsurface UHI also mirrors that of the surface UHI, with the warmest ground temperatures found beneath pavements and built-up areas, and the coolest beneath green spaces (Ferguson and Woodbury, 2007). Local heat sources can also produce intense local effects, where proximity to features such as landfills or heated swimming pools produced anomalies of 20–60°C in subterranean temperatures (Menberg *et al.*, 2013).

Contrasting Urban and Global Warming

Urban warming is primarily a surface-driven phenomenon. Anthropogenic modification of the land surface alters its thermal and radiative properties and drives changes in surface temperature, which are then propagated to other atmospheric and subsurface layers. In contrast, global warming is first an atmospheric phenomenon (with some contribution from global land-use change). Trace amounts of greenhouse gas in the atmosphere absorb outgoing longwave radiation (heat emitted by the earth's surface that would otherwise escape into space) and retain this energy within the atmosphere. As of 2018, this effect had increased earth's energy budget by about 3.1 W/m^2 compared to pre-industrial levels (Butler and Montzka, 2019).

The resulting warming is analysed primarily in terms of air temperature. For those attempting to synthesize urban and global climate literature, we note that the climate change literature often refers to air temperature at 2 m as 'surface temperature' or 'surface air temperature', whereas values comparable to urban surface temperature are called 'land surface skin temperature', 'land surface temperature' or 'ground surface temperature' (Beltrami, 2002; Jin and Dickinson, 2010).

Despite the distinctly different origins of urban and global warming, there are several broad parallels between the present-day canopy UHI and the expected global warming of air temperatures. The typical magnitude of canopy UHIs across the globe (around 1–3°C on average) is similar to the range of projected warming in global air temperatures by the end of the century, relative to a pre-industrial baseline (0.3–4.8°C, depending on emissions trajectories) (IPCC, 2014). Both urban and projected global warming are of greater magnitude at higher latitudes (Wienert and Kuttler, 2005; IPCC, 2013, Section 2.2). One broad air temperature trend that diverges between urban and global warming is seasonality: global climate models consistently predict more intense warming in winter in northern high latitudes (Cohen *et al.*, 2012), whereas the canopy UHI tends to be most intense in the summer.

It is more difficult to evaluate the potential parallels between urban and global warming for surface and subterranean temperatures. Whereas surface temperatures have a major role in the UHI literature, with surfaces heated more strongly than air (Stoll and Brazel, 1992; Shiflett *et al.*, 2017), land surface temperature changes are generally not reported in the climate change literature, except with respect to changes in the constituent components of the surface energy balance. One possible reason for this relative inattention is simply that, on a global scale, the trends in surface temperature are highly correlated with the near-surface air temperature trends (Jin and Dickinson, 2010; Good *et al.*, 2017).

Studies that have drawn biological parallels between urban and global warming have typically done so using the currency of air temperature (e.g. Ziska *et al.*, 2003; Youngsteadt *et al.*, 2015). From the most general view of global trends, this review broadly supports such a comparison (despite seasonal differences). However, the comparison on the basis of air temperature could be misleading if uniquely urban patterns of subterranean and surface warming modulate biotic responses differently than air temperature alone. Other sources of error in the comparison of urban and global warming include tremendous variation within and among cities in the intensity and timing of warming; methodological problems with measuring each type of UHI; and the geographic variation and uncertainty in projected climate change. Attempts to use urban warming to draw inferences about global warming should proceed with caution and, when possible, with local validation of the comparison. As described below, mechanistic approaches will improve confidence in the thermal basis of observed biological correlates of urban warming.

Ectotherms and Warming: Questions Cities Could Answer

Temperature is an essential driver of biological activity at the molecular level, and physiological changes at this scale have the potential to cascade through levels of organization – from molecular to organismal, population, community and ecosystem scales (Helmuth *et al.*, 2005; Kingsolver, 2009). Predicting the consequences of such changes is a major challenge in ecological research. Here we highlight key ecological questions about effects of warming on ectotherms, to which the study of urban ecosystems is poised to make strong contributions (Lahr *et al.*, 2018). In each of these questions, we note that a disadvantage of using urban study systems is that the habitats themselves usually differ from non-urban habitats in many ways other than climate – such as amount of fragmentation or pollution. Careful methods choices and complementary experimental approaches are needed to attribute biotic effects to temperature rather than confounding factors.

What is the role of evolutionary adaptation in biological responses to anthropogenic warming?

If species can adapt rapidly to warming, extinction rates and ecological disruptions may be buffered against climate change. Opportunities to examine real-time thermal adaptation to climate change in wild populations are limited and rely on the rare opportunity to resurrect living propagules from historically cool time periods (Geerts *et al.*, 2015). Urban warming gradients provide a complementary opportunity to examine evolutionary and plastic responses to recent warming, taking advantage of spatial rather than temporal sampling designs. Elevational and latitudinal studies also provide opportunities to examine thermal adaptation (Elmendorf *et al.*, 2015). However, they may better represent adaptation to long-standing climates rather than to the rapid trends caused by urbanization and global warming. Several studies have documented thermal adaptation in cities, with examples in soil fungi (McLean *et al.*, 2005), clover plants (Thompson *et al.*, 2016), aquatic crustaceans (Brans *et al.*, 2017) and ants (Diamond *et al.*, 2017). Those in relatively young cities imply rapid evolution. Acorn ants in Cleveland, Ohio, for example, may have evolved higher heat tolerance in fewer than 20 generations (Diamond *et al.*, 2017). These results demonstrate the potential of cities to assay the roles of thermal adaptation, acclimation and behavioural compensation in response to warming.

How does global change alter species interactions?

Complex networks of interactions such as herbivory, predation, parasitism and pollination maintain biological diversity and mediate ecosystem services (Tylianakis *et al.*, 2008). The stability of these interactions in the face of climate change is difficult to predict. In some cases, effects can be documented via short-term experimental warming (Barton and Schmitz, 2009) or long-term observations of ecological interactions, such as mutualism (Miller-Struttmann *et al.*, 2015). Urban areas provide study systems that complement long-term observations and experimental warming. Cities encompass habitats that have been warmed for many decades, where species may participate in a more complete web of interactions than can be reproduced experimentally, and where temperatures may already exceed the range of historical observations. One urban study, for example, demonstrated that warming changes the outcome of a parasitoid–herbivore interaction due to phenological mismatch, contributing to enhanced rates of herbivory in urban hotspots (Meineke *et al.*, 2014). Another suggested that urban warming favours the development of mosquito vectors, altering the human–mosquito–dengue interaction such that urban hotspots experienced elevated disease risk (Araujo *et al.*, 2015).

Where on the planet are organisms most vulnerable to warming?

Across latitudes, ectotherms' physiological heat tolerance varies much less than environmental temperatures (Sunday *et al.*, 2011). Tropical ectotherms tend to be thermal specialists adapted to temperatures that are relatively stable throughout the year, with narrow safety margins of a few °C, or less, before the onset of heat stress and fitness decline (Deutsch *et al.*, 2008; Huey *et al.*, 2009). Although global warming is most intense in northern high latitudes, ectotherms may be more vulnerable to warming in the tropics.

Cities, if treated as warming experiments distributed across latitudes, may provide insights into how biotic effects of warming vary against differing background climates. Urban studies have begun to test the prediction that ectotherms in already-warm climates are most vulnerable to warming, but to date have done so only across limited latitudinal gradients in higher-latitude temperate zones. These studies have indeed detected more positive responses to warming, and greater potential for evolution of increased heat tolerance, among high-latitude ectotherms (Youngsteadt *et al.*, 2017; Diamond *et al.*, 2018).

Validation

Few studies have had the opportunity to directly and quantitatively compare biotic responses to urban warming with responses to historical or recent warming outside of cities. In one such study, scale insects on red maple trees reached their highest densities in the warmest parts of Raleigh, North Carolina, and in warm historical time periods in the south-eastern USA. But their density increased more steeply per °C air temperature – and had a six-fold greater peak density – in urban habitats than in natural habitats at similar temperatures (Youngsteadt *et al.*, 2015). Similarly, a comparison of birch-flowering phenology using urban and historical datasets in Germany found that trees advanced flowering by 4.4 days per °C increase in urban air temperature, compared to only 1.9 days per °C in a historical dataset that spanned a similar temperature range (Jochner *et al.*, 2013). These results emphasize that, on their own, urban settings could not be used to directly predict response to warming in other habitats. The potential explanations for amplified biotic responses to urban warming are many, speculative and non-mutually exclusive. One clear problem to be addressed, however, is the use of air temperature as the currency of comparison.

Towards a Biophysical Urban Ecology

Two truisms have been largely overlooked in the ecology of urban warming: first, body temperature is the thermal variable that drives ectotherm activity and fitness; and second, body temperature is poorly predicted by air temperature. Despite the recent growth of an exciting literature on the ecological and evolutionary effects of urban warming, there is still a weak mechanistic understanding of how the urban environment alters body temperatures and drives physiological changes that scale up through ecosystems. The subdiscipline of biophysical ecology has long recognized that defining the thermal parameters to which individual organisms respond is essential to generalizing results through time and space and making sound predictions at higher levels of biological organization (Helmuth, 1998). Here we briefly review the relationship between body temperature and fitness and the drivers of ectotherm body temperatures.

Body temperature and fitness

In ectotherms, the relationship between body temperature and fitness is often represented as a thermal fitness curve or thermal performance curve (where performance may be measured as locomotion, growth, developmental rate, etc.). Thermal performance curves are typically asymmetrical in shape, with a gradual increase up to an optimum temperature, followed by a steep decline (Fig. 8.1). The form of the relationship between whole-organism performance and temperature is remarkably similar to that of the relationship between biochemical reaction rates and temperature (Kingsolver, 2009).

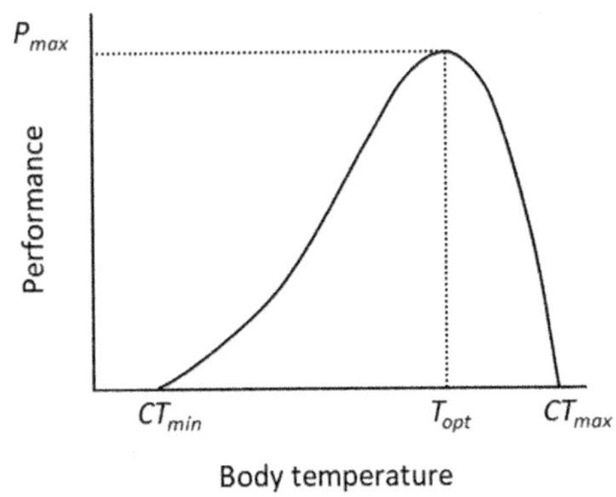

Fig. 8.1. Example thermal performance curve. The range of tolerable temperatures is bounded by the critical thermal minimum (CT_{min}) and the critical thermal maximum (CT_{max}); within this range, performance increases gradually up to an optimum temperature (T_{opt}) and then declines rapidly.

Indeed, organismal performance likely arises from biochemical and cellular processes including enzyme function, membrane integrity, and their joint effects on oxygen limitation; as well as the cost of repairing cellular machinery after a heat shock (Angilletta, 2009, but see Hofmann and Somero, 1996; Amarasekare and Savage, 2012).

Thermal performance curves bear several implications for understanding ectotherms' responses to warming. First, the width of the curve denotes the range of temperatures at which an organism can survive, with some taxa having narrow (specialized) curves, as is typical of tropical taxa, and others being more broadly adapted (Kingsolver, 2009; Huey *et al.*, 2012). Second, the non-linear and asymmetrical form of the curve means that organisms are more sensitive to temperature change when they are near or above their optima. In contrast, warming at temperatures much cooler than the optima, where the slope of the thermal performance curve is shallow, has more muted effects on performance (Kingsolver, 2009; Huey *et al.*, 2012). Finally, as a result of the non-linear, asymmetric curve, the fitness of an organism at the mean temperature is not equal to the mean fitness across fluctuating temperatures, a fact that can lead to errors when temperature is measured over inappropriate timescales. For example, Bernhardt *et al.* (2018) measured population growth rates of a marine alga under a series of constant temperatures versus fluctuating temperatures, centred on the same means. Under fluctuating conditions, population growth became negative at a temperature 4°C cooler than expected based on the mean temperature alone. Similar results have been found for insects (Paaijmans *et al.*, 2013).

Although body temperature is a key mediator of fitness, ectotherms also face costs and risks associated with maintaining an ideal body temperature, such as exposure to predation while basking, or lost productivity while waiting in cool refuges. An organism's decisions about behavioural thermoregulation may reflect these trade-offs (Huey and Stevenson, 1979).

Predictors of body temperature

Although ectotherms lack the mechanisms of thermal homeostasis found in endotherms, their body temperatures are rarely equal to the temperatures of the surrounding air or surfaces (Helmuth, 1999; Angilletta, 2009). Instead, body temperature depends on complex exchanges of energy between the organism and its environment via convection, conduction and radiation. These exchanges, in turn, depend on the properties of the animal itself: the extent to which its surface absorbs or reflects various wavelengths of radiation; the extent to which its surface conducts heat toward and away from its core; its body size, shape and orientation to the sun; and its metabolic and moisture-loss rates (Porter and Gates, 1969; Helmuth, 1998). Because body temperature depends on many climate variables, a variety of conditions within this multi-dimensional 'climate space' can yield the same body temperature. Meanwhile, the same air temperature can yield different body temperatures if combined with different radiative environments or wind speeds (Porter and Gates, 1969; Gilman *et al.*, 2006; Angilletta, 2009). This is a key point when comparing urban warming to warming in other habitats.

Urban warming is often characterized by removal and replacement of vegetation with manufactured materials. The larger heat capacities and drying potentials of these materials increase their heat storage and delay heat release. Urbanization is also often associated with reduced but spatially variable wind speeds (Kuttler, 2008), which may reduce convective heat exchange. Thus, warming along urban habitat gradients is associated with changes on multiple axes of the climate space an ectotherm experiences. Global warming, when applied to an otherwise undisturbed habitat, will alter the climate space less dramatically. The well-mixed nature of greenhouse gases means that, for a given location, shifts in air temperature are the dominant change to multivariate climate space. Although air temperature changes do alter the surface energy balance, global warming is not associated with the dramatic and spatially structured variation in heat and radiation fluxes that are found in urban areas. Other climate variables that drive urban warming, such as surface wind speed, are expected to undergo less pronounced changes globally, with large uncertainties about the sign of the change (IPCC, 2013). Because urbanization and global warming alter multivariate climate space in different ways, comparisons of biological effects of urban and global warming on the basis of air temperature alone should be interpreted with caution.

Unlike air temperature, operative temperature is a metric that integrates many of these additional predictors of body temperature. Operative temperature is the body temperature an organism would experience if it were in equilibrium with its environment, that is, if it made no effort to thermoregulate. Operative temperatures can be estimated by measurement or modelling. Measurement involves the use of physical models that match the organism of interest in surface area, aerodynamic roughness and absorbance. The models are placed in the environment where the organism occurs, or could occur. The equilibrium temperature of the model is equated to the operative temperature of the organism. If this value differs from the optimal body temperature (as measured during laboratory performance tests with live organisms), the difference may cause reduced performance, or be overcome by behavioural thermoregulation (Dzialowski, 2005). Modelling, rather than measuring, the operative temperature requires knowledge of the size, shape and surface properties of the organism, as well as microclimatological data, such as air temperature, wind speed and incidence of longwave and shortwave radiation, on a spatial scale of measurement appropriate to the organism. In some cases, researchers have used a combination of local measurements and regional weather stations to successfully model operative or body temperatures over certain timescales (Helmuth, 1999; Porter *et al.*, 2010).

When combined with information about a species' thermal performance curves, measured or modelled operative temperatures can provide insights into where, how often and when during the diurnal and seasonal cycle a habitat is suitable for activity. These results can, in turn, predict fitness and demographic processes under historical, current and future conditions (Crozier and Dwyer, 2006; Buckley, 2008; Huey *et al.*, 2009; Porter *et al.*, 2010). For example, Gilman *et al.* (2006) used a mechanistic model of heat transport through the bodies of intertidal mussels to predict mussel body temperature due to a combination of air and water temperature, wind speed, solar radiation and other microclimatic factors. At four sites spanning 14 degrees latitude, the mechanistic model was a better predictor of actual mussel body temperature than were air or water temperature alone. A simulated 1°C increase in air or water temperature raised the monthly average of maximum daily body temperature by only 0.07–0.92°C. The relationship between environmental and body temperatures differed among sites and among elevations within sites, due to differences in wind speed and other environmental variables. Because body temperature itself is the variable most closely linked to physiological stress and population growth in mussels (Helmuth and Hofmann, 2001; Somero, 2002), this study demonstrates that the use of air temperature alone would likely cause erroneous predictions of mussel population responses to climate change. Moreover, these errors would not represent a consistent bias but would vary geographically. We suspect that similar errors will arise in urban studies if air temperature is used as the sole proxy for the climate experienced by urban ectotherms, since cities differ from each other and from non-urban habitats on multiple climate axes.

Another example, in lizards, demonstrates more explicitly how body temperature can be traced through to population-level outcomes, and how doing so can produce different predictions to those derived from air temperature alone. Buckley (2008) used pre-existing information about lizards' preferred body temperatures, along with operative temperatures in habitats across their North American ranges, to define thermal constraints on foraging. Ultimately, after imposing additional constraints based on prey availability and lizard population density, this study modelled equilibrium population densities across the continent under current and future climate scenarios. Population densities, in turn, imply range limits at locations where the equilibrium population is zero. In comparison to a correlative model based on air temperatures alone, the mechanistic model was a better predictor of current ranges, and predicted more extensive northward range shifts due to climate change.

Despite the power of biophysical ecology to link environment with individual physiology, population- and community-level outcomes, these approaches have not been fully applied to urban habitats. Although observed correlations between metrics of urban warming, thermal tolerance traits and community composition (Hamblin *et al.*, 2017) imply mechanistic links across these scales, it is still unknown how ectotherm body temperatures actually vary across urban thermal gradients, making results difficult to generalize to other cities or non-urban habitats.

We are aware of only a single, very recent study that has taken a detailed biophysical approach to urban ectotherm ecology (Battles and Kolbe, 2019). This study, located in the metropolitan area of Miami, Florida, measured both operative temperatures (at randomly selected locations within lizard habitats) and body temperatures of free-living brown anoles (*Anolis sagrei*) and crested anoles (*Anolis cristatellus*) in urban and natural sites. The authors also compared these values to preferred and optimal temperatures for the same lizard populations assessed in the laboratory. In urban habitats, both operative and body temperatures were warmer than in natural habitats, an effect attributed to increased solar radiation under open urban tree canopies as opposed to closed canopies in natural areas. For *A. sagrei*, the urban operative and body temperatures were more often within the preferred and optimal range, indicating that this species needed to spend less time basking in urban areas than in natural areas, potentially lowering the cost of thermoregulation and improving performance. For *A. cristatellus*, urbanization did not increase access to optimal body temperatures. Instead, up to half the operative temperatures were too warm for this species in urban sites, perhaps contributing to the relative dominance of *A. sagrei* over *A. cristatellus* in urban habitats.

By linking environmental conditions to the relative costs of maintaining preferred body temperatures, this study of urban lizards laid a foundation for tracing the effects of urban thermal environments through individual, population, and community scales. In this case, although urban habitats in Miami did produce operative and body temperature distributions warmer than those in adjacent natural areas, the urban operative and body temperatures did not exceed those observed in natural habitats elsewhere in the species' ranges (Huey *et al.*, 2009; Gunderson and Leal, 2012). In some cases, however, urban habitats will create operative temperature distributions outside the range observed in natural habitats. In these situations, they could also provide insights into the magnitude and geographic distribution of responses to future climate change, in the meaningful currency of body temperatures.

Summary and Conclusion

This chapter has reviewed the diurnal, seasonal and geographic patterns of urban warming, a process that places cities in some ways 'ahead of the global warming curve' (Lahr *et al.*, 2018). Although the amplitude, diurnal cycle and geographic patterns of urban air temperature warming broadly correspond to patterns of global warming, we also highlight qualitative differences. Urban heat islands are characterized not only by warming night-time air temperatures but also by increased penetration of solar radiation compared to forested areas, by intense daytime warming of anthropogenic surfaces, and by time-lagged warming of the soil and subsurface. These processes will alter the thermal equilibria of urban ectotherms in ways not predicted by models that consider air temperature alone. Although a growing urban literature highlights the potential of cities to aid in the development and testing of important hypotheses about biotic effects of warming, the mechanistic understanding of warming remains simplistic and incomplete. The methods of biophysical ecology emphasize links between the environment and body temperature, with subsequent effects on individual fitness, demography and population processes. Integrating these approaches into urban ecology will improve the mechanistic understanding of the role of urban thermal environments in shaping the ecology and evolution of urban organisms, and will allow results to be more meaningfully compared, and potentially generalized, to other habitats and to future cities.

Disclaimer

Any use of trade, product, or firm names is for descriptive purposes only and does not imply endorsement by the US government.

References

Amarasekare, P. and Savage, V. (2012) A framework for elucidating the temperature dependence of fitness. *The American Naturalist* 179(2), 178–191. DOI: 10.1086/663677.

Angilletta, M.J. (2009) *Thermal Adaptation: A Theoretical and Empirical Synthesis*. Oxford University Press, Oxford, UK.

Araujo, R.V., Albertini, M.R., Costa-da-Silva, A.L., Suesdek, L., Franceschi, N.C.S. *et al.* (2015) São Paulo urban heat islands have a higher incidence of dengue than other urban areas. *The Brazilian Journal of Infectious Diseases* 19(2), 146–155. DOI: 10.1016/j.bjid.2014.10.004.

Arnfield, A.J. (2003) Two decades of urban climate research: a review of turbulence, exchanges of energy and water, and the urban heat island. *International Journal of Climatology* 23(1), 1–26. DOI: 10.1002/joc.859.

Barton, B.T. and Schmitz, O.J. (2009) Experimental warming transforms multiple predator effects in a grassland food web. *Ecology Letters* 12, 1317–1325. DOI: 10.1111/j.1461-0248.2009.01386.x.

Battles, A.C. and Kolbe, J.J. (2019) Miami heat: urban heat islands influence the thermal suitability of habitats for ectotherms. *Global Change Biology* 25(2), 562–576. DOI: 10.1111/gcb.14509.

Beltrami, H. (2002) Climate from borehole data: energy fluxes and temperatures since 1500. *Geophysical Research Letters* 29(23), 26-1–2626. DOI: 10.1029/2002GL015702.

Bernhardt, J.R., Sunday, J.M., Thompson, P.L. and O'Connor, M.I. (2018) Nonlinear averaging of thermal experience predicts population growth rates in a thermally variable environment. *Proceedings of the Royal Society B: Biological Sciences* 285(1886), 20181076. DOI: 10.1098/rspb.2018.1076.

Bowler, D.E., Buyung-Ali, L., Knight, T.M. and Pullin, A.S. (2010) Urban greening to cool towns and cities: a systematic review of the empirical evidence. *Landscape and Urban Planning* 97(3), 147–155. DOI: 10.1016/j.landurbplan.2010.05.006.

Brans, K.I., Jansen, M., Vanoverbeke, J., Tüzün, N., Stoks, R. *et al.* (2017) The heat is on: genetic adaptation to urbanization mediated by thermal tolerance and body size. *Global Change Biology* 23(12), 5218–5227. DOI: 10.1111/gcb.13784.

Buckley, L. B. (2008) Linking traits to energetics and population dynamics to predict lizard ranges in changing environments. *The American Naturalist* 171(1), E1–E19. DOI: 10.1086/523949.

Butler, J.H. and Montzka, S.A. (2019) *The NOAA Annual Greenhouse Gas Index (AGGI*. NOAA Earth System Research Laboratory, Boulder, CO. Available at: https://www.esrl.noaa.gov/gmd/aggi/aggi.html

Carreiro, M.M. and Tripler, C.E. (2005) Forest remnants along urban-rural gradients: examining their potential for global change research. *Ecosystems* 8(5), 568–582. DOI: 10.1007/s10021-003-0172-6.

Chow, W.T.L. and Roth, M. (2006) Temporal dynamics of the urban heat island of Singapore. *International Journal of Climatology* 26(15), 2243–2260. DOI: 10.1002/joc.1364.

Chu, C., Bartlett, M., Wang, Y., He, F., Weiner, J. *et al.* (2016) Does climate directly influence NPP globally? *Global Change Biology* 22(1), 12–24. DOI: 10.1111/gcb.13079.

Clinton, N. and Gong, P. (2013) MODIS detected surface urban heat islands and sinks: global locations and controls. *Remote Sensing of Environment* 134, 294–304. DOI: 10.1016/j.rse.2013.03.008.

Cohen, J.L., Furtado, J.C., Barlow, M., Alexeev, V.A. and Cherry, J.E. (2012) Asymmetric seasonal temperature trends. *Geophysical Research Letters* 39(4), L04705. DOI: 10.1029/2011GL050582.

Crozier, L. and Dwyer, G. (2006) Combining population-dynamic and ecophysiological models to predict climate-induced insect range shifts. *The American Naturalist* 167(6), 853–866. DOI: 10.1086/504848.

Deutsch, C.A., Tewksbury, J.J., Huey, R.B., Sheldon, K.S., Ghalambor, C.K. *et al.* (2008) Impacts of climate warming on terrestrial ectotherms across latitude. *Proceedings of the National Academy of Sciences* 105(18), 6668–6672. DOI: 10.1073/pnas.0709472105.

Diamond, S.E., Chick, L., Perez, A., Strickler, S.A. and Martin, R.A. (2017) Rapid evolution of ant thermal tolerance across an urban-rural temperature cline. *Biological Journal of the Linnean Society* 121(2), 248–257. DOI: 10.1093/biolinnean/blw047.

Diamond, S.E., Chick, L.D., Perez, A., Strickler, S.A. and Martin, R.A. (2018) Evolution of thermal tolerance and its fitness consequences: parallel and non-parallel responses to urban heat islands across three cities. *Proceedings of the Royal Society B: Biological Sciences* 285(1882), 20180036. DOI: 10.1098/rspb.2018.0036.

Dzialowski, E.M. (2005) Use of operative temperature and standard operative temperature models in thermal biology. *Journal of Thermal Biology* 30(4), 317–334. DOI: 10.1016/j.jtherbio.2005.01.005.

Elmendorf, S.C., Henry, G.H.R., Hollister, R.D., Fosaa, A.M., Gould, W.A. *et al.* (2015) Experiment, monitoring, and gradient methods used to infer climate change effects on plant communities yield consistent patterns. *Proceedings of the National Academy of Sciences* 112(2), 448–452. DOI: 10.1073/pnas.1410088112.

Farrell, C., Szota, C. and Arndt, S.K. (2015) Urban plantings: 'living laboratories' for climate change response. *Trends in Plant Science* 20(10), 597–599. DOI: 10.1016/j.tplants.2015.08.006.

Fenner, D., Meier, F., Scherer, D. and Polze, A. (2014) Spatial and temporal air temperature variability in Berlin, Germany, during the years 2001–2010. *Urban Climate* 10, 308–331. DOI: 10.1016/j.uclim.2014.02.004.

Ferguson, G. and Woodbury, A.D. (2007) Urban heat island in the subsurface. *Geophysical Research Letters* 34(23), L23713. DOI: 10.1029/2007GL032324.

Fortuniak, K., Kłysik, K. and Wibig, J. (2006) Urban–rural contrasts of meteorological parameters in Łódź. *Theoretical and Applied Climatology* 84(1-3), 91–101. DOI: 10.1007/s00704-005-0147-y.

Geerts, A.N., Vanoverbeke, J., Vanschoenwinkel, B., Van Doorslaer, W., Feuchtmayr, H. *et al.* (2015) Rapid evolution of thermal tolerance in the water flea *Daphnia*. *Nature Climate Change* 5(7), 665–668. DOI: 10.1038/nclimate2628.

Gilman, S.E., Wethey, D.S. and Helmuth, B. (2006) Variation in the sensitivity of organismal body temperature to climate change over local and geographic scales. *Proceedings of the National Academy of Sciences* 103(25), 9560–9565. DOI: 10.1073/pnas.0510992103.

Good, E.J., Ghent, D.J., Bulgin, C.E. and Remedios, J.J. (2017) A spatiotemporal analysis of the relationship between near-surface air temperature and satellite land surface temperatures using 17 years of data from the ATSR series. *Journal of Geophysical Research: Atmospheres* 122(17), 9185–9210. DOI: 10.1002/2017JD026880.

Grimm, N.B., Faeth, S.H., Golubiewski, N.E., Redman, C.L., Wu, J. *et al.* (2008) Global change and the ecology of cities. *Science* 319(5864), 756–760. DOI: 10.1126/science.1150195.

Gunderson, A.R. and Leal, M. (2012) Geographic variation in vulnerability to climate warming in a tropical Caribbean lizard. *Functional Ecology* 26(4), 783–793. DOI: 10.1111/j.1365-2435.2012.01987.x.

Hall, S.J., Learned, J., Ruddell, B., Larson, K.L., Cavender-Bares, J. *et al.* (2016) Convergence of microclimate in residential landscapes across diverse cities in the United States. *Landscape Ecology* 31(1), 101–117. DOI: 10.1007/s10980-015-0297-y.

Hamblin, A.L., Youngsteadt, E., López-Uribe, M.M. and Frank, S.D. (2017) Physiological thermal limits predict differential responses of bees to urban heat-island effects. *Biology Letters* 13(6), 20170125. DOI: 10.1098/rsbl.2017.0125.

Helmuth, B.S.T. (1998) Intertidal mussel microclimates: predicting the body temperature of a sessile invertebrate. *Ecological Monographs* 68(1), 51–74. DOI: 10.1890/0012-9615(1998)068[0051:IMMPTB]2.0.CO;2.

Helmuth, B. (1999) Thermal biology of rocky intertidal mussels: quantifying body temperatures using climatological data. *Ecology* 80(1), 15–34. DOI: 10.1890/0012-9658(1999)080[0015:TBORIM]2.0.CO;2.

Helmuth, B.S.T. and Hofmann, G.E. (2001) Microhabitats, thermal heterogeneity, and patterns of physiological stress in the rocky intertidal zone. *The Biological Bulletin* 201(3), 374–384. DOI: 10.2307/1543615.

Helmuth, B., Kingsolver, J.G. and Carrington, E. (2005) Biophysics, physiological ecology, and climate change: does mechanism matter? *Annual Review of Physiology* 67(1), 177–201. DOI: 10.1146/annurev.physiol.67.040403.105027.

Hofmann, G.E. and Somero, G.N. (1996) Interspecific variation in thermal denaturation of proteins in the congeneric *mussels Mytilus trossulus and M. galloprovincialis:* evidence from the heat-shock response and protein ubiquitination. *Marine Biology* 126(1), 65–75. DOI: 10.1007/BF00571378.

Howard, L. (1833) *The Climate of London*. Harvey and Darton, London.

Huang, G., Zhou, W. and Cadenasso, M.L. (2011) Is everyone hot in the City? spatial pattern of land surface temperatures, land cover and neighborhood socioeconomic characteristics in Baltimore, MD. *Journal of Environmental Management* 92(7), 1753–1759. DOI: 10.1016/j.jenvman.2011.02.006.

Huey, R.B., Deutsch, C.A., Tewksbury, J.J., Vitt, L.J., Hertz, P.E. *et al.* (2009) Why tropical forest lizards are vulnerable to climate warming. *Proceedings of the Royal Society B: Biological Sciences* 276(1664), 1939–1948. DOI: 10.1098/rspb.2008.1957.

Huey, R.B., Kearney, M.R., Krockenberger, A., Holtum, J.A.M., Jess, M. *et al.* (2012) Predicting organismal vulnerability to climate warming: roles of behaviour, physiology and adaptation. *Philosophical Transactions of the Royal Society B: Biological Sciences* 367(1596), 1665–1679. DOI: 10.1098/rstb.2012.0005.

Huey, R.B. and Stevenson, R.D. (1979) Integrating thermal physiology and ecology of ectotherms: a discussion of approaches. *American Zoologist* 19(1), 357–366. DOI: 10.1093/icb/19.1.357.

IPCC (2013) *Climate Change 2013: The Physical Science Basis*. Cambridge University Press, Cambridge, UK.

IPCC (2014) *Climate Change 2014: Synthesis Report*. IPCC, Geneva, Switzerland.

Imhoff, M.L., Zhang, P., Wolfe, R.E. and Bounoua, L. (2010) Remote sensing of the urban heat island effect across biomes in the continental USA. *Remote Sensing of Environment* 114(3), 504–513. DOI: 10.1016/j.rse.2009.10.008.

Jin, M. and Dickinson, R.E. (2010) Land surface skin temperature climatology: benefitting from the strengths of satellite observations. *Environmental Research Letters* 5(4), 044004. DOI: 10.1088/1748-9326/5/4/044004.

Jin, M., Dickinson, R.E. and Zhang, D. (2005) The footprint of urban areas on global climate as characterized by MODIS. *Journal of Climate* 18(10), 1551–1565. DOI: 10.1175/JCLI3334.1.

Jochner, S., Caffarra, A. and Menzel, A. (2013) Can spatial data substitute temporal data in phenological modelling? A survey using birch flowering. *Tree Physiology* 33(12), 1256–1268.

Kingsolver, J. G. (2009) The well-temperatured biologist. *The American Naturalist* 174(6), 755–768. DOI: 10.1086/648310.

Kuttler, W. (2008) The urban climate – basic and applied aspects. In: Marzluff, J.M., Shulenberger, E., Endlicher, W., Alberti, M., Bradley, G., *et al.* (eds) *Urban Ecology*. Springer, Boston, Massachusetts.

Lahr, E.C., Dunn, R.R. and Frank, S.D. (2018) Getting ahead of the curve: cities as surrogates for global change. *Proceedings of the Royal Society B: Biological Sciences* 285(1882), 20180643. DOI: 10.1098/rspb.2018.0643.

McLean, M.A., Angilletta, M.J. and Williams, K.S. (2005) If you can't stand the heat, stay out of the city: Thermal reaction norms of chitinolytic fungi in an urban heat island. *Journal of Thermal Biology* 30(5), 384–391. DOI: 10.1016/j.jtherbio.2005.03.002.

Meineke, E.K., Dunn, R.R. and Frank, S.D. (2014) Early pest development and loss of biological control are associated with urban warming. *Biology Letters* 10(11), 20140586. DOI: 10.1098/rsbl.2014.0586.

Menberg, K., Bayer, P., Zosseder, K., Rumohr, S. and Blum, P. (2013) Subsurface urban heat islands in German cities. *Science of the Total Environment* 442, 123–133. DOI: 10.1016/j.scitotenv.2012.10.043.

Miller-Struttmann, N.E., Geib, J.C., Franklin, J.D., Kevan, P.G., Holdo, R.M. *et al.* (2015) Functional mismatch in a bumble bee pollination mutualism under climate change. *Science* 349(6255), 1541–1544. DOI: 10.1126/science.aab0868.

Mills, G. (2008) Luke Howard and the climate of London. *Weather* 63(6), 153–157. DOI: 10.1002/wea.195.

Oke, T.R. (1973) City size and the urban heat island. *Atmospheric Environment* 7(8), 769–779. DOI: 10.1016/0004-6981(73)90140-6.

Oke, T.R., Johnson, G.T., Steyn, D.G. and Watson, I.D. (1991) Simulation of surface urban heat islands under ideal conditions at night. Part 2: diagnosis of causation. *Boundary-Layer Meteorology* 56(4), 339–358. DOI: 10.1007/BF00119211.

Oke, T.R., Mills, G. and Voogt, J. (2017) *Urban Climates*. Cambridge University Press, Cambridge, UK.

Paaijmans, K.P., Heinig, R.L., Seliga, R.A., Blanford, J.I., Blanford, S. *et al.* (2013) Temperature variation makes ectotherms more sensitive to climate change. *Global Change Biology* 19(8), 2373–2380. DOI: 10.1111/gcb.12240.

Peng, S., Piao, S., Ciais, P., Friedlingstein, P., Ottle, C. *et al.* (2012) Surface urban heat island across 419 global big cities. *Environmental Science & Technology* 46(2), 696–703. DOI: 10.1021/es2030438.

Pereira Filho, A.J. and Karam, H.A. (2016) Estimation of long term low resolution surface urban heat island intensities for tropical cities using MODIS remote sensing data. *Urban Climate* 17, 32–66.

Phelan, P.E., Kaloush, K., Miner, M., Golden, J., Phelan, B. *et al.* (2015) Urban heat island: mechanisms, implications, and possible remedies. *Annual Review of Environment and Resources* 40(1), 285–307. DOI: 10.1146/annurev-environ-102014-021155.

Pincebourde, S., Murdock, C.C., Vickers, M. and Sears, M.W. (2016) Fine-scale microclimatic variation can shape the responses of organisms to global change in both natural and urban environments. *Integrative and Comparative Biology* 56(1), 45–61. DOI: 10.1093/icb/icw016.

Pollack, H.N., Huang, S. and Shen, P.-Y. (1998) Climate change record in subsurface temperatures: a global perspective. *Science* 282(5387), 279–281. DOI: 10.1126/science.282.5387.279.

Porter, W. P., Ostrowski, S. and Williams, J. B. (2010) Modeling animal landscapes. *Physiological and Biochemical Zoology* 83(5), 705–712. DOI: 10.1086/656181.

Porter, W.P. and Gates, D.M. (1969) Thermodynamic equilibria of animals with environment. *Ecological Monographs* 39(3), 227–244. DOI: 10.2307/1948545.

Qian, H. and Ricklefs, R.E. (2008) Global concordance in diversity patterns of vascular plants and terrestrial vertebrates. *Ecology Letters* 11(6), 547–553. DOI: 10.1111/j.1461-0248.2008.01168.x.

Roth, M. (2007) Review of urban climate research in (sub)tropical regions. *International Journal of Climatology* 27(14), 1859–1873. DOI: 10.1002/joc.1591.

Santamouris, M. (2015) Analyzing the heat island magnitude and characteristics in one hundred Asian and Australian cities and regions. *Science of the Total Environment* 512-513, 582–598. DOI: 10.1016/j.scitotenv.2015.01.060.

Shi, B., Tang, C.-S., Gao, L., Liu, C. and Wang, B.-J. (2012) Observation and analysis of the urban heat island effect on soil in Nanjing, China. *Environmental Earth Sciences* 67(1), 215–229. DOI: 10.1007/s12665-011-1501-2.

Shiflett, S.A., Liang, L.L., Crum, S.M., Feyisa, G.L., Wang, J. *et al.* (2017) Variation in the urban vegetation, surface temperature, air temperature nexus. *Science of the Total Environment* 579, 495–505. DOI: 10.1016/j.scitotenv.2016.11.069.

Somero, G.N. (2002) Thermal physiology and vertical zonation of intertidal animals: optima, limits, and costs of living. *Integrative and Comparative Biology* 42(4), 780–789. DOI: 10.1093/icb/42.4.780.

Stewart, I.D. (2011) A systematic review and scientific critique of methodology in modern urban heat island literature. *International Journal of Climatology* 31(2), 200–217. DOI: 10.1002/joc.2141.

Stewart, I.D. and Oke, T.R. (2012) Local climate zones for urban temperature studies. *Bulletin of the American Meteorological Society* 93(12), 1879–1900. DOI: 10.1175/BAMS-D-11-00019.1.

Stoll, M.J. and Brazel, A.J. (1992) Surface-air temperature relationships in the urban environment of Phoenix, Arizona. *Physical Geography* 13(2), 160–179. DOI: 10.1080/02723646.1992.10642451.

Suazo, J.M.A., Vasques, R.J.A., Rojas, J.L.F. and Karam, H.A. (2019) Estimación de Isla de Calor Urbana superficial en El area Metropolitana de Iquitos/Peru. *Anuario do Instituto de Geociencias* 42, 135–145.

Sunday, J.M., Bates, A.E. and Dulvy, N.K. (2011) Global analysis of thermal tolerance and latitude in ectotherms. *Proceedings of the Royal Society B: Biological Sciences* 278(1713), 1823–1830. DOI: 10.1098/rspb.2010.1295.

Terando, A.J., Youngsteadt, E., Meineke, E.K. and Prado, S.G. (2017) Ad hoc instrumentation methods in ecological studies produce highly biased temperature measurements. *Ecology and Evolution* 7(23), 9890–9904. DOI: 10.1002/ece3.3499.

Thompson, K.A., Renaudin, M. and Johnson, M.T.J. (2016) Urbanization drives the evolution of parallel clines in plant populations. *Proceedings of the Royal Society B: Biological Sciences* 283(1845), 20162180. DOI: 10.1098/rspb.2016.2180.

Tran, H., Uchihama, D., Ochi, S. and Yasuoka, Y. (2006) Assessment with satellite data of the urban heat island effects in Asian mega cities. *International Journal of Applied Earth Observation and Geoinformation* 8(1), 34–48. DOI: 10.1016/j.jag.2005.05.003.

Tylianakis, J.M., Didham, R.K., Bascompte, J. and Wardle, D.A. (2008) Global change and species interactions in terrestrial ecosystems. *Ecology Letters* 11(12), 1351–1363. DOI: 10.1111/j.1461-0248.2008.01250.x.

Wienert, U. and Kuttler, W. (2005) The dependence of the urban heat island intensity on latitude–a statistical approach. *Meteorologische Zeitschrift* 14(5), 677–686. DOI: 10.1127/0941-2948/2005/0069.

Youngsteadt, E., Dale, A.G., Terando, A.J., Dunn, R.R. and Frank, S.D. (2015) Do cities simulate climate change? A comparison of herbivore response to urban and global warming. *Global Change Biology* 21(1), 97–105. DOI: 10.1111/gcb.12692.

Youngsteadt, E., Ernst, A.F., Dunn, R.R. and Frank, S.D. (2017) Responses of arthropod populations to warming depend on latitude: evidence from urban heat islands. *Global Change Biology* 23(4), 1436–1447. DOI: 10.1111/gcb.13550.

Zhan, W., Ju, W., Hai, S., Ferguson, G., Quan, J. *et al.* (2014) Satellite-derived subsurface urban heat island. *Environmental Science & Technology* 48(20), 12134–12140. DOI: 10.1021/es5021185.

Zhang, P., Imhoff, M.L., Wolfe, R.E. and Bounoua, L. (2010) Characterizing urban heat islands of global settlements using MODIS and nighttime lights products. *Canadian Journal of Remote Sensing* 36(3), 185–196. DOI: 10.5589/m10-039.

Zhao, L., Lee, X., Smith, R.B. and Oleson, K. (2014) Strong contributions of local background climate to urban heat islands. *Nature* 511(7508), 216–219. DOI: 10.1038/nature13462.

Zhou, B., Rybski, D. and Kropp, J.P. (2013) On the statistics of urban heat island intensity. *Geophysical Research Letters* 40(20), 5486–5491. DOI: 10.1002/2013GL057320.

Zhou, B., Rybski, D. and Kropp, J.P. (2017) The role of city size and urban form in the surface urban heat island. *Scientific Reports* 7(1), 4791. DOI: 10.1038/s41598-017-04242-2.

Ziska, L.H., Gebhard, D.E., Frenz, D.A., Faulkner, S., Singer, B.D. *et al.* (2003) Cities as harbingers of climate change: common ragweed, urbanization, and public health. *Journal of Allergy and Clinical Immunology* 111(2), 290–295. DOI: 10.1067/mai.2003.53.

9 The Green Cities Movement: Progress and Future Challenges

John G. Kelcey*

Prologue

In writing this chapter, all I can do is to commit my thoughts to paper in the hope that something intelligent will spill onto the page, whether it enrages the reader or not; it is for him or her to determine whether to read on, or not and if the former, to question it, as every reader of everything should do. What is a green city and who determines whether it is or is not and on what criteria? The chapter is divided into three parts (topped and tailed by a prologue and an epilogue). Part 1 discusses the importance of language; Part 2 (How have we got to where we are?) considers the green environment in its widest terms; Part 3, which is biased towards the role of the biological sciences, describes the challenges (What of the future?).

This author does not like the use of abbreviations but, regrettably, it has become essential for practical reasons to abbreviate Green City(ies) Movement(s) to GCM. The GCMs have been subdivided into those concerned with the abiotic environment and those concerned with the biotic environment; they are distinguished in the text with the prefixes 'a', b' or 'a/b'.

Over the millennia, the creation and expansion of cities have undergone major changes in philosophy and practice within and between continents and countries. The urban environment has been and remains one of stark contrasts; the pinnacle of human achievement on the one hand and human deprivation on the other. Compare the affluent districts of Rio de Janeiro with the Favela in Rocina; Mumbai with Hollywood. The environmental quality of cities has been of concern for four millennia, mainly in relation to poverty, disease and urban cramming. A dispassionate assessment indicates that there has been little progress in achieving a high-quality environment throughout any city – spend two or three days walking around any of them and weigh up the proportion of the good and the bad.

Regrettably, the chapter is mainly concerned with the situations in western and parts of eastern Europe; the author is ashamedly ignorant of a/bGCM outside Europe, save for a limited knowledge of some environmental issues in parts of Asia, the Americas and Oceania. However, the application of general knowledge and experience suggests that the environmental issues in cities throughout the world are similar.

There is a tension running through the chapter – science and technology versus the humanities. It is a simplistic rerun of the 'Two Cultures'; a vitriolic debate in the 1960s between C.P. Snow and F.R. Leavis, in Britain. The former believed that the solutions to the human predicament were technological while Leavis argued that they lay in the humanities.

The Importance of Language

All disciplines have developed their own language (in pejorative terms, jargon) in order to

*johnkelcey@hotmail.com

establish sub-tribal boundaries, to provide a sense of identity and psychological protection – no more so than when it comes to environmental matters. It is pertinent to start with the definition of the three key words in the title of this chapter as given in the *Collins English Dictionary*: 'green' is given eight meanings, including area 'covered with grass, plants, trees' and 'of or concerned with conservation and improvement of the environment used in a political context'. Nine meanings are ascribed to 'movement' of which one comprises two sub-meanings – a group of people with a common ideology; and the organized action and campaigning of such a group. 'Cities' is given four meanings, including 'a large town' and 'a collective noun for an assemblage of people'. All three definitions have been interpreted and reinterpreted to suit particular interest groups to the extent that they have become of little value. For example, 'green' was originally used as an adjective (to describe the natural environment); it is now used as a noun ('I am a green'); the collective noun for just about everything; and a verb ('to green'). Then there is the futile exercise of defining a 'city' in quantitative terms. In this chapter, 'city' is used as a synonym for an area dominated by buildings and related developments.

Because 'green' has become a political, media, advertising/promotional synonym for the environment *sensu lato*, for the purpose of this chapter GCM has been divided into two aspects, the biotic (to do with natural history – the province of biologists) and abiotic (to do with pollution, noise, buildings, traffic etc. – the province of architects, accountants, lawyers, engineers, urban designers, chemists and many others). (The former appears as bGCM and the latter as aGCM or a/b for both.)

By and large, those involved in the aGCM have professional institutions and a high level of understanding, respect and credibility for each other. Regrettably, this does not extend to bGCM, mainly because of a combination of the lack of professional recognition and understanding and the strong association with pressure groups and environmental extremism. In general terms, the professions (as a whole) are deeply suspicious of the objectives and motivations of the bGCM. For example, a highly qualified and experienced ornithologist castigated an environmental assessment because he had no knowledge or understanding of most of the technical terms used by the planners, engineers etc. Conversely, the professions make assumptions that they are not entitled to make – an engineer included a clause in a contract requiring the protection of the 'flora and fauna'; when asked for clarification he replied, 'trees and shrubs'.

The professions fail to understand the differences between ecology and wildlife conservation, why a taxon has several different scientific and common names, why non-native taxa are *persona non grata*, why creating or translocating habitats (especially the latter) are considered to be heretical and why wildlife conservationists dislike people and are obsessed with nature reserves. The same principle applies to terms such as eutrophication, hydroseres, rarity, species richness, relative abundance, etc. On the other hand, biologists involved in the development processes have little understanding of the 'jargon' used by the other professions – for example VOs, PD, Triples, Arisings, Drotts, Caterpillars and Bills of Quantity. If progress is to be made in correctly determining the impact of developments, and creating and maintaining a high-quality urban environment, it is essential for the professions to learn and understand each other's languages.

Some terms are frequently, inconsistently or incorrectly used as synonyms by all of the professions and the a/bGCM; for example, Open Space, Green Space and Amenity Space. Open space may be devoid, or more or less devoid, of any vegetation. Wildlife, green, biological and other names are used synonymously for 'corridors' (also called conduits), although they all have different (if overlapping) functions with different physical and biological characteristics. The professions are baffled when wildlife conservationists and ecologists argue *inter alia*, that the linear integrity of corridors is sacrosanct when it is evident that they are discontinuous and the scientific validity of the theory has yet to be proven.

How Have We Got to Where We Are?

To begin at the beginning, with four essential books: Geoffrey and Susan Jellicoe's *The Landscape of Man*; *6000 Years of Housing* by Norbert

Schoenauer; *Modern Environmentalism* by David Pepper, published in 1996; *Environmentalism – A Global History*by Ramachandra Guha, published in 2000. Interestingly, none of them is a biologist.

All cities were built *de novo*, usually on trade routes, such as adjacent to rivers. It appears that the golden age of the creation of cities in Europe, Asia and north Africa was between the fourth millennium BC and the 15th century, with short, localized bursts in the 18th and 19th centuries in Canada, USA, Australia and New Zealand. Since that time most urban development has been related to the expansion of existing cities or reconstruction following destruction. A comprehensive account of the GCM that have, or are likely to have, occurred in the past 6000 years cannot be contained in this chapter; consequently, the 'movements' are considered in a series of 'stepping stones' of some of the more important stages.

It took more than 50,000 years from the emergence of hominids out of Africa into the Middle East before enterprising *Homo sapiens* perceived the benefits of establishing permanent settlements, initially in Mesopotamia. Other cities began to be built and expanded as people gravitated to them, while others colonized Europe, north Africa, Asia and the rest of the world. That the Sumerians were building 'great' cities in the mid-centuries of the fourth millennium BC suggests that urban development started much earlier, maybe in the fifth millennium BC. Prosperous urban civilizations are reported from Egypt about 3100 BC and from the Punjab probably towards the end of the fourth millennium BC. The first known designed landscapes and evidence of horticulture occurred in the fourth millennium BC, presumably contemporaneously with the building of the cities. Uruk (built in Mesopotamia in the third millennium BC) is described as comprising one third city, one third garden and one third fields. Although Babylon was also established in the third millennium BC, the Hanging Garden (maybe the first roof garden), which cascaded from a building approx. 23 m high, was not created until the middle of the first millennium BC. The Phoenicians founded many coastal cities in the second millennium BC, Cadiz created in the early part of the first millennium BC.

The next stage is the zenith of classical Greek urban development (which was extensive in time and space) starting *c*.900 BC and lasting to *c*.600 BC, about 4000 years after the construction of the first known city. Athens had become a significant city during the 8th century BC. During the 4th century, Hippodamus (498–408 BC), a polymath in today's terms and the pioneer of urban planning in Europe, redesigned the reconstruction of Miletus and Athens (among other cities) following their destruction by the Persians, in 494 BC and 480 BC. Many later philosophers and political scientists including Plato (428–348 BC) (*The Dialogues* (including *The Republic*)) and Aristotle (384–322 BC) (*The Politics*) wrote about cities, mainly in respect of their political and social structures and the values of people rather than the landscape and natural history. However, it is fair to state that the presence of plants and animals and the quality of the environment in cities is entirely dependent on political and social structures and values.

Most, if not all, the Greek philosophers and polymaths were born, lived and died in cities, which suggests that they are likely to have been knowledgeable about, interested in and appreciated plants and animals. This assumption is supported by four example. First, Aristotle and his friend Theophrastus (371–287 BC) made extensive studies of natural history – Aristotle in relation to zoology and Theophrastus in relation to botany (*Historia Plantarum*). Second, it is known that during these times gardens were being created for aesthetic purposes and the production of fruit, medicinal herbs and spices and to attract bees. Third, John Raven (1914 - 1980), a former classics scholar in the University of Cambridge and an excellent botanist, found 800 plant species being referred to in Greek literature. It is assumed that some form of the bGCM had been established and was growing. Fourth, the extensive use of images of plants and animals in decorations, for example ceramics.

Starting during the middle of the first millennium BC (or thereabouts) the planning and development of cities moved imperceptibly from the Greeks to the Romans (the Greek-Roman Transition). Rome was created in 753 BC and was followed by the creation of many more cities as the Romans colonized and occupied a large part of Europe and elsewhere. It is likely that most, if not all, the cities and even villas in rural areas were associated with the creation of gardens, the planting of trees and general landscaping.

During the second century BC, Marcus Porcius Cato the Elder (234–149 BC) was writing about agriculture and related matters; at the turn of the century Marcus Terentius Varro (116–28 BC), writing his *Treatise Rerum Rusticarum* (37 BC) on similar subjects as well as architecture. In the first century BC Virgil (Vergil) (70–19 BC) published *The Georgics*, a poetic account of agricultural management. Agriculture is a branch of natural history and in those days was not distinguished from it. The large number of plant and animal species listed in both books, and the references to gardens, indicate a considerable awareness and knowledge of them, suggesting that they were not confined to rural areas and that many (maybe all) were likely to occur in cities. In his book *The Ten Books of Architecture*, which was also written during the first century BC, the Roman architect Vitruvius set out 12 criteria for identifying a site for the development of a city.

It appears that the book was timely because during the first century AD and subsequently, extensive areas of Rome were frequently destroyed by fire. One (or more) of the fires is reported to have been attributable to Nero; in AD 64, a fire destroyed or seriously damaged ten of the 14 districts of the city. Whether started deliberately or otherwise, it was beneficial in removing poor-quality housing, disease and providing opportunities for a comprehensive programme of rebuilding, as with many cities before and after. In the same century, Pliny the Elder (AD 23–79) produced his 37 volume *magnum opus*, *Historia Naturalis*. Books III–VI consider cities; Books VIII–XI are concerned with animals (including reptiles, birds and insects) while Books XII–XXV are concerned with plants (including trees, crops and garden plants), animals (including vertebrates and invertebrates) and gardens. The reconstruction of the extensive gardens associated with the villa of his nephew Pliny the Younger (AD 61–113) indicates the use of a large number of plant species and the animal species attracted by them. It is reasonable to assume that similar situations existed in cities.

By the third century AD, Rome had expanded to include green spaces while wealthy citizens established gardens and many parks used plants collected throughout Europe. The literature reports the occurrence of many animal species, presumably attracted by the trees, shrubs and herbs, or brought with them. Rome suffered a serious decline in its population from a peak of 1.5 million to 50,000 by the mid-fifth century AD. The 80% (or thereabouts) reduction would have resulted in opportunities to increase the amount of green space. Was it exploited by the a/bGCM of the day? We will never know.

Similar catastrophic events were to occur in London, Copenhagen and many other cities throughout the world during the ensuing centuries, resulting in similar opportunities for redevelopment and environmental improvements, although the populations at the time would not have seen it that way. These are examples of the sequence of creation–destruction–creation that has occurred since time immemorial and is repeated in the arts, religion (Christmas and Easter of Christianity), astrophysics (see the writings of Fred Hoyle) and the fundamental principle of biology – sex (creation) and death (destruction) which the Irish poet W.B. Yeats said are the only matters that should concern artists.

Following the demise of the Roman Empire there was a cessation of urban development followed by a massive renaissance during the Middle Ages. In the late 13th century, the British King Edward I ordered 24 towns to elect men who knew how best to design new towns. It assumed that the developments were associated with some form of landscaping while religious houses established physic gardens. The hunting grounds and palaces with extensive gardens established by the aristocracy (and to a more modest standard, by merchants) during the 13th–16th centuries, relatively close to cities, were eventually incorporated into them. The pattern continued throughout the Renaissance and Baroque periods when the affluent created elaborate gardens containing many different habitats, plants and probably by default, animals. It is likely that the aristocracy and merchants were stimulated by the visual aspects of their estates and motivated by 'one-upmanship'.

In his book *Utopia*, published in 1516, the lawyer Thomas More described 'Utopia' as a large island comprising 54 towns all of the same size and design and a minimum distance of 24 miles (39 km) apart, set in agricultural land. The 'urban green issues' *per se* being confined to the provision of large gardens with lawns and the growing of fruit trees and flowers. More describes the inhabitants of his Utopia as keen gardeners who enjoy the fertility and beauty of their gardens and hold competitions of the best-kept gardens. As with the Greeks and Romans, he was primarily concerned with social issues – behaviour, politics, health, food production, commerce and recreation.

The beginning of the 16th century saw the colonization of South America and elsewhere by Europeans, a practice that Europeans inflicted on many countries during the ensuing centuries in pursuit of resources without any humanity, ethics, morality or consideration for the quality of the environment of the cities they destroyed. At the beginning of the 18th century Peter the Great founded St Petersburg by draining an extensive area of marshland adjacent to the Gulf of Finland creating canals, boulevards, parks and other 'natural' features as well as superb architecture. Was he an environmental vandal or a visionary?

In the mid-19th century, Napoleon III approved the demolition of a large part of Paris, mainly to facilitate the movement of troops; it provided his architect Georges-Eugene Hausmann with an opportunity to create boulevards, parks and gardens. As with previous emperors, monarchs and other dictators, could and should, they have claimed to be protagonists of the bGCM? It is interesting to speculate on the reactions of a biotic green lobby (had one existed) to these and previous events and what would have resulted if a formal environmental impact assessment had been carried out?

The late 19th and early 20th centuries were bridged by large 'clusters' of pioneers and innovators in town planning and architecture (including the landscape) – a golden age to equal ancient Greece. They include Frederick Olmsted (1822–1903), William Morris (1834–1896), who founded the Arts and Crafts Movement in England in the 1880s, Gertrude Jekyll (1843–1932), Ebenezer Howard (1850–1928), Patrick Geddes (1854–1932), Frank Lloyd Wright (1867–1959), Edward Lutyens (1869–1944), Mies van der Rohe (1886–1969) and Le Corbusier (1887–1965).

Although most had no formal training in town planning, architecture or the biological sciences, they were major influences in urban development worldwide. Two British examples illustrate the point. First, Geddes was a botanist and zoologist (with strong views about specialization) before becoming a sociologist and concerned with the poverty, disease, pollution and human behaviour he first experienced in Edinburgh and which he took practical steps to improve before planning new urban developments in Scotland, India and what is now Israel.

Second, Howard was at various times a farmer, journalist and clerk, who became concerned with the poor living and working conditions and the need for radical social reform. Howard's approach was firmly based on a capitalist system including commercial and economic considerations. The second edition of his seminal work was published in 1902 as *Garden Cities of Tomorrow*. Howard's ideas were manifested by the creation of two garden cities, Letchworth and Welwyn (both in Hertfordshire, England) in 1903 and 1920, respectively. Two other extremely important issues occupied Howard's thoughts: first, the continued movement of people into cities, which he tried to stop and reverse; and second, that garden cities could not be built via the democratic process; in other words, 'If you want something done, democracy is not the way to achieve it' (a/bGCM please note). Regrettably, a dispassionate and thorough assessment of whether Letchworth and Welwyn and all the British new towns that were to follow have achieved and maintained the aspirations set out for them has not been carried out, except for the ecology of Milton Keynes, which the author monitored over ten years.

Olmsted was the major influence on urban landscape design in the USA in the late 19th century responsible for designing many important projects, especially in relation to parks. For example, he was instrumental in the creation of Central Park in New York, connecting isolated parks with green space and shared with Geddes a dislike of the grid system on which cities had been built for millennia. However, Olmsted proposed to develop a substantial area of marsh in Boston into a park. It is interesting to speculate what the bGCM would have thought of that compared with their current worship of him.

The turn of the 19th century also saw the emergence of two political movements, which considered the solutions to the health, poverty, working and living conditions in cities in political/revolutionary terms, although they differed in the approach and the means. Friedrich Engels and Karl Marx followed one course while the

Third Reich followed another – both missed their way and developed into two opposing extremes of political theories.

In 1888 the Lever brothers started to build Port Sunlight (England), a model village (the design was influenced by William Morris) to accommodate the workers from their soap factory. Five years later the Cadbury family started to build Bournville, a model village on the outskirts of Birmingham, England, to alleviate the 'evils' of the conditions in which their workers were living. Another major contributor to the history and understanding of the planning of cities from the earliest times to the mid-20th century was Lewis Mumford (1895–1990) who started his career as a sociologist before abandoning it to become a historian and journalist. He was greatly influenced by Geddes and like him, was neither a planner nor an architect, although he wrote two standard works on urban development, *The Culture of Cities* (1938) and *The City in History* (1961); both give little consideration to biotic green issues.

The wars of 1914–18 and 1939–45 resulted in the destruction of many cities, mainly in Europe, including Dresden and parts of London. As in previous millennia, these events provided considerable opportunities to redesign the whole or parts of the cities that were affected. Where were the bGCM? The approach to the restoration of the destroyed areas varied within and between countries. For example, following 1945, three types of redevelopment were adopted in Germany: (i) reconstruction based on the old city plans; (ii) the construction of a new modern city with wide streets and large housing areas; and (iii) in the German Democratic Republic (Eastern Germany) many of the cities did not have the money for a complete reconstruction; consequently, the damaged buildings were demolished and the land converted to green space rather than housing, with high-rise flats being built around the fringes.

In order to replace the loss of housing employment, areas destroyed by warfare slum areas in the major cities and to provide for population growth, the British government, influenced by Howard's Garden City Model, established what became known as the New Towns Movement, which resulted in the creation of 32 new towns throughout Britain between 1946 and 1967, although all but the last are based on the expansion of existing urban developments. The government accepted Howard's thesis that new towns could not be built via the normal democratic processes and therefore established development corporations with draconian powers to 'get on with and complete the job within 20 years'. It reinforces the view that projects can only be achieved by 'dictators' of one form or another. The major feature of the planning and design of the new towns was the creation of a high-quality environment, essential housing and employment areas set in extensive landscaping, including the planting of a large number of trees complemented by parks and incidental green space. The master plans were prepared by planning consultants and to a lesser extent, engineers, without any signs of the a/bGCM, save for foresters and landscape architects.

In 1951, the German government established the Garden Festival Movement, ostensibly to finance the restoration of derelict industrial land to provide parks and incidental green spaces. It turned out to be a means of servicing and improving unused (i.e. derelict/waste) land for industrial development. The model was adopted by other countries (mainly The Netherlands and Britain) before disappearing without trace and relatively little sign of a green legacy. Andrew Theoka's book *Grounds for Review – the Garden Festival in Urban Planning and Design*, provides an excellent assessment of the now defunct movement.

It is instructive (in terms of the a/bGCM) to provide a brief consideration of the planning, design and construction of two capital cities that were built during the late 20th century, namely Brasilia (built between 1956 and 1960) and Abuja (designed, including McHarg's Practice, in the 1970s and mainly constructed in the 1980s). In both cases the environmental considerations appear to relate mainly to culture, buildings, cars and recreation, with tree belts being used to screen the apartment blocks from each other.

Eugene Odum's classic work *The Fundamentals of Ecology*, published in 1959, contains a chapter on human ecology, which he claims brings together ecology, sociology and anthropology and contains references to ecological studies of cities in the 1920s, 30s and 50s. He includes a quote from a review of the ecology of American cities made by Calvin Schmid in 1950: 'The ecological structure of the large American city conforms to a consistent and regular pattern in which the socio-economic status of the population is a dominant feature.' Odum then goes on to comment that ecological principles have important applications in city planning, a plea that has been largely ignored by ecologists, planners and the other professions involved.

Even by the mid-20th century environmental concerns in cities were still largely confined to areas associated with poverty, working and living conditions, health and pollution. By the mid-1960s, some scientists, landscape architects and ordinary people were becoming more aware of and concerned about specific environmental issues, such as air pollution and wildlife conservation, giving rise to what are now called 'single-issue focus groups'. Politicians (and their officials), planners and sociologists (among many others) were aghast, unable to cope with the issues but more importantly, the solutions. Consequently, such people were labelled as radical socialist extremists and cranks bent on unsettling the status quo who had to be stopped at all costs.

The cause was and remains a fundamental conflict between people with different values and aspirations. The following three approaches illustrate the point, which is mainly related to cognitive psychology:

1. The bGCM prefer urban green spaces to be mainly 'wildernesses' dominated by long grass, scrub and dense tree planting, which attract many species of birds, mammals and invertebrates.
2. The police, politicians and many other people prefer short, mown grass because it is neat and tidy, does not attract litter and does not give an impression of neglect, is 'pest'-free, provides good sightlines, is not a safety and security hazard and is easier and cheaper to maintain.
3. The preference of most people lies between the two extremes; they like parks and parkland with individual trees or groups of trees and shrubs that have attractive shapes, flowers, foliage and fruits set in short, mown grass (lawns). The taxa, which are mainly non-native (at least in most of Europe), are strongly disliked by the bGCM who assert that they support significantly fewer animals when compared with the larger number associated with native species, which are generally much less visually attractive. The theory has yet to be tested by rigorous scientific studies.

Unfortunately, instead of forming a united band, the bGCM became highly fragmented, dissipating their energies, resources and influence and therefore their effectiveness. They became infected by the specialization that had infected their predecessors; sub-tribes were spawned giving birth to sub-sub-tribes and sub-sub-sub-tribes and sadly, to eco-terrorists, such that the police have had to deploy undercover officers to investigate their activities. The epidemic of specialization has had and is having, a serious detrimental effect on achieving a high-quality environment in cities.

The 1960s saw the emergence of three 'modern' pioneers of different aspects of the bGCM – the academic scientists Herbert Sukopp and Maciej Luniak in Germany and Poland, respectively and the academic and practising landscape architect Ian McHarg in the USA. The researches of Sukopp, Luniak and their colleagues have been instrumental in encouraging the study of urban ecosystems throughout the world. The comprehensive surveys of the plants and animals of Warsaw, carried out in the 1980s the later surveys of Skaryszewski Park (50 ha) that were carried out by 40 scientists in 2014–15 and the surveys of Pole Mokotowski Park (60 ha) by about 30 scientists in 2016, are particularly impressive.

McHarg's seminal work *Design with Nature* has had a major international influence on landscape design and has spawned similar publications in other countries, including Ian Laurie's *Nature in Cities*, published in 1979. McHarg had a much easier task than Sukopp and Luniak; understanding the structure and function of the urban ecosystem is much more difficult, complex, geographically restrictive and time-consuming than landscape design, which has no geographical constraints and only requires an office, a drawing board and a flair for design. In addition, politicians, the professions and people can comprehend the visual landscape while failing to understand its structure and function.

Another major contributor to the a/bGMC was the Australian academic and polymath George Seddon, who at various times held professorships in English, Geology, Environmental Science and History and Philosophy of Science, eventually becoming the Director of Environmental Studies and the Dean of Architecture and Planning in the University of Melbourne. As so many of his predecessors in the urban development business, he did not have any formal training in biology, planning or architecture. He and his colleagues prepared many reports for the planning and design of cities, including an open-space system for Canberra in 1977 and the Middle-Ring Suburb in Melbourne in 1979.

Green events were also moving in China. In the 1970s, several new towns were built in the New Territories in Hong Kong during the time it was leased to Britain. This was followed by the creation of 310 National Garden Cities throughout China between 1992 and 2016. Although many landscape architects were employed in the New Territories, it is not known (to this author) whether they or any of the 310 Garden Cities were influenced by the a/bGCM and/or employed ecologists. It is also instructive to examine the environmental consequences of three major events that produced markedly different results and which none of the a/bGCM appear to have played a part.

An earthquake that affected Bucureşti in 1977 provided Nicolae Ceauşescu, the then president of Romania, with an excuse to demolish part of the centre of the city in order to build a monumental 'palace' set in extensive parkland. The demolition of 7 km² of houses, shops, factories and other buildings and the displacement of about 40,000 people started in 1982, construction in 1984 with practical completion in 1997. There are 100 ha of parkland within 1 km radius of what is now known as the Palace of Parliament. The creation of an enormous building set in such a large area of green space in the heart of a major city could only be achieved by a megalomaniac dictator, as indeed has been the case in the past. It is an objective that the bGCM could not have hoped to achieve and probably dared not even suggest. It raises the questions of whether Ceauşescu should be considered to be a major contributor to it and whether the present a/bGCM would applaud his achievement; in environmental terms, which is preferable, pre-1982 or post-1997?

As a consequence of general political uprisings in central and eastern Europe, the Berlin Wall, which included a strip of land up to 100 m wide, was demolished. During the 30 years of its existence, the green space between the walls had become of considerable interest for wildlife, arguably the best protected nature reserve in the world. The demolition provided exciting opportunities for the retention of areas of wildlife importance and opportunities for the creation of parks and incidental green space. Regrettably, the opportunities were not exploited, the land being used for dense commercial, high-rise developments, which involved, among others, the British architect Richard Rogers, whose thoughts on urban development are discussed later in this chapter. Where were the a/bGCM? In contrast to what happened in Potsdamer Platz, there was a popular uprising when it was proposed to develop 18 ha of railway land no longer being used; the objections were so great that the proposal was dropped and the area designated a nature reserve (Natur-Park Sugelande).

The contribution that the various international agencies have made to the a/bGCM and their influence (if any) on the quality of the urban environment are subjects that need to be examined, although only briefly in this chapter. There are far too many of them to consider individually; consequently, three have been chosen for illustrative purposes.

1. In the early 1990s, the UN Development Programme and the UN Centre for Human Settlements (Habitat) established a project to examine the environmental degradation in the metropolitan areas in three countries including Brazil. The latter project included a study of Rio de Janeiro by the World Bank, which was published in 1993.
2. The IUCN established a Specialist Group on Cities and Protected Areas in 2005 with the aim of increasing and managing protected areas in cities and encouraging communications with institutions such as museums and zoos.
3. The Convention on Biological Diversity, which was established in 1992, was to examine the quality of the urban ecosystem. In 2007, the Executive Secretary of the

> Convention on Biological Diversity stated that the battle for life on earth will be won or lost in the urban area.

The effectiveness of these organizations can be tested by one question: 'Have they made a significant practical contribution to the quality of the environment of cities? Put another way, have the environment and people of cities benefitted substantially from their existence? The answer to the question is no; this author awaits Utopia or Armageddon with interest.

During the latter part of the 20th century, wildlife conservation became the dominant environmental issue in rural and eventually, urban areas, a quasi-evangelical religious bGCM, whose main objectives were the maintenance of the status quo. It was and remains a matter of self-interest; the wildlife conservation branch of the bGCM grew exponentially and became a 'godsend' to people who feared that the value of their houses would fall dramatically if the unused land (so-called derelict, wasteland, brownfield sites) they overlooked was to be developed. Suddenly, house-owners become obsessively interested in wildlife, giving rise to a major conflict (wildlife versus housing for the underprivileged). The resolution of the issues became and still is, an adversarial process when it should be a dispassionate examination of what is the most appropriate use of resources for the benefit of the environment and people as a whole. Many of these issues are discussed by the political scientist Frank Fisher in his interesting and perceptive book *Citizens, Experts and the Environment*.

Also, in the early 1980s, the bGCM seemed to be on the verge of a breakthrough in relation to urban development – a glimmer of hope. Shortly after, she was appointed Chief Planning Officer for Liverpool, Audrey Lees established an ecological unit. Unfortunately, it did not survive long and faded away in 1981 when she became the Controller of Planning in the Greater London Council, where she established another unit; but that also disappeared when Margaret Thatcher (the then British prime minister) abolished the Council. The unit metamorphosed into the London Ecology Unit, which became an urban wildlife conservation pressure group funded by most of the London boroughs to oppose development proposals, until it was subsumed into the Greater London Authority to provide advice on strategic issues.

The histrionic and generally ill-informed views and actions of wildlife conservation pressure groups in a democratic society raises many serious issues in relation to what should be an objective and dispassionate approach – although they appear to be more effective. For instance, is it morally right and acceptable for the bGCM to use its influence on politicians and planning authorities to impose constraints (generally based on inadequate data and analysis) on the use of land and other resources that they do not own, bearing in mind that they could buy the land? A major difficulty is that the scientific inputs are not robust. For example, the biodiversity of a unit of land, whether 1.0 m² or 100 km², is not and never will be, known, while rarity cannot be quantified. The other major evaluation criteria are sociological and psychological, which are highly subjective and cannot be scientifically tested. In other words, people differ in their values and aspirations, which are determined by their socio-economic status. People living in poor-quality environments are not going to be concerned about wildlife but will be very concerned about other environmental factors such as noise, litter and pollution.

The bGCM has sought to impose arbitrary standards on the minimum amount of green space that should be provided. For example, in England, the government has accepted the Urban Green Space Standards devised by what is now Natural England (its publicly unaccountable statutory wildlife advisor), namely:

1. 1.20 ha per 100 people.
2. 1 × 500ha area within 10 km of every dwelling.
3. 1 × 100 ha area within 5 km of every dwelling.
4. 1 × 20 ha area within 2 km of every dwelling.

Many questions arise as to the scientific or other basis of these criteria because Natural England has no remit or expertise in town planning, cognitive psychology, economics or indeed many aspects of the biological sciences. Commonsense vindicates the assertion made by Ebenezer Howard at the beginning of the last century that high-quality urban development and renewal cannot be achieved democratically in any city throughout the world, especially where it involves extensive demolition of existing houses and employment areas – and at what cost in human and financial terms?

Two important but conflicting questions emerge: by objecting to it, is the bGCM responsible for the shortage of social housing in most cities? and, conversely, how detrimental has development been to the quality of the urban environment in general or wildlife conservation in particular? What price a home for the homeless versus what price a home for plants and animals?

The environmental situation in cities at the end of the 20th century has merged imperceptibly into the first decades of the 21st century, with no significant 'improvements' in sight. Thus, what follows is an assessment of what the a/bGCM should do in order to create high-quality environments in existing cities and those that may be expanded or built *de novo*. There is no doubt that many readers will have disagreed with the prognosis and will continue to do so with current prescriptions.

What of the Future?

The a/bGCM need to consider the relationship between two fundamental and sometimes conflicting issues – science versus the humanities, a rerun of the Snow versus Leavis debate. The author understands and largely agrees with both sides and concludes that neither can provide the solutions alone. The challenges will vary within and between nations, depending on the politics, economics and to a lesser extent, the values of the people, even in a democracy. Consequently, what follows is in the generality, disjointed and presented in no particular order.

Those who wish to pursue the environmental aspects of the planning, design, construction and management of cities are faced with many choices, including:

1. to continue the status quo, as numerous, fragmented pressure groups often are with conflicting ideologies and pursuing negative objectives;
2. to continue with the pursuit of small-scale projects including the establishment of small nature reserves, urban farms and allotments (which usually have educational and/or sociological objectives);
3. to act as specialist advisors in public authorities or commercial organizations and always subservient to the professions;
4. to pusue academic research (pure or applied); or
5. to follow their ambition to climb the professional ladder, to become a Chief Planning Officer, Director of Parks or Chief Executive, i.e. the boss.

The critical issue facing the 21st-century a/bGCM is urban cramming versus rural depopulation, which was first identified in the fourth millennium BC and has been increasing ever since with no signs of easing and exacerbated by government policies. There are three fundamental causes: economics (cities are rich, rural areas are poor); cities are more efficient in terms of the use of resources (although they may be totally dependent on rural areas); they are also the main home of culture (museums, art galleries, concert halls, sports stadia, rock concerts, cinemas and so on).

An important consequence is the issue of supply and demand; there is a 'surplus' of relatively cheap agricultural land in rural areas while there is a desperate shortage of land available in cities, resulting in land values being very high. There are at least five options for resolving the urban housing/green space issues:

1. Governments should stop movements into cities and reverse the trend – by repatriation, voluntary or forced.
2. The redevelopment of all available previously built-upon land (known as brownfield sites), although the development costs may be prohibitively expensive and therefore not commercially viable because of the presence of hazardous substances. The issue turns on land values.
3. 'Going up' – the construction of extremely high-rise apartments, offices etc.
4. Expanding into rural areas by the expansion of existing villages or construction of new developments, or both. Here lies a dilemma; in many cases the wildlife value of sites (large and small) in cities is of much greater

importance than that in rural areas, especially areas dominated by arable land.

5. Take steps to reduce birth rates. The corollary is to increase infant mortality and decrease life expectancy significantly. Are the bGCM up to advocating such solutions?

The better solutions are a combination of options 1, 4 and 5 but they are also likely to be the least politically acceptable. The most politically acceptable solutions are likely to be options 2, 3 and part of 4. Even these will result in vitriolic arguments between various interest groups. As alluded to earlier in this chapter, the enthusiasm of the wildlife conservation component of the bGCM (in both the public and private sectors) for continually seeking to persuade politicians to adopt restrictive policies and organizing negative opposition to all activities that disturb the status quo is not helpful in determining the solution, which can only be achieved by rational, dispassionate and objective analysis.

Another major issue and one that the a/bGCM neglect at their peril (although they have managed it so far) is the abolition of slums (*villa misserias*) and their replacement, *in situ*, by urban cramming or *de novo* development in rural areas. Which does the bGCM prefer, assuming it does not consider the status quo to be acceptable? But, there are two opposing views as to whether or not the abolition of slums is entirely desirable. First, a genuine concern that their replacement will have serious adverse consequences in fragmenting the social structure and destroying communities. Second, which is more cynical, is that slums are caused by people and that at least some people will turn new developments into slums quite quickly – regrettably, as it happens, there are examples in Britain and central and eastern Europe and it is reasonable to assume in many other countries also. It raises the philosophical question of whether politicians and sociologists should impose their values and aspirations on others and fail (if not refuse) to consult the people about what they would like.

Yet another major challenge facing a/bGCM and one that needs an urgent radical solution is 'specialization', an issue that concerned Geddes (among others) and which has become exponentially worse. For millennia, the humanities and sciences were as one but over the centuries they became separated. In the past, knowledge and interest in the sciences were broad-based with few specialists; now the pyramid has become inverted so that there are too many specialists and too few people with a broad knowledge. In writing this chapter, the author came across a discipline called 'palaeo-ethnobotany' – whatever next? Bring back natural history – all is forgiven. The problem is exacerbated because the top of the pyramids (inverted or otherwise) cannot, or are reluctant to, communicate with and understand the bottom. It is not that specialization is wrong in principle, only that there is too much of it. In some respects it is a matter of personal values, the wish to know a little about a lot (the author's approach) or a lot about a little, which has been the approach of biologists for far too long. The latter can be justified in terms of some aspects of research but not in relation to understanding the urban environment and the implementation of 'greening' where the preference is for knowing a little about a lot (the phrase should read 'quite a lot'). The specialist medic versus the GP is a reasonable analogy. Despite what Odum and others have written about the unifying concepts in ecology, in terms of urban development it is too specialized. Imagine the response of those involved in the design and construction of a new city or the environmental improvement of an existing one being told that they must employ at least one mammologist, herpetologist, lepidopterist, coleopterist, conchologist, ornithologist, algologist, bryologist, arboriculturist/silviculturalist, fisheries biologist, paleo-ethnobotanist, mycologist, plant taxonomist and freshwater, woodland and grassland ecologists, a soil scientist and probably at least another ten other disciplines and even some of their many sub-categories. The current solution to not knowing is to find out, usually within 24 hours, five working days at most. Solutions are not handed out on a plate. The theme runs through this chapter.

If the bGCM (biologists as a whole) are to establish a high degree of credibility, be accepted by the professions and be effective in achieving their objectives, they need to operate from a broad base of academic and practical expertise. Politicians and the professions wish to have advice and practical help from one person, not 20 or 30 ! On the other hand, the operators of plant and equipment have more respect if the person directing them knows what plant and equipment can and cannot do.

A radical change is needed to the tertiary education system if it is to produce the expertise that is essential if biologists and other scientists are going to have a direct influential and positive role in all aspects of the environment of cities. The following, which is modelled on the education of the medical professions, is only one of many possibilities. It involves the establishment of 'undergraduate' courses of six years of academic training in a wide range of disciplines selected from the sciences and humanities; for example botany, zoology, forestry, horticulture, genetics, chemistry, public health, medical science, sociology, cognitive psychology, philosophy, visual arts, landscape design, property valuation, geography, languages, anthropology and various aspects of architecture, engineering and law. This would be followed by, say, five years of professional practice (on the job) before becoming fully qualified, after which the person would be allowed to specialize.

Ecological research of urban ecosystems (to which this and the next paragraph are restricted) also need to undergo a massive transformation, first by dividing them into pure and applied. The need for pure research programmes is beyond question – a phrase that was once anathema to this author and one he never thought he would write. The conversion resulted from a conversation with a drainage engineer about an urban designer with whacky ideas. His response to the criticisms was: 'Such people are needed because if only one of many projects is realized, his employment will have been worthwhile'. A similar sentiment is expressed by Ramachandran in his book *The Tell-Tale Brain* – the need for scientists to speculate.

Notwithstanding the previous paragraph, priority should still be given to applied research, which also requires a radical shake-up. To put it kindly, the current disparate and unco-ordinated subjects and studies of the urban environment as reported in journals and books are far from helpful. Even the collection and analysis of the thousands of papers and books on the ecology of cities that have beenpublished are unlikely to assist putting the jigsaw of urban ecosystems together. It becomes even worse if the research in relation to sociology, engineering, urban design etc. are included, which they should be. Sadly, ecological researchers think they know what the disciplines want when they have little, if any, experience and knowledge of what is actually needed and the time in which it is needed. On the other hand, the professions rarely perceive what they need and should know; they need to be told. It is regrettable that ecologists and other disciplines involved in urban development never read, let alone are able to understand, papers and books written by the different professions, when they should do so.

The sad fact is that despite the pioneering and excellent research by Herbert Sukopp, Maciej Luniak and their colleagues and their disciples, very little is known about the structure and function of the urban ecosystem (past and present) of any city in any country. Without such information it is impossible for the bGCM to make any significant contribution to urban development. The solutions lie in a six-pronged approach, which are considered below, in no particular order:

1. Comprehensive ten-year research programmes following the Chinese economic plans that were started in the 1950s and the Ecopolis Programme in Pushchino (Russia).
2. The renaissance of the original claim for ecology, namely as a unifying concept.
3. Much improved communications and understanding, as discussed elsewhere in this chapter.
4. Production of (non-PR) reports on the state of the environment of cities, as is done in relation to economics and public health among others.
5. The establishment and substantial funding of an Institute for Urban Ecology in every country with a remit to be a thinktank, to commission and undertake research on ALL aspects of the urban environment, to provide a forum for national and international communications and to co-ordinate research between nations. It is not possible to compare within any ecological zone, let alone between two or more.
6. Monitoring changes in the structure and function of the urban ecosystem over time. We have no idea of what changes occur over any given period – 10, 25, 50, 100 years. Without this information it is not possible to

manage the wise use of natural resources effectively for people as a whole.

While governments are willing to provide the long-term funding of long-term research programmes in relation to space exploration, theoretical and astro-physics and the origins of the universe, with a few small-scale *ad hoc* exceptions they are unwilling to fund much-needed research in the structure and function of urban ecosystems.

As discussed earlier, the bGCM and the professions involved in urban development have not shown much enthusiasm to communicate with and understand each other technically or in terms of their respective aspirations – bridges need to be built, which can be achieved in many ways starting with individuals taking a broader approach (inherent or learned). The publishers and editors of some journals and books concerned with the different aspects of cities should include contributions from other disciplines in the sciences and humanities. The same principle applies to conferences, some of which have become incestuous. In short, there should be opportunities for the different disciplines to have their approaches questioned; challenging the status quo is a highly desirable activity.

The final example of mechanisms to encourage inter-disciplinary communications is the establishment of organizations with a remit to do just that – another uphill struggle. The Professional Institutions Council for Conservation was established in Britain in 1972 at the behest of the Duke of Edinburgh to provide a forum for the exchange of views and the promotion of understanding between 16 professional institutes and 42 quasi-professional organizations. The silence at its meetings became deafening and the antagonism palpable. It lost its purpose, became moribund and in the early 1990s faded away. Nevertheless, such organizations are urgently needed – yet another formidable task for the a/bGCM.

There are many considerable conflicts within the bGCM that need to be resolved. There is serious mutual antagonism between wildlife conservationists, foresters, horticulturalists and landscape architects, among others, all of whom have different approaches. For example, foresters like to increase the diversity of woodlands by planting non-native species while landscape architects have a considerable enthusiasm for planting non-native trees and shrubs anywhere and everywhere, even on sites of wildlife importance. In his 1981 book *The Ecology of Urban Habitats*, Oliver Gilbert (pre-empting Fred Pearce – see later) states that Japanese knotweed (*Fallopia japonica)* along the river banks of Sheffield is beneficial in mitigating water erosion, is valuable as a food source for many insect taxa and provides the canopy for woodland herbs where there are no trees. On the other hand, there is pressure from wildlife conservationists to remove it because it out-competes or suppresses herbaceous vegetation.

Many people who consider themselves part of the bGCM like to feed animals such as feral pigeons (*Columba livia f. domestica*), grey squirrels (*Sciurus carolinensis*) and seagulls (*Larus* spp.) and object in the strongest possible terms to being stopped, while others complain about the nuisance and public health hazard they represent. A similar situation applies to the feeding of ducks with bread and the consequential hyper-eutrophication of water bodies. There are vociferous objections from the bGCM about removal of dead, dying and diseased trees because they support fungi, insects and other organisms – science becomes *persona non grata*. Ironically, some parts of the bGCM encourage the feeding of birds in gardens, especially during the winter, which distorts the urban bird populations by giving a small number of species a considerable competitive advantage.

The internal conflicts continue in the management of nature reserves and other areas for wildlife. Invertebrate zoologists dislike birds because they eat insects while botanists dislike insects because they damage plants but there are even more philosophical issues that need to be resolved. For example, the main objective of wildlife conservation is to maintain the status quo and prevent the dynamics of biological processes, presumably on the basis that the known is preferable to the unknown – certainty versus uncertainty. A lot of the tenets of the modern wildlife conservation movement are scrutinized by the journalist Fred Pearce in his excellent, much-needed and thought-provoking book, *The New Wild – Why Invasive Species Will Be Nature's Salvation*. He makes many valid and important points such as drawing attention to the distinction between ecologists and the nature conservation movement and undertaking a critical assessment of the dangerous mythologies that have been applied as to how nature works including native versus non-native species and the premise that 'native' ecosystems will collapse when invaded by foreign species, which he states (correctly) has not happened so far and implies is unlikely to happen, a view shared by this author. He concludes that 21st-century conservation should not be about preserving nature in aspic or trying to recreate the past – 'the natural world is not a zoo', which nature reserves are. It comes as no surprise that Pearce has been labelled a heretic, which has been applied with considerable vigour to this author for much the same reasons and for advocating the creation of habitats, the translocation of plants and animals that would otherwise perish when a site is developed and for daring to question the approach, motives and activities of the wildlife conservation movement.

Epilogue

As a general rule, most, if not all, of the concerns and what little is done about them, are related to capital cities with other cities receiving little, if any, attention. Of course, many wildlife conservationists and city councils will disagree with the analyses described in this chapter and point to many green projects that have been undertaken in recent years. To an extent the author agrees but it is highly specialized and on a very small scale – grossly inadequate in relation to the scope and magnitude of the problems that need to be resolved.

In his 1995 BBC Reith Lecture '*Cities for a Small Planet*' the internationally renowned British architect Richard Rogers considered the ecological problems of cities to be traffic, highways, communications, noise, slum clearance, waste disposal, sociological and energy conservation. Biological issues did not get a mention. He implies, if not asserts, that the solutions are more or less exclusively the province of architects. Rogers identifies six attributes of the sustainable city without comprehending that cities are not sustainable but 'black holes' with an insatiable appetite for sucking in renewable and non-renewable natural resources and spewing out various forms of pollution and waste. The failure is confirmed in Attribute 4, in which he considers an ecological city to have a circular metabolism which gives as much to the environment as it takes out. In Attribute 6 he states that a beautiful city is where art, architecture and landscape move the spirit. However, Rogers gives no consideration to the green environment of cities and the role of biologists.

Rogers advocates the demolition of buildings that have passed their sell-by date and their replacement with developments that are more energy-efficient and have other environmental benefits. It is a principle that can be extrapolated to a whole or part of a city – following Nero and Napoleon III, among others. This implies that additional land will need to be provided in or outside the city to allow this to happen. The principle can be extrapolated to the whole of a city, which raises the question of whether or not there is a case for demolishing and rebuilding cities on say a 50–100-year cycle – a conundrum that should exercise the minds of the a/bGCM.

How architects get away with asserting that they have exclusive rights to urban development is a travesty. The bGCM should be up in arms, they are not but why not? Biologists have allowed themselves to become seriously marginalized; for five millennia town planning has been the province *inter alia* of philosophers, architects, dictators, journalists and administrators with no expertise in the natural sciences. Regrettably, there are very few (if any) ecologists employed in those departments of central and local governments concerned with urban development and the management of cities, let alone being chief executives.

Nevertheless, Rogers is right, fundamentally; cities are about sensuality in all its forms. However, the priority that is given to each of the senses individually or in combination is a matter for individuals and the population corporately – the province of the neurologist and cognitive psychologist. It is the brain where the appreciation of aesthetics, values and issues of morality and ethics occurs (see Ramachandran's *The Tell-Tale Brain*). For example, as to whether an area, habitat or species is of such value to society as a whole to overwhelm the provision of housing for those living in slums or who are homeless and many other accompanying environmental benefits, these are matters that a/bGCM must grapple with.

To reiterate a point made earlier in this chapter and which cannot be repeated often enough, it is an international and national disgrace that we know and politicians care more about, outer space and the origins of life than we do about the structure and function of urban ecosystems in which 70% (and increasing) of the world's population lives, 50% of it in slums. Governments prefer to pour in billions of dollars, euros, pounds and other currencies for political kudos rather than the welfare of people and the quality of the urban environment. It is, at best, irresponsible. Outer space with no people versus cities with lots of people. The a/b GCM are silent. Why?

The greatest achievements of the aGCM have emerged from significant public and political pressure in relation to the improvements in air and water quality, the disposal of solid waste and to a much more limited extent, the suppression of excessive noise. In short, the public perceived the problems, got on the bus as it passed the door and demanded that the aGCM solve them.

Nevertheless, all is not well and there are many serious environmental problems to overcome, including air pollution. It is essential to convert combined sewers to separate systems, which is a very expensive exercise. However, environmental improvements may not always be beneficial. For example, improving water quality may reduce the productivity of an estuary resulting in a substantial reduction in the waterfowl population. The situation in respect of noise is even more interesting; the effect of noise levels is determined on a mathematical scale only. The dBs of a dawn chorus can be significantly higher than noise generated by mechanical sources. It results in conflicts within and between individuals; a naturalist may be opposed to, say, a particular level of traffic-generated noise but enjoy a much higher level produced by birds. The reverse also occurs – some people will tolerate traffic noise but not that produced by birds or livestock.

There is another aspect of cognitive psychology that the a/bGCM should examine in some detail: Why is it that people who live in cities in the northern hemisphere are attracted to and enthused about large mammals such as tigers, whales and elephants but are indifferent to and often dislike, some of the mammals they see in their locality? (although this author suspects that the residents of cities in which polar bears, tigers or elephants occur are not particularly fond of their presence).

It will have become more than apparent to the reader that three major issues emerge from this chapter. First, that the most influential people involved in urban development are dictators of one form or another, or 'professionals' with no formal training in town planning or the biological sciences. But does it matter, especially if biologists have opted out? Second, despite the huge number of reports of various kinds that are published by international organizations, national governments, numerous agencies, academics and voluntary organizations during many decades, very little of substance has been achieved in the way of understanding urban ecosystems and improving the environment of cities. Third, many (may be most) organizations that governments have established with the remit to improve the quality of the urban environment have disappeared, having achieved very little of permanent value. My detractors will consider this assessment to be negative and somewhat cynical but examine what has actually happened, qualitatively and quantitatively, in cities and the understanding of them.

As this chapter comes to a close it is important to ask whether the bGCM has made a significant contribution to achieving, in practical terms, a high-quality environment for the benefit of people as a whole in any existing or new city. The answer appears to be no, although it has raised public and therefore political awareness but with two diametrically opposed results. The bGCM seems incapable of taking practical steps to resolve environmental issues, preferring to hold demonstrations accompanied by ill-informed histrionics, which have tended to antagonize people and politicians. On the other hand, the cult of the personality has had significant positive effect; for example, at the end of a recent television programme on the oceans, the naturalist David Attenborough made a short comment on plastic pollution. The problem has existed for decades without any action being taken. Attenborough's comment immediately ignited a massive international campaign against the use of plastics and action to clean up the marine environment.

This chapter ends with another challenge, namely the design of a city. The bGCM should follow the examples of More, Howard and many others and plan and design (with an estimate of the total cost, consideration of the implications for construction and a construction programme) a new city for, say, a million people on a theoretical greenfield site and the upgrading of an existing city with a poor-quality environment to one with a high-quality environment and then submit their utopias for public examination.

10 Light Pollution: How Urban Conditions Influence Ecological Interactions

Iriana Zuria[1], Ignacio Castellanos[1*] and Pedro Barbosa[2]
[1]Universidad Autonoma del Estado de Hidalgo, Hidalgo, Mexico; [2]Department of Entomology, University of Maryland, College Park, Maryand

Urbanization poses many challenges to plants and animals. Many natural or human-induced factors, sometimes referred to as ecological drivers, directly or indirectly cause significant changes in ecosystem processes, and consequently in urban and non-urban ecosystems. However, the same drivers can unequivocally have very distinct influences in urban and non-urban ecosystems. For example, changes in climate associated with the heat island effect in urban environments can modify plant-flowering and insect herbivore developmental times, which in turn can disrupt pollination and predator–prey interactions. Similarly, the intensity and extent of variables such as noise, artificial light at night, and a variety of ecological traits and forces can, to a significant degree, influence what plants are available in urban environments and thus to herbivores. The problem is that the intensity and mix of colours emitted by artificial light sources do not change with the time of day, severely limiting the ability of artificial light to provide the necessary cues for synchronizing ecological events and interactions. Further, they can limit the growth and survival of the plants, as well as the nature and extent of resources use by extant herbivore species. Clearly, it would be unwieldy, if not impossible, to discuss the differences among all such factors in urban versus non-urban habitats. Thus, in this chapter we focus on what we term light pollution. We define the latter as the disruptive effects of the brightening of the night sky caused by street lights and other human-made sources.

Urbanization has resulted in a worldwide increase of artificial light at night (ALAN, Fig. 10.1). It is considered pollution because it alters the natural lighting levels, disturbing living organisms, altering the utilization of resources and flow of information in urban ecosystems, which can be significant in that it can reach areas that are hundreds of kilometres from its source (Longcore and Rich, 2004; Gaston *et al.*, 2013; Falchi *et al.*, 2016). Light pollution has been considered one of the most pervasive forms of environmental alteration (Cinzano *et al.*, 2001), which can have effects on terrestrial and aquatic ecosystems (Davies *et al.*, 2014), altering behaviour, physiology and survival of individuals of many species, and the species

*Corresponding author: ignacioe@uaeh.edu.mx

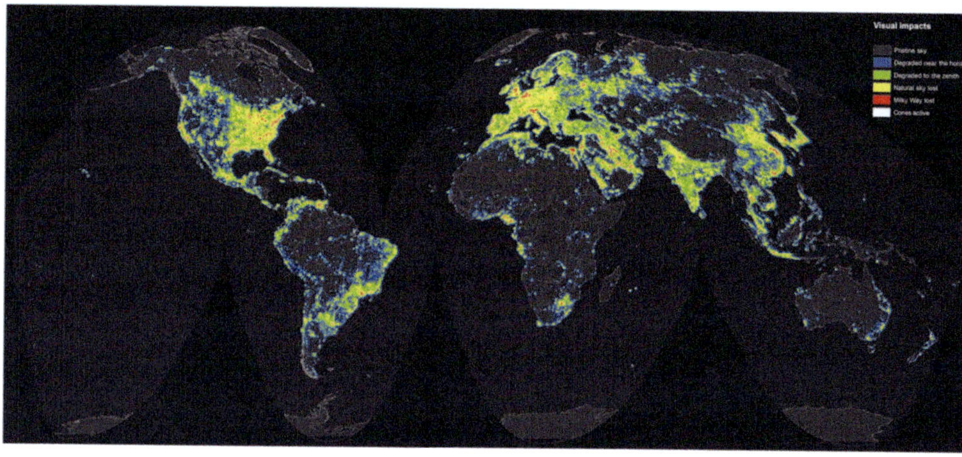

Fig. 10.1. Map of light pollution's visual impact on the night sky. (From Falchi *et al.*, 2016)

composition in an area. Further, other processes may be affected, such as primary productivity and the partitioning of temporal niches, which in turn can change population size and affect ecological interactions; changes that ultimately can cascade throughout the ecosystem, potentially exerting strong selection pressures on organisms living exposed to ALAN (Gaston *et al.*, 2013, Gaston *et al.*, 2015; Hopkins *et al.*, 2018).

Within urban and suburban environments organisms are locally exposed to artificial lightscapes which include isolated light sources, linearly lit features (i.e., roads illuminated by street lights), areas illuminated mainly by skyglow, and areas with different spectral compositions (Gaston *et al.*, 2015). Clearly, the nature and intensity of light depends on the type of light, proximity of the light sources, shadows created by buildings, topography and vegetation near light sources. Thus, light intensity and its spectral composition is highly heterogeneous within short distances (Bennie *et al.*, 2014). The source of ALAN also can be from moving sources such as road vehicle headlights (which typically are relatively high intensity light sources) (Gaston and Holt, 2018). Although some animals can move through the lightscape to reduce exposition (Gaston *et al.*, 2015), very little is known about how these complex lightscapes affect individuals, the dynamics of their populations, and ecological interactions.

Light pollution has been investigated and reviewed mainly with regard to its effects on individual organisms of a variety of species or particular groups, i.e. the susceptibility of populations, communities and ecosystems to an illuminated environment (Gaston *et al.*, 2015; Schroer and Hölker, 2016), including plants (e.g. Bennie *et al.*, 2015; Ffrench-Constant *et al.*, 2016), invertebrates (e.g. Seymoure, 2018; Owens and Lewis, 2018), insects (Desouhant *et al.*, 2019) and bats (e.g. Stone *et al.*, 2015; Rowse *et al.*, 2016), among others. However, until recently, the effects of ALAN in changing interactions between species and impacting communities were rarely analysed (Davies *et al.*, 2012; Plummer *et al.*, 2016; Sanders and Gaston, 2018).

Species interactions have evolved in response to natural lighting regimes, so any changes in behaviour or physiology as a result of living exposed to ALAN can affect interactions. It has been shown that trophic interactions can be influenced by ALAN; for instance, predator–prey interactions are often determined by light, as many prey species have evolved to be active at night to avoid diurnal predators or have particular behaviours at night to reduce risk of predation (Riley *et al.*, 2013). Although some studies have failed to find an effect of light pollution on predator–prey interactions (e.g. Grenis *et al.*, 2015; Firebaugh and Haynes, 2019), many studies have found negative effects. One of the most studied predator–prey interactions is the one that involves insectivorous bats and their prey. Many of these studies have focused on moths, which, besides being prey, also have other important functional roles as pollinators and in food webs as herbivores.

Moth behaviour is altered by night lighting. For example, moths subjected to ALAN spend less time feeding than moths in darkness, contributing to moth population declines in strongly illuminated areas, and affecting critical ecosystem functions (van Langevelde *et al.*, 2017). Also, ALAN can interfere with anti-bat defensive behaviours in moths. Thus, moth consumption by bats is low in unlit conditions, but increases six-fold under lit conditions despite a decrease in relative moth abundance (Minnaar *et al.*, 2015). The increase in moth consumption is a result of decreased moth defensive behaviour induced by light. LED street lights also reduce the anti-predator behaviour in moths, thus shifting the balance in favour of their bat predators (Wakefield *et al.*, 2015). Differences in prey consumption by bats in lit versus unlit sites can be species-specific, given that light affects the foraging behaviour of bat species in different ways (Cravens *et al.*, 2018). Similarly, artificial light can alter the foraging behaviour of fruit-eating bats. A reduction in nocturnal seed-dispersal activity can occur in lit areas, resulting in a lower likelihood that fruits will be harvested when plants are illuminated (Lewanzik and Voigt, 2014).

Artificial night light can significantly alter trophic interactions involving visually dependent species. Illuminated webs of the orb weaver spider (*Nephila pilipes*) had lower rates of moth interception than unlit webs (Yuen and Bonebrake, 2017). Thus, one should understand that environmental factors such as urban artificial light can complicate simple predictions about species interactions. Similarly, the abundance of certain species of slugs (*Arionidae*) increases in illuminated sites, possibly because of reduced predation and/or increased food quality in the form of carcasses of insects initially attracted by the light. As slugs in many parts of Europe are important pest species, the increase of slugs under ALAN can have important economic consequences (van Grunsven *et al.*, 2018).

Even in aquatic systems, ALAN can have significant effects on trophic interactions (Davies *et al.*, 2014; Zapata *et al.*, 2019). For example, exposure to ALAN along migratory routes may potentially impede the navigation that leads to successful migration of salmonids to natal habitats. Such natal habitats are critical because experiences with natal habitats can shape an individual's habitat preferences (Davis and Stamps, 2004). ALAN can also increase nocturnal activity of predominantly diurnal predatory fishes, increasing predation pressure on nocturnal fish (Becker *et al.*, 2013; Manfrin *et al.*, 2017), and can have negative effects on marine keystone predators, potentially modifying entire community structures (Manríquez *et al.*, 2019).

Increasing intensity of ALAN, as compared to moonlight, and the broadening of artificial light spectra, provide greater opportunities for predatory species to find and recognize their prey (Davies *et al.*, 2014). Localized population declines, or changes in community composition, can occur as prey species experience intensified predation pressure. The increased abundance of aquatic prey in areas illuminated by artificial light at night can lead to a dietary shift among consumers (Manfrin *et al.*, 2018) and can influence the behaviour of animals such that the balance of inter-specific interactions involved in community structuring may be affected (Underwood *et al.*, 2017). It has also been demonstrated that moderate-to-high levels of ALAN can alter invertebrate community composition in aquatic riparian food webs; cause changes favouring primarily predators and detritivores; and alter the flows of energy between aquatic and terrestrial systems (Sullivan *et al.*, 2019).

Plant–herbivore interactions are also affected by living exposed to ALAN. Street lights alter plant toughness and the size of larvae that feed on those leaves (Grenis and Murphy, 2019). Similarly, white LED lighting can significantly enhance the impact of grazing gastropods on epilithic microphytobenthos (i.e. on unicellular eukaryotic algae and cyanobacteria growing on the surface of rocks), so that consumers compensate for the positive effect of night lighting on primary producers (Maggi and Benedetti-Cecchi, 2018).

Some vertebrate and invertebrate predators in aquatic ecosystems can benefit from artificial lighting. The addition of ALAN on a bridge results in increased predation upon migrating Pacific salmon (*Oncorhnchus* spp.) by harbour seals (*Phoca vitulina*) (Yurk and Trites, 2000). Similarly, visual foragers increase their foraging effort in illuminated areas, i.e. areas with ALAN (Santos *et al.*, 2010). Exposure to artificial lights (yellow, orange and red lights) also can increase the number of ghost crabs that prey upon loggerhead turtles' (*Caretta caretta* L.) eggs and hatchlings, and the yellow light in particular produced a significant change in aggressive and prey-searching behaviours, suggesting that predation risk on loggerhead turtle nests by crabs may be greater under this light (Silva *et al.* (2017). Besides, exposure to artificial light decreased nesting attempts of loggerhead sea turtles by at least 20%, and disoriented the turtles. Living exposed to ALAN also increases the time that turtles spend on the nesting process, and light significantly disrupted behaviour. Other studies have also demonstrated that different types of lights affect predator–prey interactions in different ways (e.g. Tałanda *et al.*, 2018).

Urbanization can affect host–parasite interactions (Bradley and Altizer, 2007). For example, both parasite prevalence and intensity can increase in urban compared to rural individuals (Jiménez-Peñuela *et al.*, 2019; Majewska *et al.*, 2019; Thawley *et al.*, 2019), and may influence which host species are able to exploit urban habitats (Delgado-V. and French, 2012). However, the mechanisms responsible for differences in parasite loads between urban and rural hosts are not clear. Although a growing body of evidence links the presence of urban pollutants such as ALAN with physiological effects like depressed immune systems and increased risk of diseases (Navara and Nelson, 2007; Bedrosian *et al.*, 2011; Durrant *et al.*, 2020), no study has determined if ALAN influences the relationship between parasites and hosts along sites with different degrees of urbanization.

The impacts of artificial lighting can cascade through food webs. Both, top-down and bottom-up effects of ALAN have been investigated. For example, although no evidence for top-down control of a pea aphid in a grassland community was observed, evidence was provided for bottom-up effects mediated through the impact of light on the leguminous food plant (Bennie *et al.*, 2015). Both effects of living communities, exposed to ALAN, were found in a plant–aphid–parasitoid system. Light reduced the abundance of aphids as a consequence of reductions in bean plant biomass. In addition, parasitoids were negatively affected by the light treatment as a result of reduced host numbers and possible behavioural changes among plants, aphids and parasitoids (Sanders *et al.*, 2015). Predatory behaviour of fish increased under ALAN and as a consequence the sessile invertebrate assemblage structure was changed (Bolton *et al.*, 2017). Additionally, the intensity of light can have differential effects on organisms. For example, artificial light can increase the efficiency of parasitioid wasps attacking aphids, resulting in twice the parasitism rate of aphids under low light levels compared to that of unlit controls (Sanders *et al.*, 2018). Nevertheless, at higher light levels, parasitoid wasps spent more time away from the aphid host plants, thus diminishing this increased efficiency and resulting in increased aphid numbers under increased light intensities.

In general, artificial light at night is a threat to pollination because it disrupts nocturnal pollination networks and thus has negative consequences for plant reproductive success (MacGregor *et al.*, 2015; Knop *et al.*, 2017). Some have suggested that moths are very efficient pollinators and they can be severely affected by ALAN, impacting pollination interactions between moths and plants and resulting in significant reductions in pollen transport (MacGregor *et al.*, 2017). Indeed, in artificially illuminated plant-pollinator communities, nocturnal visits to plants can be reduced by 62% compared to unilluminated areas, resulting in an overall 13% reduction in fruit set of focal plants (Knop *et al.*, 2017). Further complicating the task of predicting the impact of ALAN on plant-pollinator communities is the fact that lamp type, lighting regime and distance from the light source all significantly affect different aspects of pollination biology (Macgregor *et al.*, 2019).

Unfortunately, most studies consider urbanization stressors singularly, e.g. they examine the effects of light independently of other factors like noise or heat pollution, even though these are frequently correlated (Votsi *et al.*, 2017) and multiple factors may be having an impact. However, the effects of single stressors can differ dramatically from the effects of a combination of urban stressors. For example, light and noise pollution can interact to disrupt host–parasite interactions between a frog (*Engystomops pustulosus*) and frog-biting midges (*Corethrella* spp.) in ways that differ if only light or only noise were being considered. Indeed, both light and noise pollution disrupt this host–parasite interaction and highlight the importance of considering interactions among species and each type of pollutant or urbanization stressor in order to accurately assess the impacts of urbanization on ecological communities due to each stressor. Thus, one could investigate and conclude that frog abundance was not significantly correlated with urbanization. However, if one considers multiple urban stressors, one would conclude that midges were sensitive to light and noise pollution, and that increased light intensity significantly reduced midge abundance at low noise levels, while at high noise intensity there were no midges, regardless of light level. Similarly, the combined effects of night warming and light pollution on predator–prey interactions can differ from the effects of singular factors (Miller *et al.*, 2017). Neither night-time warming nor light pollution changed the suppression of aphids by ladybeetles that forage in darkness. But for more-visual predators, warming and light had non-additive effects in which, together, they caused much lower aphid abundances. In addition, one stressor may reduce the negative effect of another stressor. For example, ALAN can reduce the toxicity of heavy metals on the plant litter decomposition process in aquatic ecosystems because light exposure increases litter biodegradability (Pu *et al.*, 2019). More research is needed that seeks to consider interactions among species and different type of pollutants acting on interacting species in order to accurately assess the impacts of urbanization on ecological communities.

Some management options have been proposed to reduce the ecological consequences of ALAN (Bradley and Altizer, 2007; Gaston *et al.*, 2012) and designed to prevent areas from being artificially lit, to limit the duration of lighting, to reduce the trespass of lighting into areas that are not intended to be lit, and to change the intensity and the spectral composition of lighting. However, the principal problem is how to maximize the benefits of artificial night-time lighting while limiting the ecological costs (Gaston *et al.*, 2015).

Recently, the rapid global increase in artificial light at night has been proposed to be a new threat to global biodiversity and ecosystem services. Light pollution has been investigated and reviewed mainly with regard to its effects on individual organisms of a variety of species or particular groups. In this chapter we focused on the effects of ALAN in changing interactions between species and its impact on communities and we identified some gaps in our knowledge. For example, there is a need for experiments where ALAN alone is experimentally manipulated while all other environmental stressors remain constant. Also, studies in ecosystems other than temperate are needed and there is also a need for studies with a wider breadth of taxonomic groups in response to ALAN in natural habitats. Finally, it is promising that our understanding of the effects of ALAN will increase promptly as attested by the rapidly increasing rate of published works on the subject.

References

Becker, A., Whitfield, A.K., Cowley, P.D., Järnegren, J. and Naesje, T.F. (2013) Potential effects of artificial light associated with anthropogenic infrastructure on the abundance and foraging behaviour of estuary-associated fishes. *Journal of Applied Ecology* 50(1), 43–50. DOI: 10.1111/1365-2664.12024.

Bedrosian, T.A., Fonken, L.K., Walton, J.C. and Nelson, R.J. (2011) Chronic exposure to dim light at night suppresses immune responses in Siberian hamsters. *Biology Letters* 7(3), 468–471. DOI: 10.1098/rsbl.2010.1108.

Bennie, J., Davies, T.W., Inger, R. and Gaston, K.J. (2014) Mapping artificial lightscapes for ecological studies. *Methods in Ecology and Evolution* 5(6), 534–540. DOI: 10.1111/2041-210X.12182.

Bennie, J., Davies, T.W., Cruse, D., Inger, R. and Gaston, K.J. (2015) Cascading effects of artificial light at night: resource-mediated control of herbivores in a grassland ecosystem. *Philosophical Transactions of the Royal Society B: Biological Sciences* 370(1667), 20140131. DOI: 10.1098/rstb.2014.0131.

Bolton, D., Mayer-Pinto, M., Clark, G.F., Dafforn, K.A., Brassil, W.A. *et al.* (2017) Coastal urban lighting has ecological consequences for multiple trophic levels under the sea. *Science of the Total Environment* 576, 1–9. DOI: 10.1016/j.scitotenv.2016.10.037.

Bradley, C.A. and Altizer, S. (2007) Urbanization and the ecology of wildlife diseases. *Trends in Ecology & Evolution* 22(2), 95–102. DOI: 10.1016/j.tree.2006.11.001.

Cinzano, P., Falchi, F. and Elvidge, C.D. (2001) The first world atlas of the artificial night sky brightness. *Monthly Notices of the Royal Astronomical Society* 328(3), 689–707. DOI: 10.1046/j.1365-8711.2001.04882.x.

Cravens, Z.M., Brown, V.A., Divoll, T.J. and Boyles, J.G. (2018) Illuminating prey selection in an insectivorous bat community exposed to artificial light at night. *Journal of Applied Ecology* 55(2), 705–713. DOI: 10.1111/1365-2664.13036.

Davies, T.W., Bennie, J. and Gaston, K.J. (2012) Street lighting changes the composition of invertebrate communities. *Biology Letters* 8(5), 764–767. DOI: 10.1098/rsbl.2012.0216.

Davies, T.W., Duffy, J.P., Bennie, J. and Gaston, K.J. (2014) The nature, extent, and ecological implications of marine light pollution. *Frontiers in Ecology and the Environment* 12(6), 347–355. DOI: 10.1890/130281.

Davis, J.M. and Stamps, J.A. (2004) The effect of natal experience on habitat preferences. *Trends in Ecology & Evolution* 19(8), 411–416. DOI: 10.1016/j.tree.2004.04.006.

Delgado-V., C.A. and French, K. (2012) Parasite–bird interactions in urban areas: current evidence and emerging questions. *Landscape and Urban Planning* 105(1-2), 5–14. DOI: 10.1016/j.landurbplan.2011.12.019.

Desouhant, E., Gomes, E., Mondy, N. and Amat, I. (2019) Mechanistic, ecological, and evolutionary consequences of artificial light at night for insects: review and prospective. *Entomologia Experimentalis et Applicata* 167(1), 37–58. DOI: 10.1111/eea.12754.

Durrant, J., Green, M.P. and Jones, T.M. (2020) Dim artificial light at night reduces the cellular immune response of the black field cricket, *Teleogryllus commodus*. *Insect science* 27(3), 571–582. DOI: 10.1111/1744-7917.12665.

Falchi, F., Cinzano, P., Duriscoe, D., Kyba, C.C.M., Elvidge, C.D. *et al.* (2016) The new world atlas of artificial night sky brightness. *Science Advances* 2(6), e1600377. DOI: 10.1126/sciadv.1600377.

ffrench-Constant, R.H., Somers-Yeates, R., Bennie, J., Economou, T., Hodgson, D. *et al.* (2016) Light pollution is associated with earlier tree budburst across the United Kingdom. *Proceedings of the Royal Society B: Biological Sciences* 283(1833), 20160813. DOI: 10.1098/rspb.2016.0813.

Firebaugh, A. and Haynes, K.J. (2019) Light pollution may create demographic traps for nocturnal insects. *Basic and Applied Ecology* 34, 118–125. DOI: 10.1016/j.baae.2018.07.005.

Gaston, K.J., Davies, T.W., Bennie, J. and Hopkins, J. (2012) Reducing the ecological consequences of night-time light pollution: options and developments. *Journal of Applied Ecology* 49(6), 1256–1266. DOI: 10.1111/j.1365-2664.2012.02212.x.

Gaston, K.J., Bennie, J., Davies, T.W. and Hopkins, J. (2013) The ecological impacts of nighttime light pollution: a mechanistic appraisal. *Biological Reviews* 88(4), 912–927. DOI: 10.1111/brv.12036.

Gaston, K.J., Visser, M.E. and Hölker, F. (2015) The biological impacts of artificial light at night: the research challenge. *Philosophical Transactions of the Royal Society B: Biological Sciences* 370(1667), 20140133. DOI: 10.1098/rstb.2014.0133.

Gaston, K.J. and Holt, L.A. (2018) Nature, extent and ecological implications of night-time light from road vehicles. *Journal of Applied Ecology* 55(5), 2296–2307. DOI: 10.1111/1365-2664.13157.

Grenis, K., Tjossem, B. and Murphy, S.M. (2015) Predation of larval Lepidoptera in habitat fragments varies spatially and temporally but is not affected by light pollution. *Journal of Insect Conservation* 19(3), 559–566. DOI: 10.1007/s10841-015-9777-2.

Grenis, K. and Murphy, S.M. (2019) Direct and indirect effects of light pollution on the performance of an herbivorous insect. *Insect science* 26(4), 770–776. DOI: 10.1111/1744-7917.12574.

Hopkins, G.R., Gaston, K.J., Visser, M.E., Elgar, M.A. and Jones, T.M. (2018) Artificial light at night as a driver of evolution across urban-rural landscapes. *Frontiers in Ecology and the Environment* 16(8), 472–479. DOI: 10.1002/fee.1828.

Jiménez-Peñuela, J., Ferraguti, M., Martínez-de la Puente, J., Soriguer, R. and Figuerola, J. (2019) Urbanization and blood parasite infections affect the body condition of wild birds. *Science of the Total Environment* 651(Pt 2), 3015–3022. DOI: 10.1016/j.scitotenv.2018.10.203.

Knop, E., Zoller, L., Ryser, R., Gerpe, C., Hörler, M. *et al.* (2017) Artificial light at night as a new threat to pollination. *Nature* 548(7666), 206–209. DOI: 10.1038/nature23288.

Lewanzik, D. and Voigt, C.C. (2014) Artificial light puts ecosystem services of frugivorous bats at risk. *Journal of Applied Ecology* 51(2), 388–394. DOI: 10.1111/1365-2664.12206.

Longcore, T. and Rich, C. (2004) Ecological light pollution. *Frontiers in Ecology and the Environment* 2(4), 191–198. DOI: 10.1890/1540-9295(2004)002[0191:ELP]2.0.CO;2.

MacGregor, C.J., Pocock, M.J.O., Fox, R. and Evans, D.M. (2015) Pollination by nocturnal Lepidoptera, and the effects of light pollution: a review. *Ecological Entomology* 40(3), 187–198. DOI: 10.1111/een.12174.

MacGregor, C.J., Evans, D.M., Fox, R. and Pocock, M.J.O. (2017) The dark side of street lighting: impacts on moths and evidence for the disruption of nocturnal pollen transport. *Global Change Biology* 23(2), 697–707. DOI: 10.1111/gcb.13371.

MacGregor, C.J., Pocock, M.J.O., Fox, R. and Evans, D.M. (2019) Effects of street lighting technologies on the success and quality of pollination in a nocturnally pollinated plant. *Ecosphere* 10(1), 1–16. DOI: 10.1002/ecs2.2550.

Maggi, E. and Benedetti-Cecchi, L. (2018) Trophic compensation stabilizes marine primary producers exposed to artificial light at night. *Marine Ecology Progress Series* 606, 1–5. DOI: 10.3354/meps12769.

Majewska, A.A., Satterfield, D.A., Harrison, R.B., Altizer, S. and Hepinstall-Cymerman, J. (2019) Urbanization predicts infection risk by a protozoan parasite in non-migratory populations of monarch butterflies from the southern coastal U.S. and Hawaii. *Landscape Ecology* 34(3), 649–661. DOI: 10.1007/s10980-019-00799-7.

Manfrin, A., Singer, G., Larsen, S., Weiß, N., van Grunsven, R.H.A. *et al.* (2017) Artificial light at night affects organism flux across ecosystem boundaries and drives community structure in the recipient ecosystem. *Frontiers in Environmental Science* 5, 61. DOI: 10.3389/fenvs.2017.00061.

Manfrin, A., Lehmann, D., van Grunsven, R.H.A., Larsen, S., Syväranta, J. *et al.* (2018) Dietary changes in predators and scavengers in a nocturnally illuminated riparian ecosystem. *Oikos* 127(7), 960–969. DOI: 10.1111/oik.04696.

Manríquez, P.H., Jara, M.E., Diaz, M.I., Quijón, P.A., Widdicombe, S. *et al.* (2019) Artificial light pollution influences behavioral and physiological traits in a keystone predator species, *Concholepas concholepas*. *Science of the Total Environment* 661, 543–552. DOI: 10.1016/j.scitotenv.2019.01.157.

Miller, C.R., Barton, B.T., Zhu, L., Radeloff, V.C., Oliver, K.M. *et al.* (2017) Combined effects of night warming and light pollution on predator-prey interactions. *Proceedings of the Royal Society B: Biological Sciences* 284(1864), 20171195. DOI: 10.1098/rspb.2017.1195.

Minnaar, C., Boyles, J.G., Minnaar, I.A., Sole, C.L. and McKechnie, A.E. (2015) Stacking the odds: light pollution may shift the balance in an ancient predator-prey arms race. *Journal of Applied Ecology* 52(2), 522–531. DOI: 10.1111/1365-2664.12381.

Navara, K.J. and Nelson, R.J. (2007) The dark side of light at night: physiological, epidemiological, and ecological consequences. *Journal of Pineal Research* 43(3), 215–224. DOI: 10.1111/j.1600-079X.2007.00473.x.

Owens, A.C.S. and Lewis, S.M. (2018) The impact of artificial light at night on nocturnal insects: a review and synthesis. *Ecology and Evolution* 8(22), 11337–11358. DOI: 10.1002/ece3.4557.

Plummer, K.E., Hale, J.D., O'Callaghan, M.J.O., Sadler, J.P. and Siriwardena, G.M. (2016) Investigating the impact of street lighting changes on garden moth communities. *Journal of Urban Ecology* 2016, 1–10.

Pu, G., Zeng, D., Mo, L., Liao, J. and Chen, X. (2019) Artificial light at night alleviates the negative effect of Pb on freshwater ecosystems. *International Journal of Molecular Sciences* 20(6), 1343. DOI: 10.3390/ijms20061343.

Riley, W.D., Davison, P.I., Maxwell, D.L. and Bendall, B. (2013) Street lighting delays and disrupts the dispersal of Atlantic salmon (*Salmo salar*) fry. *Biological Conservation* 158, 140–146. DOI: 10.1016/j.biocon.2012.09.022.

Rowse, E.G., Lewanzik, D., Stone, E.L., Harris, S. and Jones, G. (2016) Dark matters: the effects of artificial lighting on bats. In: Voigt, C.C. and Kingston, T. (eds) *Bats in the Anthropocene: Conservation of Bats in a Changing World*. Springer, London, pp. 187–214.

Sanders, D. and Gaston, K.J. (2018) How ecological communities respond to artificial light at night. *Journal of Experimental Zoology Part A: Ecological and Integrative Physiology* 329(8-9), 394–400. DOI: 10.1002/jez.2157.

Sanders, D., Kehoe, R., Tiley, K., Bennie, J., Cruse, D. *et al.* (2015) Artificial nighttime light changes aphid-parasitoid population dynamics. *Scientific Reports* 5, 15232. DOI: 10.1038/srep15232.

Sanders, D., Kehoe, R., Cruse, D., van Veen, F.J.F. and Gaston, K.J. (2018) Low levels of artificial light at night strengthen top-down control in insect food web. *Current Biology* 28(15), 2474–2478. DOI: 10.1016/j.cub.2018.05.078.

Santos, C.D., Miranda, A.C., Granadeiro, J.P., Lourenço, P.M., Saraiva, S. *et al.* (2010) Effects of artificial illumination on the nocturnal foraging of waders. *Acta Oecologica* 36(2), 166–172. DOI: 10.1016/j.actao.2009.11.008.

Schroer, S. and Hölker, F. (2016) Impact of lighting on flora and fauna. In: Karlicek, R., Sun, C.C., Zissis, G. and Ma, R. (eds) *Handbook of Advanced Lighting Technology*. Springer, London, pp. 1–33.

Seymoure, B.M. (2018) Enlightening butterfly conservation efforts: the importance of natural lighting for butterfly behavioral ecology and conservation. *Insects* 9(1), 22–25. DOI: 10.3390/insects9010022.

Silva, E., Marco, A., da Graça, J., Pérez, H., Abella, E. *et al.* (2017) Light pollution affects nesting behavior of loggerhead turtles and predation risk of nests and hatchlings. *Journal of Photochemistry and Photobiology B: Biology* 173, 240–249. DOI: 10.1016/j.jphotobiol.2017.06.006.

Stone, E.L., Harris, S. and Jones, G. (2015) Impacts of artificial lighting on bats: a review of challenges and solutions. *Mammalian Biology* 80(3), 213–219. DOI: 10.1016/j.mambio.2015.02.004.

Sullivan, S.M.P., Hossler, K. and Meyer, L.A. (2019) Artificial lighting at night alters aquatic-riparian invertebrate food webs. *Ecological Applications* 29(1), e01821. DOI: 10.1002/eap.1821.

Tałanda, J., Maszczyk, P. and Babkiewicz, E. (2018) The reaction distance of a planktivorous fish (*Scardinius erythrophthalmus*) and the evasiveness of its prey (*Daphnia pulex* × *pulicaria*) under different artificial light spectra. *Limnology* 19(3), 311–319. DOI: 10.1007/s10201-018-0548-0.

Thawley, C.J., Moniz, H.A., Merritt, A.J., Battles, A.C., Michaelides, S.N. *et al.* (2019) Urbanization affects body size and parasitism but not thermal preferences in *Anolis* lizards. *Journal of Urban Ecology* 5(1), 1–9. DOI: 10.1093/jue/juy031.

Underwood, C.N., Davies, T.W. and Queirós, A.M. (2017) Artificial light at night alters trophic interactions of intertidal invertebrates. *Journal of Animal Ecology* 86(4), 781–789. DOI: 10.1111/1365-2656.12670.

van Grunsven, R.H.A., Jähnichen, D., Grubisic, M. and Hölker, F. (2018) Slugs (Arionidae) benefit from nocturnal artificial illumination. *Journal of Experimental Zoology Part A: Ecological and Integrative Physiology* 329(8-9), 429–433. DOI: 10.1002/jez.2170.

van Langevelde, F., van Grunsven, R.H.A., Veenendaal, E.M. and Fijen, T.P.M. (2017) Artificial night lighting inhibits feeding in moths. *Biology Letters* 13(3), 20160874–20160877. DOI: 10.1098/rsbl.2016.0874.

Votsi, N.-E.P., Kallimanis, A.S. and Pantis, I.D. (2017) An environmental index of noise and light pollution at EU by spatial correlation of quiet and unlit areas. *Environmental Pollution* 221, 459–469. DOI: 10.1016/j.envpol.2016.12.015.

Wakefield, A., Stone, E.L., Jones, G. and Harris, S. (2015) Light-emitting diode street lights reduce last-ditch evasive manoeuvres by moths to bat echolocation calls. *Royal Society Open Science* 2(8), 150291. DOI: 10.1098/rsos.150291.

Yuen, S.W. and Bonebrake, T.C. (2017) Artificial night light alters nocturnal prey interception outcomes for morphologically variable spiders. *PeerJ* 5, e4070. DOI: 10.7717/peerj.4070.

Yurk, H. and Trites, A.W. (2000) Experimental attempts to reduce predation by Harbor Seals on out-migrating juvenile salmonids. *Transactions of the American Fisheries Society* 129(6), 1360–1366. DOI: 10.1577/1548-8659(2000)129<1360:EATRPB>2.0.CO;2.

Zapata, M.J., Sullivan, S.M.P. and Gray, S.M. (2019) Artificial lighting at night in estuaries: implications from individuals to ecosystems. *Estuaries and Coasts* 42(2), 309–330. DOI: 10.1007/s12237-018-0479-3.

11 Citizen Science in Urban Habitats

Amanda E. Sorensen* and Rebecca C. Jordan
Department of Community Sustainability, College of Agriculture and Natural Resources, Michigan State University

Abstract

Citizen science has been broadly defined as the inclusion of non-scientific expert members of the public in the authentic practices and processes of scientific inquiry. In the coming decades, as cities and urbans areas around the world expand, they will continue to face mounting environmental pressure and degradation. For researchers to collect data effectively in urban landscapes, inclusion of the local community is paramount. Urban landscapes in the USA, however, are not places popular for public participation in environmental research. Here we explore the creation of a citizen science project, the Baltimore Mosquito Study, within an urban community that is comprised of individuals who have been under-represented in citizen science (i.e. predominately African-Americans, many of whom are struggling economically). While we suggest greater research be emphasized exploring the mutual benefits between citizen science, science and the community, we also have learned from our case study a number of practices which we felt were essential to project development and success. First, when developing outreach materials, we found it was important to frame these materials to particular individuals with high efficacy and a strong connection to their community. The second practice was developing other programmatic strategies that leverage stewards' environmental knowledge and civic awareness, and might take the form of mobilizing stewards as communication leaders. Finally, it is important to be clear what data are necessary to address the questions posed. It is critical for all parties involved to understand the rationale and limitations of the data collection efforts. These core practices can help foster greater trust between scientists and urban communities who have been historically sidelined in academic scientific research.

Introduction

Citizen science has been broadly defined as the inclusion of non-scientific expert members of the public in the authentic practices and processes of scientific inquiry (Bonney *et al.*, 2009; Silvertown, 2009; Jordan *et al.*, 2011). While the professionalization of science is a relatively recent phenomenon, citizen science has a robust history with earliest examples of citizen science projects emerging in the 1900s with the Christmas Bird Count, run by the National Audubon Society in the USA (see www.audubon.org for further history). In today's context, citizen science projects can range in scope, focal area and level of participant engagement. Citizen science projects are usually defined by the level of participant engagement, with Bonney *et al.* (2009) conceptualizing three levels

*Corresponding author: amandaesorensen@gmail.com

of citizen science. At the lowest-engagement end of the spectrum are contributory citizen science projects. Contributory projects are a top-down model of engagement where professional researchers lead all aspects of the scientific process and members of the public perform a data collection role only. These projects are often created when a researcher is looking to engage a certain number of people to help gather data that the researcher otherwise could not. The middle-engagement level is collaborative projects, where members of the public are typically not involved at the project's onset but are involved in data collection and analysis, and results dissemination. These projects prove critically important when the data are of use beyond the scientific community. Toward the greatest-engagement end of the spectrum are co-created citizen science projects. These projects are true partnerships between local citizens and scientists in which members of the public are included in all the steps of scientific inquiry, from defining research questions to data collection and analysis, and disseminating results of the research. These types of projects have long been successful in preservation and restoration efforts, and, more recently, have been critical to aiding communities who seek a greater understanding of environmental use impacts. One recent and salient example of co-created projects are those partnerships between residents and scientists in Flint, Michigan, in response to the Flint Water crisis that started in 2014. Here, residents in Flint partnered with water-quality scientists to collect data on local water quality and advocate for change with those data to political leaders after the municipal drinking water source was changed to an unsuitable source and residents began to notice health problems (see Roy and Edwards, 2019 for origin and timeline of Flint Water crisis). In the Flint Water crisis, and many other co-created projects, citizen scientists initiate data collection efforts in response to local, systemic environmental problems the community is facing and the professional research scientists are brought in as partners to support and bolster the efforts already being conducted by the citizen scientists. Since this initial effort to develop a framework of citizen science project typologies, many other scholars have contributed new or refined existing frameworks for characterizing citizen science projects by participant engagement to better capture the range and scope of citizen science activities (Danielsen *et al.*, 2009; Wiggins and Crowston, 2011; Shirk *et al.*, 2012; Haklay, 2013). Many of these frameworks seek to better understand the outcomes for the communities that participate.

Citizen science may confer other benefits on communities including environmental democracy (in which there is meaningful public participation in the procedures and approaches that provide access to information on land and natural resource decisions, environmental issues such as conservation and biodiversity, as well as participation in the decision-making process and the application of environmental protection laws or compensation) (Cooper *et al.*, 2007; Conrad and Hilchey, 2011; Mueller *et al.*, 2012; Larese-Casanova and Prysby, 2018; Ballard *et al.*, 2018). Studies have also shown positive outcomes for citizen science participants in terms of civic engagement (Dunlap, 1992; Marcinkowski, 1993; Nerbonne and Nelson, 2004), increased scientific literacy (Bonney *et al.*, 2009; Silvertown, 2009; Dickinson *et al.*, 2010), and scientific thinking and learning (Trumbull *et al.*, 2000; Evans *et al.*, 2005; Phillips *et al.*, 2018). Additionally, projects have found that participants increase their scientific knowledge about the specific focal issues or taxa throughout the project (Jordan *et al.*, 2011). Whereas many professionals using citizen science intentionally include educational outcomes for participants as a goal of the project (Crall *et al.*, 2013), citizen scientists frequently show knowledge gains by simply engaging in the project even if it was not an explicit goal (Nerbonne and Nelson, 2004). The beneficial outcomes from participating in citizen science has been shown to occur across projects with different disciplinary focus (environmental, astronomical, biomedical etc.) and participant level (Raddick *et al.*, 2013). Finally, citizen science has the potential to democratize science (USEPA (U.S. Environmental Protection Agency), 2018) by making the scientific process participatory and by including many stakeholders in engendering knowledge, the dissemination of information/knowledge, and being included in the actions or processes involved in making decisions (McCormick, 2007; Jordan *et al.*, 2017). In this chapter we consider the urban audience and share a case study that links benefits of citizen science to urban communities. Further, we place special consideration on the potential for projects like the one we describe to help underserved groups find a platform on which to advocate.

Citizen Science Projects and Benefits to Research in Urban Areas

In the coming decades, as cities and urban areas around the world expand (Sadik, 1999; Grimm *et al.*, 2008), they will continue to face mounting environmental pressure and degradation (Blanco *et al.*, 2009). Indeed, there has been a developing body of literature exploring the unique environmental and ecological characteristics of cities (Sukopp *et al.*, 1995; Grimm *et al.*, 2000). A major challenge in developing a comprehensive scientific understanding of phenomena in cities is the spatially and temporally heterogenous nature of the urban landscape (Machlis *et al.*, 1997). Further, scholars have characterized cities as having a unique ecological nature, emergent from the biophysical and socio-economic interactions that are inherent in cities (Alberti *et al.*, 2003). This uniqueness makes scientific data collection at the various scales (temporal, spatial, social) challenging. For researchers to collect data across these scales effectively, inclusion of the local community is paramount. Urban landscapes in the USA, however, are not places popular for public participation in environmental research. In the USA, non-Hispanic whites, though a national majority, have been a minority of the population in most cities and urban areas since 2000 (PEW Social Trends, 2018). Additionally, demographics in urban areas tend to skew towards younger age groups, with suburban and rural areas seeing more rapid ageing of their population demographics (PEW Social Trends, 2018). While there has not been a formal comprehensive analysis of the demographic representation of citizen science projects across the USA, the studies that do exist suggest concerning trends of lack of diversity and representation in citizen science. A recent report on diversity in citizen science has found that demographics of participants in citizen science tend to skew predominately toward non-Hispanic white, well-educated (Pandya and Dibner, 2018) and older individuals (Merenlender *et al.*, 2016). While we only cite demographic trends in the USA, some of these demographic trends of citizen science participants are also observed in studies of international citizen science projects (e.g. the Galaxy Zoo project, Raddick *et al.*, 2013). Unfortunately, citizen science in the USA does not truly reflect the broader demographics of the country (Pandya, 2012) nor are citizen science projects (with some notable exceptions) particularly prevalent in cities.

The Baltimore Mosquito Study

The Baltimore Mosquito Study (hereafter BMS) provided an opportunity to create a citizen science project. This citizen science project included under-represented individuals, i.e. predominantly African-Americans, many of whom were struggling economically. The BMS was a scientist-initiated project designed to investigate the social and ecological drivers, and the interplay between them, of an invasive nuisance mosquito in the city of Baltimore, Maryland. The mosquito, commonly known as the Asian tiger mosquito (scientific name *Aedes albopictus* (Skuse)), is an invasive mosquito species found throughout the southern and eastern USA, coming from China in the late 1980s (Kraemer *et al.*, 2015). While the Asian tiger mosquito is only of moderate concern as a disease transmitter, it has the potential to carry and transmit diseases such as Zika, Chikungunya, Dengue, Eastern Equine Encephalitis and West Nile and Yellow Fever (Kraemer *et al.*, 2015). They typically have a limited geographic distribution or are rarely identified in US travellers. However, although some of these diseases do not occur in the US mainland, they may occur in Caribbean locations such as Puerto Rico, which is a commonwealth of the USA. Because of this potential, there is concern among experts that the Asian tiger mosquito could play a role in significant disease outbreaks across the temperate areas of the USA where it exists (Manore *et al.*, 2017).

Regardless of its potential as a disease vector, the Asian tiger mosquito is still a major nuisance (Moore and Mitchell, 1997). Because it is a biting pest it significantly reduces outdoor physical activity of children and reduces the quality of life of individuals living in communities where this mosquito is present (Worobey *et al.*, 2013; Halasa *et al.*, 2014). Further, it is difficult to keep Asian tiger mosquito populations under control using traditional mosquito control methods due to its ability to thrive using small, transient, water-holding containers (e.g. rubbish, tyres, potted plants, unmaintained buildings and infrastructure) for larval development (Unlu *et al.*, 2011; LaDeau *et al.*, 2013). In other words, these mosquitoes thrive in urban areas, with current and potential threat to community well-being, with few management strategies available to effectively manage them (https://mda.maryland. gov/plants-pests/Pages/asian_tiger_mosquito_md.aspx). Prior studies have found collecting good data on the Asian tiger mosquito is challenging in cities as the habitat they are found in tends to be privately owned or inaccessible spaces, and they occur in variable patches (Unlu *et al.*, 2011; LaDeau *et al.*, 2013; Becker *et al.*, 2014). Citizen science, then, was identified at project onset as a means to better collect data at the varying spatial, temporal and social scales in which the Asian tiger mosquito exists in the urban landscape.

As it was initially conceived, the BMS was to be a contributory project where individuals gathered data on potential and actual mosquito habitat as well as on mosquito nuisance. While these efforts indicated that individuals were able to gather reliable mosquito data (Jordan *et al.*, 2017), many individuals in the community who were invited did not elect to participate (Jordan *et al.*, 2019). Those that did participate tended to report an interest in the science or, more generally, wanted to help, which is not unlike the motivations found in other citizen science programmes (Phillips and Gates, 2011). It is interesting to note that in our project, among those individuals who did participate in the citizen science, several individuals showed, early on, an increase in their sense of agency (i.e. the personal belief that one's actions could be meaningful and have an impact on the problem). This increase in sense of agency manifested as individuals reporting to the research team that they felt they personally had the ability to improve the environment, and further that their actions would have an impact (Jordan *et al.*, 2019). However, it may be a chicken-and-egg situation, where it is challenging to disentangle the role a person's sense of agency plays in influencing their motivations to contribute to citizen science, or whether that sense of agency is developed as a result of their participation in citizen science. As we cannot often survey those individuals who do not choose to participate in citizen science, the role of an individual's sense of agency in driving participation is not well understood. It could be that, for individuals, a pre-existing high sense of agency drives their motivation to participate in citizen science, or, conversely, the act of committing to participate engenders a sense of agency.

As the BMS project progressed, a small group of participants expressed an interest in becoming actively involved in improving the local environment, such as reducing rubbish and litter and beautifying the local green spaces, but felt that their voice would not be heard among Baltimore City's municipal decision makers. Working as a small group and following the Civic Ecology Practices described in Krasny *et al.* (2014), we worked with the participants to suggest ideas for tangible small projects and to find ways to collect and present evidence for why such ideas may or may not work to achieve the desired goals. During these conversations, the absence of evidence around the impacts of these environmental issues led the group to consider projects that were co-created. Thus, scientists and community members discussed issues related to nuisance species and what we would need to know in order to help direct management practices.

In some cases, participant 'ways of knowing' did not align with the norms of scientific data collection. In this context, we mean 'ways of knowing' as the general methods through which individual knowledge becomes apparent. An example is Photovoice, which is a methodology in which participants collect visual examples that demonstrate the basis for their belief and perceived knowledge about an issue. However, the methodology used by participants in the collection of visual evidence does not always align with the principles of scientific data collection if one were abiding by a traditional experimental analysis (e.g. one that would require random sampling, the inclusion of controls, and standardized method of data collection). But, in many ways, the images produced through Photovoice drew local municipal and media attention, which can add legitimacy to a group whose voice had been going previously unheard. In other ways, the information gathered by the public can align well with scientific norms. For example, in the BMS project, participants wanted to know whether beautifying rubbish receptacles would have an effect on illegal dumping rates around such receptacles. In this case, participants wanted to know whether the beautified rubbish bins and dumpsters would lead to reduced dumping and littering rates around the receptacles, thereby reducing potential larval habitat. To explore this, participants set up experimental manipulations of different receptacles, creating a research protocol that included treatment groups, a data collection protocol, and experimental 'control receptacles' that were not beautified. Many of the participants felt that we, as researchers, were essential to add further endorsement to the data gathered, but it was unclear as to whether it was the inclusion of experts or the more formalized representation of information (e.g. preparation and use of line graphs and bar charts often used in formal science endeavours) that gave the sense of increased validity. While the contributory citizen science project ultimately indicated that rubbish was not the major source of the nuisance mosquito population, all of us involved in the project saw its negative effects on the residents who encounter it daily. This particular finding about the rubbish from the research was particularly disappointing for the participants who were hoping to have further evidence of potential public health threats related to the illegal dumping in their city. With this we saw an opportunity for individuals to own both the responsibility that comes with data collection and the need to determine how to use these data as tools for municipal action and community justice. Because of a series of federal funding decisions and staff rotation at critical organizations, the BMS did not fully transition into a co-created project. Individuals, however, were able to form a larger organization that participates in ongoing advocacy around environmental and social injustices.

Discussion

Citizen science is a scientific research endeavour conducted in wholly, or in part, by amateur or non-professional scientists and often has occurred, or evolved, disconnected from the concerns related to, or the study of, environmental justice. Environmental justice work inherently prioritizes the role of communities of colour in defining environmental science questions for the purpose of improving health and environmental self-determination. Citizen science may also engage traditionally marginalized communities in data collection and advocacy in the context of environmental and social justice (Corburn, 2005). Instead of being viewed largely as recipients of science, with focused outreach, marginalized people also can be stakeholders, participants or even equal partners (Heaney *et al.*, 2007; Dhillon, 2017) in the generation and use of scientific evidence. The field of environmental justice has long been attentive to questions of who participates in science, and citizen science has begun to move in this direction as well (Pandya, 2012). In addition, scholars and activists argue that shedding light on and addressing environmental injustice cannot be done solely by institutions, but necessarily requires community participation (Carr, 2004). Building on the experiences and knowledge of the environmental justice field, citizen science could provide opportunities for communities to define, investigate and address pressing environmental justice problems in collaboration with professional scientists. To do so would require scientists to actively seek voices from individuals in the places and institutions that are not often invited to the decision-making process but who often have a high stake in the outcomes.

However, there are certainly challenges when broadening the umbrella of who leads and directs scientific research endeavours. When reflecting on challenges, in the context of our project as described in this chapter, one of the BMS researchers described what he believed to be a lack of alignment with perceived community needs. He noted that some of the community members he worked with suggested to him that the project was another example of community members being 'over-studied' by scientists, or being treated like an 'object' in research rather than true collaborators in it. This researcher described that when he reported at community meetings about the progress of the project, community members often saw the research findings as irrelevant to their community development goals. If, however, he discussed the mosquito research findings in the context of broader issues like vacancy, dumping or health risks associated with mosquitoes, community members saw greater value in the research. This illuminates the potential challenge for researchers when community questions do not necessarily align with access to academic or research institutional funding, or outcome priorities (e.g. publications, seed data for new funding opportunities, and reports). While we suggest that greater research be placed into the mutual benefits between citizen science and the community, we also have learned, from our case study, that there were a number of practices that were essential to project development and success. First, when developing outreach materials, we found it was important to frame these materials to particular individuals with high sense of agency (i.e. a person who believes that their actions could be meaningful and have an impact on the problem) and a strong connection to their community. In this way, our project attracted critical thought leaders who could help draw in necessary support. If, however, project managers are looking to diversify the pool of citizen scientists outside of those who are already motivated, new recruitment strategies need to be developed and tested. For recruitment messaging to reach those less-engaged populations, these messages need to 'meet people where they are' and resonate with their own cultural experiences, knowledge base and interests. One example of this is highlighted by Johnson *et al.* (2018), in which *Trees Count!*, a street tree census project created by the New York City Department of Parks & Recreation, partnered with the Afropunk Festival, an arts festival held in New York City and cities around the world, to give away free festival tickets as an incentive to attract a new demographic of participants to the tree census effort.

A second practice we identified from our project was developing other programmatic strategies that leverage citizen scientist's environmental knowledge and civic awareness. Such strategies may help mobilize citizen scientists towards becoming communication leaders, i.e. individuals that could translate outcomes and advocate for the projects to other community groups (e.g. religious organizations, schools, sports groups). From the collective agency perspective (i.e. the shared belief in the collaborative power to produce desired results; Bandura, 2000), creating discrete, actionable steps throughout the project that the participants can accomplish, and emphasizing how these steps translate to broader community benefits, could reinforce the participants' sense of agency. This would be particularly important in the event of research setbacks, such that, in the case of BMS, difficult or disappointing research outcomes, such as the lack of connection between rubbish and mosquitoes and struggles in leveraging the data to advocate for changes with municipal leaders, would not be as disappointing to participants. Finally, it is important to be clear on what data are necessary to address the questions posed by the research team, the questions posed by the community, and where there is, if at all, any overlap. By doing so, participants can decide on when and how to participate. In the BMS project we approached many of our discussions with community members at the research proposal stage. This level of community involvement was critical to forming the trust of our group within a community that felt disenfranchised in many ways (see discussion in Jordan *et al.*, 2019). With trust, it was possible to recover from accidental mis-steps that could have caused the community members to abandon our work (e.g. media reports about recent community engagement issues with science in Flint, Michigan). In this way, we were in a place to set up an adaptive and co-learning environment. We must caution, however, that when scientific and community needs are at odds, researchers may find themselves needing to make difficult choices about moving forward with the research.

References

Alberti, M., Marzluff, J.M., Shulenberger, E., Bradley, G., Ryan, C. *et al.* (2003) Integrating humans into ecology: opportunities and challenges for studying urban ecosystems. *BioScience* 53(12), 1169–1179. DOI: 10.1641/0006-3568(2003)053[1169:IHIEOA]2.0.CO;2.

Ballard, H.L., Harris, E.M. and Dixon, C.G. (2018) Science identity and agency in community and citizen science: evidence & potential. Commissioned for the Committee on Designing Citizen Science to Support Science Learning, National Academy of Sciences, Engineering, and Medicine. Available at: https://sites.nationalacademies.org/cs/groups/dbassesite/documents/ webpage/dbasse_189606.pdf

Bandura, A. (2000) Exercise of human agency through collective efficacy. *Current Directions in Psychological Science* 9(3), 75–78. DOI: 10.1111/1467-8721.00064.

Becker, B., Leisnham, P.T. and LaDeau, S.L. (2014) A tale of two city blocks: differences in immature and adult mosquito abundances between socioeconomically different urban blocks in Baltimore (Maryland, USA). *International Journal of Environmental Research and Public Health* 11(3), 3256–3270. DOI: 10.3390/ijerph110303256.

Blanco, H., Alberti, M., Forsyth, A., Krizek, K.J., Rodríguez, D.A. *et al.* (2009) Hot, congested, crowded and diverse: emerging research agendas in planning. *Progress in Planning* 71(4), 153–205. DOI: 10.1016/j.progress.2009.03.001.

Bonney, R., Ballard, H., Jordan, R., McCallie, E. and Phillips, T. (2009) Public participation in scientific research: defining the field and assessing its potential for informal science education. A CAISE inquiry group report. Online submission. Available at: https://eric.ed.gov/?id=ED519688

Carr, A.J.L. (2004) Why do we all need community science. *Society & Natural Resources* 17, 841–849.

Conrad, C.C. and Hilchey, K.G. (2011) A review of citizen science and community-based environmental monitoring: issues and opportunities. *Environmental Monitoring and Assessment* 176(1-4), 273–291. DOI: 10.1007/s10661-010-1582-5.

Cooper, C.B., Dickinson, J., Phillips, T. and Bonney, R. (2007) Citizen science as a tool for conservation in residential ecosystems. *Ecology and Society* 12(2). DOI: 10.5751/ES-02197-120211.

Corburn, J. (2005) *Street Science: Community Knowledge and Environmental Health Justice (Urban and Industrial Environments)*. The MIT Press, Cambridge, Massachusetts.

Crall, A.W., Jordan, R., Holfelder, K., Newman, G.J., Graham, J. *et al.* (2013) The impacts of an invasive species citizen science training program on participant attitudes, behavior, and science literacy. *Public Understanding of Science* 22(6), 745–764. DOI: 10.1177/0963662511434894.

Danielsen, F., Burgess, N.D., Balmford, A., Donald, P.F., Funder, M. *et al.* (2009) Local participation in natural resource monitoring: a characterization of approaches. *Conservation Biology* 23(1), 31–42. DOI: 10.1111/j.1523-1739.2008.01063.x.

Dhillon, C.M. (2017) Using citizen science in environmental justice: participation and decision-making in a southern California waste facility siting conflict. *Local Environment* 22(12), 1479–1496. DOI: 10.1080/13549839.2017.1360263.

Dickinson, J.L., Zuckerberg, B. and Bonter, D.N. (2010) Citizen science as an ecological research tool: challenges and benefits. *Annual Review of Ecology, Evolution, and Systematics* 41(1), 149–172. DOI: 10.1146/annurev-ecolsys-102209-144636.

Dunlap, R. (1992) Trends in public opinion toward environmental issues: 1965–1990. In: Dunlap, R. and Mertig, A. (eds) *American Environmentalism: The US Environmental Movement, 1970–1990*. Taylor and Francis, Philadelphia, Pennsylvania, pp. 89–116.

Evans, C., Abrams, E., Reitsma, R., Roux, K., Salmonsen, L. *et al.* (2005) The neighborhood Nestwatch program: participant outcomes of a citizen-science ecological research project. *Conservation Biology* 19(3), 589–594. DOI: 10.1111/j.1523-1739.2005.00s01.x.

Grimm, N.B., Grove, J.G., Pickett, S.T. and Redman, C.L. (2000) Integrated approaches to long-term studies of urban ecological systems: Urban ecological systems present multiple challenges to ecologists - pervasive human impact and extreme heterogeneity of cities, and the need to integrate social and ecological approaches, concepts, and theory. *BioScience* 50, 571–584.

Grimm, N.B., Faeth, S.H., Golubiewski, N.E., Redman, C.L., Wu, J. *et al.* (2008) Global change and the ecology of cities. *Science* 319(5864), 756–760. DOI: 10.1126/science.1150195.

Haklay, M. (2013) Citizen science and volunteered geographic information: overview and typology of participation. In: Sui, D., Elwood, S. and Goodchild, M.F. (eds) *Crowdsourcing Geographic Knowledge*. Springer, pp. 105–122.

Halasa, Y.A., Shepard, D.S., Fonseca, D.M., Farajollahi, A., Healy, S. *et al.* (2014) Quantifying the impact of mosquitoes on quality of life and enjoyment of yard and porch activities in New Jersey. *PLOS ONE* 9(3), e89221. DOI: 10.1371/journal.pone.0089221.

Heaney, C.D., Wilson, S.M. and Wilson, O.R. (2007) The West End Revitalization Association's community-owned and -managed research model: development, implementation, and action. *Progress in Community Health Partnerships: Research, Education, and Action* 1(4), 339–349. DOI: 10.1353/cpr.2007.0037.

Johnson, M.L., Campbell, L.K., Svendsen, E.S. and Silva, P. (2018) Why count trees?: volunteer motivations, experiences, and civic engagement in tree monitoring efforts. *Arboriculture and Urban Forestry* 44, 59–72.

Jordan, R.C., Gray, S.A., Howe, D.V., Brooks, W.R. and Ehrenfeld, J.G. (2011) Knowledge gain and behavioral change in citizen-science programs. *Conservation Biology* 25(6), 1148–1154. DOI: 10.1111/j.1523-1739.2011.01745.x.

Jordan, R.C., Sorensen, A.E. and Ladeau, S. (2017) Citizen science as a tool for mosquito control. *Journal of the American Mosquito Control Association* 33(3), 241–245. DOI: 10.2987/17-6644R.1.

Jordan, R.C., Sorensen, A.E., Biehler, D., Wilson, S. and LaDeau, S. (2019) Citizen science and civic ecology: merging paths to stewardship. *Journal of Environmental Studies and Sciences* 9(1), 133–143. DOI: 10.1007/s13412-018-0521-6.

Kraemer, M.U.G., Sinka, M.E., Duda, K.A., Mylne, A.Q.N., Shearer, F.M. *et al.* (2015) The global distribution of the arbovirus vectors *Aedes aegypti* and *Ae. albopictus*. *eLife* 4, e08347. DOI: 10.7554/eLife.08347.

Krasny, M.E., Russ, A., Tidball, K.G. and Elmqvist, T. (2014) Civic ecology practices: participatory approaches to generating and measuring ecosystem services in cities. *Ecosystem Services* 7, 177–186. DOI: 10.1016/j.ecoser.2013.11.002.

LaDeau, S.L., Leisnham, P.T., Biehler, D. and Bodner, D. (2013) Higher mosquito production in low-income neighborhoods of Baltimore and Washington, DC: understanding ecological drivers and mosquito-borne disease risk in temperate cities. *International Journal of Environmental Research and Public Health* 10(4), 1505–1526. DOI: 10.3390/ijerph10041505.

Larese-Casanova, M. and Prysby, M. (2018) Engaging people in nature stewardship through master naturalist programs. *Human–Wildlife Interactions* 12.

Machlis, G.E., Force, J.E. and Burch, W.R. (1997) The human ecosystem. Part I: the human ecosystem as an organizing concept in ecosystem management. *Society & Natural Resources* 10(4), 347–367. DOI: 10.1080/08941929709381034.

Manore, C.A., Ostfeld, R.S., Agusto, F.B., Gaff, H. and LaDeau, S.L. (2017) Defining the risk of Zika and Chikungunya virus transmission in human population centers of the eastern United States. *PLOS Neglected Tropical Diseases* 11(1), e0005255. DOI: 10.1371/journal.pntd.0005255.

Marcinkowski, T. (1993) Assessment in environmental education. In: Wilke, R.J. (ed.) *Environmental Education Teacher Resource Handbook*. Corwin Press, Thousand Oaks, California, pp. 143–197.

McCormick, S. (2007) Democratizing science movements: a new framework for mobilization and contestation. *Social Studies of Science* 37, 609–623.

Merenlender, A.M., Crall, A.W., Drill, S., Prysby, M. and Ballard, H. (2016) Evaluating environmental education, citizen science, and stewardship through naturalist programs. *Conservation Biology* 30(6), 1255–1265. DOI: 10.1111/cobi.12737.

Moore, C.G. and Mitchell, C.J. (1997) *Aedes albopictus* in the United States: ten-year presence and public health implications. *Emerging Infectious Diseases* 3, 329–334. DOI: 10.3201/eid0303.970309.

Mueller, M., Tippins, D. and Bryan, L. (2012) The future of citizen science. *Democracy and Education 20*

Nerbonne, J.F. and Nelson, K.C. (2004) Volunteer macroinvertebrate monitoring in the United States: resource mobilization and comparative state structures. *Society & Natural Resources* 17(9), 817–839. DOI: 10.1080/08941920490493837.

Pandya, R.E. (2012) A framework for engaging diverse communities in citizen science in the US. *Frontiers in Ecology and the Environment* 10(6), 314–317. DOI: 10.1890/120007.

Pandya, R. and Dibner, K.A. (eds) (2018) *Learning Through Citizen Science: Enhancing Opportunities by Design*. The National Academies Press. The National Academies of Sciences, Engineering, and Medicine. A Consensus Study of The National Academies Press, Washington, DC.

PEW Social Trends (2018) Demographic and economic trends in urban, suburban, and rural communities. Available at: https://www.pewsocialtrends.org/2018/05/22/demographic-and-economic-trends-in-urban-suburban-and-rural-communities/

Phillips, S.D. and Gates, T. (2011) A conceptual framework for understanding the stigmatization of children of incarcerated parents. *Journal of Child and Family Studies* 20(3), 286–294. DOI: 10.1007/s10826-010-9391-6.

Phillips, T., Porticella, N., Constas, M. and Bonney, R. (2018) A framework for articulating and measuring individual learning outcomes from participation in citizen science. *Citizen Science: Theory and Practice* 3(2), 3. DOI: 10.5334/cstp.126.

Raddick, M.J., Bracey, G., Gay, P.L., Lintott, C.J., Cardamone, C. *et al*. (2013) Galaxy zoo: motivations of citizen scientists. *Astronomy Education Review* 12, 1–41.

Roy, S. and Edwards, M. (2019) Citizen science during the Flint, Michigan federal water emergency: ethical dilemmas and lessons learned. *Citizen Science: Theory and Practice* 4(1). DOI: 10.5334/cstp.154.

Sadik, N. (1999) *State of world population 1999: 6 Billion - a time for choices*. United Nations Population Fund, New York.

Shirk, J.L., Ballard, H.L., Wilderman, C.C., Phillips, T., Wiggins, A. *et al*. (2012) Public participation in scientific research: a framework for deliberate design. *Ecology and Society* 17(2). DOI: 10.5751/ES-04705-170229.

Silvertown, J. (2009) A new dawn for citizen science. *Trends in Ecology & Evolution* 24(9), 467–471. DOI: 10.1016/j.tree.2009.03.017.

Sukopp, H., Numata, M. and Huber, A. (1995) *Urban Ecology as the Basis for Urban Planning*. SPB Academic, The Hague, The Netherlands.

Trumbull, D.J., Bonney, R., Bascom, D. and Cabral, A. (2000) Thinking scientifically during participation in a citizen-science project. *Science Education* 84(2), 265–275. DOI: 10.1002/(SICI)1098-237X(200003)84:2<265::AID-SCE7>3.0.CO;2-5.

USEPA (U.S. Environmental Protection Agency) (2018) Information to action: strengthening EPA citizen science partnerships for environmental protection. NACEPT (National Advisory Council for Environmental Policy and Technology). EPA 220-R-18-001.

Unlu, I., Farajollahi, A., Healy, S.P., Crepeau, T., Bartlett-Healy, K. *et al*. (2011) Area-wide management of *Aedes albopictus*: choice of study sites based on geospatial characteristics, socioeconomic factors and mosquito populations. *Pest Management Science* 67(8), 965–974. DOI: 10.1002/ps.2140.

Wiggins, A. and Crowston, K. (2011) From conservation to crowdsourcing: a typology of citizen science. *Proceedings of the International Conference on System Sciences – 2011* 44, 1–10.

Worobey, J., Fonseca, D.M., Espinosa, C., Healy, S. and Gaugler, R. (2013) Child outdoor physical activity is reduced by prevalence of the Asian tiger mosquito, *Aedes albopictus*. *Journal of the American Mosquito Control Association* 29(1), 78–80. DOI: 10.2987/12-6296R.1.

12 Ecology of Insects and Other Arthropods in Urban Agroecosystems

Heidi Liere[1]* and Monika Egerer[2]
[1]*Department of Environmental Studies, Seattle University, Seattle, Washinghton;*
[2]*Department of Ecology and Ecosystem Management, School of Life Sciences Weihenstephan, Technische Universität München, Munich, Germany*

Abstract

In cities around the world, agriculture in urban and peri-urban areas (i.e. areas immediately adjacent to, or surrounding, a city or urban area) provides an important source of fresh and affordable foods and offers an opportunity for urban dwellers to connect to nature and their community. Urban agroecosystems also provide important habitats in urban environments to support biodiversity and ecosystem services. The often-complex vegetation composition and structure of urban agroecosystems can further support associated biodiversity to thereby support ecosystem functions such as natural pest control and pollination. Of particular importance for these services are insects and other arthropods. As forms of urban agriculture (henceforth referred to as UA) are increasingly popular, the number of arthropod ecology studies in urban agroecosystems has increased accordingly. In this chapter we review these studies to inform how much we know about the local and landscape factors that affect these important organisms, their ecological interactions, and the ecosystem services they provide in urban agricultural systems. This knowledge has the potential to inform city planning policies that would help manage urban spaces within and around UA to boost biodiversity and maximize crucial ecosystem regulating services for sustainable local food production.

Introduction

It is estimated that by 2030, 80–90% of the global population will live in cities (United Nations, 2010; Seto *et al.*, 2012), and in the United States the 2012 census reported that more than 80% of the US population lives in urban areas (United States Census Bureau, 2020). To feed the growing urban population, food is imported daily, travelling sometimes thousands of miles (Zeeuw *et al.*, 2011), which not only increases its carbon footprint but also reduces the quality of the produce (if, for example, it is harvested before peak quality to better withstand shipment). Yet many urban residents lack sufficient access to fresh produce or adequate nutrition, and many communities lack the nutritional diversity and social connectedness provided by a fresh and local food market. In response to this need, urban agriculture (UA) has expanded dramatically, especially in low-income and underserved communities (Alig *et al.*, 2004). For the past 50 years, UA has increased by 3.6%, annually, in

*Corresponding author: liereheidi@seattleu.edu

developing countries and in the USA by more than 30% in the past 30 years (Siegner *et al.*, 2018). Although robust measurements of UA's actual effect on improving food security (i.e. having reliable access to a sufficient quantity of affordable, nutritious food) in low-income communities are still lacking (Siegner *et al.*, 2018), it has been estimated that UA provides 15–20% of the global food supply (Hodgson *et al.*, 2011, but see Clinton *et al.*, 2018). In addition to food, UA provides numerous social and environmental benefits, and comes in various forms including residential food gardens, community and allotment gardens and commercial urban farms (Table 12.1). All of these and others are, in essence, versions of urban agroecosystems.

Agroecosystems, defined as functionally coherent units of agricultural activity, include biophysical and human components and their interactions. They generally contain fewer species and are considered to be less complex compared to unmanaged ecosystems. However, urban agroecosystems are, essentially, similar, in contrast to rural agroecosystems, in that, typically, they are established in already-species-poor environments, and they tend to have higher species diversity than their surrounding areas. In fact, urban agroecosystems, like urban gardens and farms, have a high potential to provide high-quality habitats to many animal species due to the high abundance of flowering plants and prolonged growing season maintained by irrigation and fertilization (Faeth *et al.*, 2005; Colding *et al.*, 2006; Clucas *et al.*, 2018). In turn, biodiversity-mediated regulating ecosystem services, which are vital for sustainable urban food production, create a positive feedback loop (Fig. 12.1).

Social benefits of UA include, but are not limited to, city beautification, education and health, as well as community building and empowerment (Blaine *et al.*, 2010; Draper and Freedman, 2010; Williams *et al.*, 2018). For example, Tzoulas *et al.* (2007), using a synthesis of published literature as conceptual framework, proposed that green infrastructure can provide both ecosystem and human health benefits in urban areas. Many environmental and ecological benefits have been associated with UA. These include reduced heat island effect, improved air and stormwater quality, and carbon sequestration (Lovell, 2010; Kulak *et al.*, 2013; Lin *et al.*, 2015). In the face of rapid urbanization, UA and other urban green spaces thus play an important role in harbouring biodiversity within city landscapes (Faeth *et al.*, 2005; Goddard *et al.*, 2010).

A significant proportion of the biodiversity associated with UA is composed of insects and other arthropods. It is crucial to understand their ecological interactions in urban agroecosystems because, on the one hand, arthropods, especially insects, can cause substantial damage to urban crops and cultivated plants (Gregory *et al.*, 2016) but, on the other, provide important services such as pollination and pest control (Peisley *et al.*, 2015). These services have been estimated to be valued in billions of dollars in both rural agriculture (Losey and Vaughan, 2006) and urban agriculture (Clinton *et al.*, 2018). Agroecological principles, where external inputs are replaced by natural processes, have been applied to improve small-scale agriculture for many years (Altieri, 1995). The same principles can be applied to UA and these will necessarily include managing UA and the surrounding landscapes to optimize insect-mediated ecosystem services (Altieri and Nicholls, 2018). This will require a detailed knowledge about the factors that affect insect communities in urban landscapes.

Pollinator density and diversity are essential for optimal fruit set of many crop species (Klein *et al.*, 2007), while arthropod predators and parasitoids can keep crop pests below damaging levels (Letourneau *et al.*, 2009). These ecosystem services are particularly important in urban agriculture, where most of the crops that are grown depend upon bee pollination to set fruit or seed (Matteson *et al.*, 2008; Oberholtzer *et al.*, 2014), and urban gardeners and growers greatly rely on natural pest control because they may have strict restrictions on their use of chemical pesticides (Oberholtzer *et al.*, 2014). Farm and garden management practices such as crop diversification, hedgerow planting and beetle refuges can benefit both pollinators and pest control agents (Bolger *et al.*, 2000; Philpott *et al.*, 2014; Altieri and Nicholls, 2018). Even so, due to limited size, relative isolation from natural areas and constant disturbances, small-scale urban green spaces alone may not be able to retain viable populations (Goddard *et al.*, 2010; see also Chapter 6: Protecting pollinators in the urban environment). Consequently, a landscape

Table 12.1. Different typologies and descriptions of six non-mutually exclusive types of urban agricultural systems. References are examples of studies that focus on this scale of agricultural production.

Type	Description	Example studies at this scale
Community or **allotment gardens**	Small-scale, vegetatively complex and species-rich agroecosystems that are usually located in urban or semi-urban areas for food production. Usually are community managed by a group or individual plots are leased by households.	Drescher *et al*., 2006; Colding *et al*., 2006; Speak *et al*., 2015; Bell *et al*., 2016
Private or **home gardens**	Often located in suburban areas and may be the most prevalent form of urban agriculture in cities. For example, privately owned gardens cover about 22–27% of the total urban area in the UK, 36% in New Zealand, and 19.5% in Dayton, Ohio.	Mathieu *et al*., 2007; Loram *et al*., 2008
Easement gardens	Gardens often regulated by the local government but located within private or community properties. Urban easements are often established to improve water quality and erosion control, but they can include a wide array of biodiversity, including food plants, depending on management type. Gardening on verges (edges or borders) may also be done as a form of 'guerrilla gardening' where local community gardens are established on small patches of soil when few unpaved spaces are available.	Hunter and Hunter, 2008; Hunter and Brown, 2012
Rooftop gardens or **green roofs**	Vegetation on the roof of a building. It may be used to improve insulation, create local habitat, provide decorative amenity, and cultivate food plants.	Whittinghill and Rowe, 2012
Urban orchards	Tree-based food production systems that can be owned and run privately or by the community. For example, schools and hospitals are establishing fruit trees that provide crops, erosion control, shade and wildlife habitat, and produce food for the local community.	Drescher *et al*., 2006
Peri-urban agriculture	Primarily located at the outskirts of cities. Typically, these are multi-functional agricultural systems that include a large variety of activities and diversification approaches, and contribute to environmental, social and economic functions of, and for, the surrounding community.	Zasada, 2011
Market farms, commercial and non-commercial farms	Focus on commercial sale of crops grown in outdoor fields, in greenhouses or on rooftops. Generally, represents the largest scale of production (total area and size) of urban agriculture by utilizing several lots.	Lowenstein and Minor, 2018

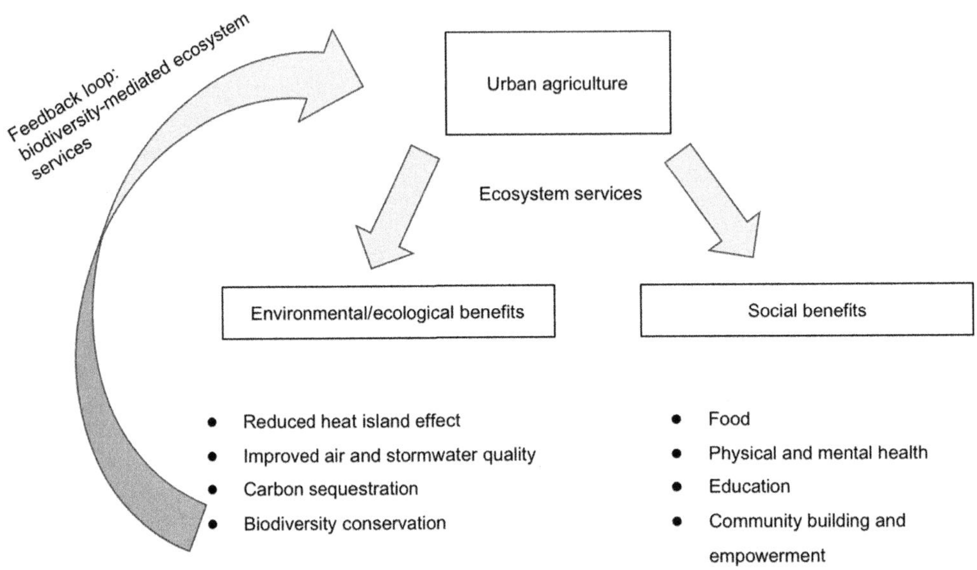

Fig. 12.1. Conceptual diagram of the links between urban agricultural systems, urban residents and associated biodiversity (i.e. the biodiversity that is not intentionally planned and managed by humans).

management approach may be necessary to maintain urban biodiversity (Beninde *et al.*, 2015) in order to fulfil UA pollination and pest control needs.

Types of Urban Agriculture

As described by Lin *et al.* (2015) and Lin and Egerer (2018), urban agroecosystems can be characterized by land tenure, management, types of food and service provision, and scale of production (Table 12.1). UA systems can harbour a rich diversity of flowering vegetables, as well as an abundance of ornamental plants that provide nectar resources for beneficial insect populations and increased species diversity (Colding *et al.*, 2006). That is, over 400 species of flowering plants were documented in community gardens surveyed in Stockholm, Sweden. Studies that directly measure the benefits of urban agroecosystems on animal species diversity, however, are relatively few and recent (Clucas *et al.*, 2018). This meta-analysis found that most of the published work on the subject is observational, biased towards a view of developed countries, and mainly focused on invertebrates (14 out of the 15 studies), predominantly pollinators (bees and butterflies) and pest control agents (ladybird beetles, ants and spiders). Indeed, the authors conclude that more studies are needed in developing countries and covering a wider range of taxa in order to generalize about the positive effect of urban agroecosystems on biodiversity. Nevertheless, the growing number of studies investigating the drivers that affect insect abundance, species richness and diversity in urban agroecosystems is already contributing to our understanding of the ecology of the system to implement agroecological gardening practices (Guitart *et al.*, 2012).

Biodiversity, Richness and Abundance Patterns

From studies in rural agricultural systems, we know that both local factors (i.e. patch effects, habitat characteristics) and landscape factors (or matrix effects such as surrounding landscape features), and the interactions among them, affect insect communities in agroecosystems (Altieri, 1999; Tscharntke *et al.*, 2005). There is growing evidence that landscape context is also important in urban systems, and that the degree to which local resource availability increases beneficial insect abundance and diversity depends on the composition of habitat types in the surrounding areas (Bennett and Gratton, 2012; Philpott *et al.*, 2014; Egerer *et al.*, 2017a). Yet the unique characteristics of cities (Cadenasso *et al.*, 2007; Pickett and Cadenasso, 2009; see also Chapter 6) may lead to different effects of landscape context on beneficial organisms than those expected from rural systems. For example, in contrast to studies in rural systems, in a study conducted in the California Central Coast, Egerer *et al.* (2017b) showed that beneficial insect abundance and diversity in community gardens do not decrease with reduced proportion of natural habitats in the surrounding landscape. Arthropod groups (ants, bees, ladybeetles, parasitoids and spiders) are influenced by landscape factors, but landscape influence differs depending on spatial scale as well as local habitat features, likely due to arthropod community interactions and life history strategies.

Pollinators respond to local factors of urban agroecosystems including habitat size, abundance, species richness and composition of plants (including crops, flowers, weeds, native and non-native plants), ground cover composition and management practices (e.g. irrigation, composting etc.). In general, bee community richness in urban agroecosystems increases with floral and plant abundance and richness across many different urban landscapes (Frankie *et al.*, 2005; Matteson *et al.*, 2008; Matteson and Langellotto, 2010; Pardee and Philpott, 2014; Lowenstein *et al.*, 2014; Bennett and Lovell, 2019). Bumblebee abundance and diversity in urban agroecosystems increases with greater flower abundance and richness (Tommasi *et al.*, 2004; Andersson *et al.*, 2007; Ahrné *et al.*, 2009), and solitary bee diversity increases with tree abundance and vegetation complexity (Smith *et al.*, 2006). Similarly, other pollinator groups, including butterflies, increase in diversity with increased floral resources. Whether native and non-native plants have a different effect on pollinators within urban agroecosystem is still unclear, with some studies suggesting positive effects (Frankie *et al.*, 2005; Pawelek *et al.*, 2009; Pardee and Philpott, 2014) while other studies found little or no effect (Matteson and Langellotto, 2011).

Natural enemies, including predators and parasitoids, also respond to local factors such as garden area, mulch cover, and floral abundance and richness (Bennett and Gratton, 2012; Burkman and Gardiner, 2014; Otoshi *et al.*, 2015; Burks and Philpott, 2017; Lowenstein and Minor, 2018). In general, larger green spaces with diverse plant communities support a greater abundance and diversity of natural enemies (Burkman and Gardiner, 2014). Lowenstein and Minor (2018) found that urban farm area and floral abundance are positively associated with all predatory groups of *Brassica* pests in UA systems in the city of Chicago, Illinois. In community gardens in California, Egerer *et al.* (2017b) found that abundance of ornamental plants and crop diversity are important for maintaining ladybird beetles, which are important predators of aphids and scale insects. The vegetation structure (e.g. taller vegetation, denser grass cover) is also important for supporting greater abundance and richness of natural enemies (e.g. beetles; Delgado de la Flor *et al.*, 2017). In addition, urban agroecosystems with greater ground cover heterogeneity (mulch and leaf-litter cover) support more active and diverse assemblages of spiders (Otoshi *et al.*, 2015) and carabid beetles (Philpott *et al.*, 2019), while soils with high organic matter, moisture and nitrogen (Grewal *et al.*, 2011) provide resources for ground-dwelling arthropod natural enemies. In sum, management of urban agricultural systems' soil and vegetation can enhance habitat availability for various arthropod natural enemies and increase their abundance and species richness (Burkman and Gardiner, 2014; Delgado de la Flor *et al.*, 2017).

Landscape connectivity, landscape diversity, proportion of impervious cover and the position along a rural-to-urban gradient are some of the landscape-scale factors that can affect ecological interactions within urban agroecosystems. Urban landscapes are similar to agricultural landscapes in regard to fragmentation and intensity of land use (Lin and Fuller, 2013). In rural agricultural landscapes, landscape-level intensification resulting in habitat loss and fragmentation can negatively impact beneficial insects, pest control and pollination (Tscharntke *et al.*, 2005). Likewise, the movement and colonization of insect and arthropod populations in urban landscapes are negatively affected by increased impervious cover and fragmentation of green spaces (McKinney, 2002, 2008; Faeth *et al.*, 2005; Williams, 2009). These factors can affect arthropods with varying degrees of dispersal capabilities, from non-flying spiders and springtails (Ponge, 2003; Sattler *et al.*, 2010) to parasitic wasps (Bennett and Gratton, 2012; Burks and Philpott, 2017) to strong fliers such as bees (Cane *et al.*, 2006; Ahrné *et al.*, 2009; Jha and Kremen, 2013).

Natural areas can provide source populations and resources to proximate human-managed habitats (Ricketts *et al.*, 2008). Accordingly, in UA systems, increasing proximity to natural areas increases the abundance and diversity of bees (Hernandez *et al.*, 2009); although bee species richness depends on the habitat position along the urban-to-rural gradient (Matteson *et al.*, 2008; Fetridge *et al.*, 2008). However, the strength, direction and scale at which landscape factors affect arthropod communities in UA system depends on the organism's life history traits and habits (Egerer *et al.*, 2017a; Bennett and Lovell, 2019). For example, Pardee and Philpott (2014) found that cavity-nesting bee abundance in community gardens is positively associated with proximity to forested habitats, while ground-nesting bees increase with proximity to wetlands. Bennett and Lovell (2019) found that large-bodied bees were positively associated with increasing amounts of impervious cover but small-bodied ones were not; instead, these responded to local floral resource variables. Furthermore, landscape effects may be dependent on the region of study. Abundance and diversity of ladybird beetles, for example, increase with urbanization in urban gardens in California, but decrease in Michigan (Egerer *et al.*, 2018d). These authors suggest that water availability in gardens as well as urbanization history could explain the divergent patterns.

Enhancing connectivity of urban agroecosystems in the urban matrix may counter some negative impacts of landscape urbanization on functional biodiversity and may be a tool to enhance regional habitat quality (Rudd *et al.*, 2002). Following this, the spatial connectivity of allotment or community gardens in a highly developed urban matrix may facilitate the movement of native bees to ultimately support pollination function (Colding *et al.*, 2006; see also Chapter 6). Thus, agroecosystems have the potential to contribute to overall green space connectivity and thus should be integrated into conservation methods for insect biodiversity.

Although the effects of many local and landscape factors on beneficial arthropods seem to be context-dependent, studies reviewed here show that floral resources within UA systems have consistent positive effects on both pollinator and natural enemy diversity and abundance. Thus, garden practitioners and urban residents should be able to enhance local habitats for beneficial insects. A common recommended strategy to enhance biological pest control in rural agriculture is to plant borders or strips of flowers early in the season to prevent pest population build-up (Uyttenbroeck, *et al.*, 2016). This practice could now be easily adopted by UA gardeners and farmers (Altieri and Nicholls, 2018).

Beyond Insect Biodiversity in Urban Agroecosystems: Functional Traits, Species Interactions and Ecosystem Function

Several studies in urban agroecosystems have investigated insect communities by evaluating metrics such as presence, abundance, species richness and diversity. The last decade has seen an increase of research that moves beyond examining biodiversity patterns and trends. These investigations in urban areas have explored trophic cascades, species interactions, functional diversity, dispersal and movement, and ecological network linkages, improving our understanding of how environmental factors affect ecosystem function and services in urban landscapes.

Functional traits and diversity

We are just beginning to identify how environmental factors affect the presence and richness of functional traits (i.e. traits that define species with regard to their ecological roles) and functional diversity of insects in urban agroecosystems. Functional diversity is of ecological importance because it, by definition, is the component of diversity that influences ecosystem dynamics, stability, productivity, nutrient balance and other aspects of ecosystem functioning (Tilman, 2001). Further, high functional trait diversity is critical for maintaining functional stability and ecosystem resilience in response to change (Walker *et al.*, 1999). Under changing and perhaps stressful conditions, such as those that might occur more readily in urban areas, ecosystem functions may still be maintained even when dominant species decline or are lost because functionally equivalent minor species are able to substitute for them. The latter species confer resilience in terms of ecosystem function (Walker *et al.*, 1999).

Although, in general, bees are ubiquitous and diverse in urban areas, it is important to understand if their species assemblages and functional diversity show specific associations with particular urban habitats. For example, in two Canadian cities, Montreal and Quebec, wild-bee communities were highly diverse but also dominated by abundant, ubiquitous and exotic species (Normandin *et al.*, 2017). Functional trait diversity, as determined based on observations of pollen specialization, pollen transportation method, tongue length, seasonal activity, sociality, nesting behaviour and intertegular distance (as a proxy for body size and bee foraging range) varied with habitat. Despite their small size, community gardens had higher functional trait diversity than urban cemeteries, and only slightly lower than that of urban parks. However, colonization potential is influenced by life history traits, such as body size and voltinism (i.e. number of generations per year) (Banaszak-Cibicka and Żmihorski, 2012). Thus, the establishment of pollinators with certain life histories may be hindered by some UA management practices. For example, widespread mulch and rock ground cover used by gardeners to retain soil moisture (Grewal *et al.*, 2011; Egerer *et al.*, 2018c) may affect nesting site availability for soil-nesting bees and thereby lower their abundance in UA systems (Quistberg *et al.*, 2016). Nevertheless, UA has a great potential to contribute to urban pollination services through the spillover of their highly functional diverse bee communities (Normandin *et al.*, 2017).

Similar to pollinators, natural enemies' functional traits also influence the degree to which they respond to UA local and landscape context. Highly mobile arthropods such as ladybird beetles (Coleoptera: Coccinellidae) may not perceive the urban matrix as a barrier to movement, and urban gardens can be inhabited by numerous native species of different sizes, and with different diets and diet breadths (Liere *et al.*, 2019). Natural enemies with low dispersal capabilities including ground-dwelling beetles (Carabidae and Staphylinidae) have been found to be more abundant in UA systems that are more connected to other green spaces in the landscape than in those with low connectivity (Vergnes *et al.*, 2012). However, in gardens with low green-space connectivity, some of these natural enemies with low dispersal capabilities can still thrive and benefit from local garden features, such as mulch woodchips and leaf litter, which provide them with shelter and refuge from predation (Philpott *et al.*, 2019). Further, UA systems are distinct microhabitats in which the combination of highly irrigated, vegetated, mulched and bare soil patches may allow for the support of organisms with a variety of life history strategies. For example, urban farms in Cleveland, Ohio, supported a high diversity and abundance of beetles with a variety of habitat preferences: xerophilous (dry/hot environment species), hygrophilous (wet environment species), open habitat affinity beetles, as well as beetle species associated with anthropogenic activity and environmental disturbance (Delgado de la Flor *et al.*, 2017). In summary, by creating highly heterogeneous habitats that attract and harbour organisms with different functional traits, UA has been shown to generate habitats with tremendous potential for the conservation of beneficial insects, including pollinator and natural enemies.

Species interactions, food webs and networks

Arthropod food-web dynamics have been widely studied in rural settings; however, anthropogenic forces may alter environmental stressors and create unique interactions in urban ecosystems (Faeth *et al.*, 2005, 2011; Shochat *et al.*, 2006), sufficient enough to cause shifts in food-web dynamics. For example, Faeth *et al.* (2005) found that altered productivity due to resource subsidies (irrigation, fertilization) in the city of Phoenix, Arizona, changed species composition, dampened season fluctuations of species abundance, and altered feeding behaviours of key animal taxa (Faeth *et al.*, 2005). They found that, in contrast to outlying areas, bird-mediated control of arthropod herbivores is a strong top-down force inside the city. These authors attribute this to the low predator-related mortality of birds in urban settings. They suggest, however, that their results are contingent on environmental conditions and that different results will likely arise in cities located in landscapes that are less water-, and resource-limited. Furthermore, it is not yet clear how ecological interactions and the drivers that affect them differ between urban and rural agroecosystems because some management practices (e.g. irrigation, planning of plant diversity, density, and composition) are similar between urban and rural agroecosystems but others are fundamentally different (e.g. chemical inputs, size) (Faeth *et al.*, 2011). Food-web approaches can improve understanding of agroecosystem functioning and response to environmental stressors (Tylianakis and Binzer, 2014), and can inform managing decisions to optimize insect-mediated ecosystem services in urban gardens and farms.

Network analysis is a powerful tool for understanding food-web and community dynamics (e.g. Bascompte *et al.*, 2003). A growing number of studies have investigated insect community networks in agricultural systems and provide suggested application of this analytical tool (Tylianakis and Binzer, 2014; for review see Woodward and Bohan, 2013). An example of network analysis in agroecosystems is the examination of interactions between plants and their seed predators to determine how they affect weed control in UK agricultural systems (Bohan *et al.*, 2011). Network analyses are also providing insights into the link between diversity within trophic levels and pest control services (for review see Woodward and Bohan, 2013).

Rapidly changing landscapes, such as urban environments, experience substantial turnover in plant, pollinator, herbivore and parasitoid species pools over time (Burkle and Alarcón, 2011; Gagic *et al.*, 2011), all of which can potentially influence network structure. For example, pollination networks along a rural–urban gradient around Paris are negatively affected by urbanization (Geslin *et al.*, 2013). In turn, these changes in the number of interactions and the interaction evenness between flower-visitors and plants, detrimentally affected the reproductive success of one functional group of plants (Geslin *et al.*, 2013). However, plant and flower-visitor community networks can be complex. Flower richness and bee richness can be higher in urban than in agricultural areas (Theodorou *et al.*, 2016). Furthermore, flower-visitor generality, which potentially leads to more consistent pollination services (Blüthgen and Klein, 2011), was higher in urban compared to agricultural areas in recent studies in Germany (Theodorou *et al.*, 2016) and in the UK (Baldock *et al.*, 2015). Thus, because moderately urbanized areas provide rich floral resources, they can positively influence bee richness and plant reproduction. Nevertheless, the prevalence of trypanosomatids such as *Crithidia bombi*, a common bumblebee parasite, can be higher in urbanized areas compared to agricultural areas (Goulson *et al.*, 2012).

An examination of multiple networks in rural agroecosystems such as natural enemy–herbivore networks, pollination networks, and seed dispersal networks suggest that different networks vary in their fragility (Pocock *et al.*, 2012). Further, because agricultural management strategies that may benefit one functional group may not, inevitably, benefit other groups, optimizing for multiple ecosystem services is not straightforward. For example, multi-network analyses show how managing and changing the abundance of one species could improve one ecosystem service, such as pollination, but at the same time it could affect the delivery of another ecosystem service, such as pest control (Pocock *et al.*, 2012). Multi-network analysis could also help identify keystone species, crucial for the delivery of ecosystem services (Woodward and Bohan, 2013). In particular, this approach could become a useful tool for urban agroecosystems because it would allow for other networks, such as social networks (focused on, for example, gardener workshops or seed exchange programmes) to be combined with ecological networks (such as those focused on natural enemy–herbivore and plant–pollinator interactions) (Woodward and Bohan, 2013). Being able to evaluate how social interactions among urban gardeners and farmers affect ecosystem services and vice versa could be enormously valuable for UA planners and managers.

Ecosystem functions and services

The provisioning of insect-mediated ecosystem services is affected by species population-level and community-level factors (Daily, 1997). For example, in rural agricultural systems, pollinator density and diversity are essential for optimal fruit set for many crops (Klein *et al.*, 2007) and predator abundance and richness may enhance pest control (Letourneau *et al.*, 2009). As presented above, many studies have examined how the diversity and abundance of beneficial insects in urban gardens are affected by both local and landscape factors. Although many of these studies point to the potential implication for ecosystem services provisioning, only a few have directly examined the cascading effects on pollination or pest control services in urban gardens (Lin *et al.*, 2015; Bennett and Lovell, 2019), and none, to our knowledge, has further linked these services to yield output. A recent global analysis, however, estimates pest control and pollination services in UA to be valued at hundreds of thousands of dollars (Clinton *et al.*, 2018).

Herbivory, plant damage and pest control

More research is needed into the relative importance of top-down effects (where predators and parasitoids control the structure or population dynamics of the ecosystem) and bottom-up effects (where plants control the increase or decrease of energy to the higher trophic levels) (Raupp *et al.*, 2010) in order to understand and be able to predict the population dynamics of herbivorous arthropods in urban environments. Higher temperatures in urbanized habitats can speed up herbivore developmental time, reducing vulnerable period to natural enemies and thus, indirectly, reducing top-down control by natural enemies (Raupp *et al.*, 2010). Abiotic factors in urban areas can also affect herbivore populations directly through bottom-up forces. For example, increases in canopy temperatures can cause higher abundance of the urban forest pest, the gloomy scale (*Melanaspis tenebricosa*), in urban trees compared to rural ones by directly increasing the fecundity and population growth of the scale insects, not by indirectly reducing natural enemy abundance or percentage parasitism (Dale and Frank, 2014). Specifically, adult female *M. tenebricosa* reproduction increases by approximately 14 eggs for every 1°C increase in temperature (Dale and Frank, 2014), which, in turn, is a consequence of increasing impervious surface (an important habitat variable reflecting increasing urbanization). In general, small arthropods with limited mobility and multiple generations on the same host plant, such as scale insects, aphids, mites and leaf miners, have a higher probability of reaching outbreak levels with increased levels of urbanization (Raupp *et al.*, 2010). Although the same could hold for arthropod assemblages in urban agroecosystems, there is as yet no evidence to support this. Biotic and abiotic factors that affect bottom-up and top-down forces in these green-spaces may be different to biotic and abiotic factors in dispersed urban trees.

Studies in urban agroecosystems have both found evidence (Egerer *et al.*, 2018b) and no evidence (Lowenstein and Minor, 2018) of bottom-up forces affecting herbivore populations in garden crops. Herbivory of *Brassica* crops, a widely cultivated crop in urban agriculture, was relatively low (less than 15%) in residential gardens, community gardens and urban farms in Chicago, Illinois (Lowenstein and Minor, 2018). The latter was apparently attributable to the attraction of parasitoids and predators to garden flowers and not by changes in host plant density or other garden characteristics. In contrast, in urban community gardens in the California Central Coast, cabbage aphid density was strongly influenced by both host plant density and by parasitism (Egerer *et al.*, 2018b). Both studies concluded that urban agriculture, of various types, can be beneficial habitats for species responsible for effective biological control.

Top-down control, as estimated with sentinel prey-removal experiments in urban agricultural systems, varies greatly from 12.8% to 100% (Gardiner *et al.*, 2014; Philpott and Bichier, 2017; Lowenstein and Minor, 2018; Morales *et al.*, 2018). Among the local factors reported to affect prey removal are garden ground-cover variables (such as mulch cover, bare ground), garden vegetation variables (such as complexity and diversity, floral abundance), and garden characteristics (size) and landscape characteristics (such as landscape diversity, percentage impervious cover, and percentage urban cover) (Philpott and Bichier, 2017; Morales *et al.*, 2018). Comparative studies have found that significant drivers of prey removal differ between cities (Morales *et al.*, 2018) and between prey taxa (Philpott and Bichier, 2017). Notably, some results in urban gardens, such as declined predation services with increases in landscape diversity, contrast sharply with patterns found in rural studies (Philpott and Bichier, 2017).

Pollination and Fruit Set

It is estimated that 92% of crops grown in urban food gardens are dependent upon bee pollination for fruit or seed set (Matteson and Langellotto, 2009). Although pollination services in urban agricultural sites are influenced by local management characteristics (Werrell *et al.*, 2009; Bennett and Lovell, 2019; Cohen *et al.*, 2020), the strength of these effects may be dependent on landscape context (Cohen *et al.*, 2020). Conspecific pollen deposition in flowers of cucumber plants is positively and significantly associated with the size of a community garden; in particular, the area of each garden dedicated to flowering plants and the area of each garden dedicated to cucumber plants (Werrell *et al.*, 2009). Here the arrangement of plants within a garden positively influenced yield in fruit- and vegetable-producing plants within urban community gardens. Of course, other factors can be important. For example, fruit set in urban agricultural sites in the city of Chicago negatively correlate with the percentage of paved areas within the sites (Bennett and Lovell, 2019). In the California Central Coast, seed number, but not fruit set, of jalapeno pepper plants was best explained by pollinator abundance and by the number of trees and shrubs within community gardens. The direction of the latter effect, however, was contingent on the proportion of natural habitat cover in the landscape (within a 2 km radius) (Cohen *et al.*, 2020).

Future Research Directions

As urban populations are growing rapidly and urban agriculture is seeing a 'renaissance' in recent decades, it is both a crucial and exciting time for research on insect ecology in urban agroecosystems. This research has the potential to inform sustainable urban food production practices and urban greening efforts. As we have presented here, several studies have revealed relationships between local and landscape factors with arthropod communities and some have found cascading effects on ecosystem functions and services. This research has already a great potential to inform managing practices within the urban agroecosystem and its surrounding landscape. We recommend more research that assesses how different insect taxa experience and, in turn, affect their urban environments so that we gain a more comprehensive picture to better support biodiversity relevant to ecosystem services and provision for sustainable food production. Specifically, we highlight five research directions that we see as valuable to understand insect ecology in urban agroecosystems and urban ecosystems broadly: (i) movement and potential spillover of insect populations; (ii) systematic comparisons between urban and rural agroecosystems; (iii) the importance of abiotic conditions in the context of land use and climate change; (iv) the utilization of new methods to ask ecological questions; and (v) the exploration of connections between humans and insects in urban agroecosystems.

Movement, connectivity and spillover

The movement of organisms across the landscape may be key for long-term population persistence in urban agroecosystems because dispersal affects colonization–extinction dynamics. Yet we know very little about the factors that influence the dispersal to and from urban agroecosystems, or what factors influence permanence or fidelity to these sites. A mark-recapture study using ladybird beetles investigated the effects of local resource manipulation and landscape composition on dispersal from community gardens (Egerer *et al.*, 2018a). This study found that landscape context is the most important driver of beetle dispersal. Thus ladybird beetles are more likely to disperse in gardens surrounded by more urban land cover (i.e. a high proportion of impervious cover). Interestingly, gardens surrounded by landscapes with more urban land cover, referred to by the authors as 'low-quality' cityscapes, had a higher diversity and abundance of ladybird beetles. They suggest that highly mobile natural enemies, such as ladybird beetles, have a high turnover rate in urban agroecosystems or in highly urbanized areas. Yet the mechanisms driving the relationship between landscape context and dispersal behaviour are still unclear. This warrants future experimental research approaches on dispersal between UA systems and surrounding areas.

To our knowledge, there are no studies that have rigorously assessed the impact of landscape or habitat-scale connectivity on insect abundance and diversity in urban agroecosystems. There is, however, indirect evidence that urban agroecosystems facilitate population spillover of beneficial insects to peri-urban farms (Langellotto *et al.*, 2018) . These conclusions were based on estimates of the size and potential foraging range of bees captured in urban agricultural systems. That is, about 30–50% of urban food-garden bee communities could spill over to pollinate adjacent agricultural crops. They also proposed a research agenda that might provide a better understanding of potentially effective strategies for managing urban habitats and providing ecosystem services across the peri-urban interface. The development of such a research agenda would require a combination of methods to quantify insect movement across urban landscapes, perhaps using approaches such as mark-recapture, molecular methods and geographic spatial analysis (which seeks to explain biological patterns and processes in terms of geographic location). Integrative and multiple approaches are needed to determine how landscape connectivity can promote dispersal and persistence in UA systems, ecosystem service multiplicity across space, and how to facilitate urban garden networks to increase ecosystem functions provided by insects.

Urban–rural agroecosystems comparisons

Arthropod communities and the ecosystem services that these communities provide are rarely systematically compared between rural and urban agroecosystems. Both systems largely differ from one another in spatial and temporal resource availability for arthropods. For example, in contrast to urban farms, urban community gardens are managed by multiple gardeners, resulting in increased local vegetation heterogeneity, floral species diversity and stability of floral availability (Pereira-Peixoto *et al.*, 2014). Although these management differences likely affect species interactions such as plant–pollinator, predator–prey and parasitoid–host, there are few comparisons of these interactions between urban and rural agriculture. But there are a few insightful exceptions (Pereira-Peixoto *et al.*, 2016). They investigated the diversity and specificity of interactions between bees and their parasitoids in urban, peri-urban and rural agricultural systems. Higher abundance of parasitoids was observed in the urban–rural interface with high habitat heterogeneity (i.e. a mix of urban, rural, semi-natural and natural elements), but bees experienced the highest parasitism rates in urban gardens. They also found that parasitoid specificity, which implies interaction 'vulnerability', was highest in the urban–rural interface systems; whereas both urban and rural systems had more generalist parasitoids, suggesting that the interactions were more stable. The results suggest that there may be trade-offs between factors that promote pollinators versus those that promote natural enemies in urban and rural agroecosystems. Consequently, more studies are needed which examine and compare insect communities and species interactions within urban, peri-urban and rural agroecosystems.

Abiotic factors in changing environments

Urban environments are characterized by pesticides, heavy metal contamination, light and noise pollution, and urban heat (Epstein, 1995; Frumkin, 2001; Grimm *et al.*, 2008; see Chapter 11). The combinations of these environmental stressors can impact insects and their interactions with other organisms (McIntyre, 2000; Gardiner and Harwood, 2017), but their effects in urban agroecosystems are largely understudied. Elucidating the relative impacts of environmental stressors including pesticides and pollution on insects and their ecological interactions in urban agroecosystems is increasingly important due to global insect decline (Sánchez-Bayo and Wyckhuys, 2019), as well as the functional roles that insects provide in agroecosystems. Here we review some of these environmental stressors and potential future directions.

Pesticides

Pesticides, which are commonly applied on urban public and private green spaces, can have a negative impact on insect biodiversity (Politi Bertoncini *et al.*, 2012) and affect community dynamics. Citywide pesticide applications to manage biting flies, for example, have caused pest outbreaks in urban trees by disrupting top-down control by natural enemies (Raupp *et al.*, 2010 and references therein). In addition to mortality, pesticide exposure can affect insect reproduction and growth (Goulson *et al.*, 2012), learning (Yang *et al.*, 2012), as well as feeding and foraging ability (Feltham *et al.*, 2014). Although many urban agroecosystems are managed organically or pesticide-free (Oberholtzer *et al.*, 2014), mobile insects with large foraging ranges will likely be exposed to pesticides in the urban matrix. This is similar to agricultural landscapes where pesticide exposure in one agroecosystem affects pollination services in another (Krupke *et al.*, 2012). Most pesticide-related research has taken place in rural agricultural systems on honey bees (*Apis mellifera*) and bumblebees (*Bombus*), warranting future work in urban landscapes and particularly urban agroecosystems where pesticide exposure, and food production and consumption, intersect. Future studies should adopt a landscape approach and environmental chemistry methods to quantify pesticide exposure on insects of varying taxonomic and functional groups with different life histories in urban agroecosystems.

Heavy metals

Chemical pollution sources in cities are numerous (Madrid *et al.*, 2002; Imperato *et al.*, 2003) and, compared to rural agricultural landscapes, contamination of heavy metals like lead, zinc and copper is significantly higher in cities (Ordóñez *et al.*, 2003; Imperato *et al.*, 2003). Heavy metals bioaccumulate in living tissues (Peterson *et al.*, 2003; Neilson and Rajakaruna, 2015; Bourioug *et al.*, 2015) and can move up the food chain from contaminated soil to plants (Peralta-Videa *et al.*, 2009), to primary and secondary consumers (Prince *et al.*, 2001; Heikens *et al.*, 2001; Peralta-Videa *et al.*, 2009). Ingesting plants grown in contaminated soil can carry human health risks (Brown and Jameton, 2000; Ljung *et al.*, 2006). However, metal concentration is generally diluted to non-risk levels when the diet also includes food from other (non-contaminated) sources, as is usually the case for humans (Sridhara Chary *et al.*, 2008). In contrast, invertebrates are constantly in close contact with the soil, have a diet restricted to contaminated sources and have thus a high risk of being negatively affected by metal pollution (Heikens *et al.*, 2001).

Heavy metals are known to accumulate in the bodies of soil decomposers such as earthworms, isopods and springtails (Heikens *et al.*, 2001), and this, potentially, has negative effects on soil decomposition rates (Giller *et al.*, 2009). Further, elevated concentration of metals in soil invertebrates can be passed on to higher trophic levels, thus affecting important pest control agents like predatory insects, spiders and birds (Heikens *et al.*, 2001). For example, spiders collected in urban agroecosystems with heavy-metal soil contamination have slower development and reproduction rates (Gardiner and Harwood, 2017). Pollinators also can be affected by heavy metals. Bees can acquire heavy metals through consuming contaminated nectar, pollen or water (Perugini *et al.*, 2011). Thus, in general, contamination can alter bees' foraging behaviour (Meindl and Ashman, 2013) and affect plant–pollinator interactions (Perugini *et al.*, 2011). Consequently, heavy metals could diminish the effective delivery of animal-mediated ecosystem services, including decomposition, pollination and pest control, which are essential for sustainable agriculture. Though studies have investigated the effect of metal contamination on individual invertebrate groups in urban areas (see Gall *et al.*, 2015 for a review) and in urban agriculture (Gardiner and Harwood, 2017), the effects on community composition and on food-web dynamics are still poorly understood in urban agroecosystems (but see Peterson *et al.*, 2003; Boyd *et al.*, 2006; Bourioug *et al.*, 2015).

Other pollutants: nutrient, thermal, light and noise pollution

Other pollutants characteristic of urban environments can affect arthropod communities and their interactions. For example, ozone and atmospheric nitrogen deposition from car combustion can alter the susceptibility and resistance of urban trees to insect herbivores (Hain, 1987; Eatough Jones *et al.*, 2004). In addition, other associated features of urban environments, including thermal, light and noise pollution, can also affect insect populations and behaviour (McIntyre, 2000). For example, prolonged warmer temperatures due to urban heat effects and increased artificial illumination may induce flight responses, thus extending arthropod foraging time and increasing dispersal likelihood (Longcore and Rich, 2004; see also Chapter 8). By speeding up developmental times, elevated temperatures in urban habitats may reduce arthropod vulnerability to natural enemies (for review see Raupp *et al.*, 2010). Yet we do not know how urban heat effects, light exposure and noise affect insects and their interactions in urban agroecosystems. Further work in this area is needed to determine the potential mechanisms of community disassembly that could affect food production.

New methods in ecological investigations

Advancements in remote sensing and molecular technologies provide new pathways to examine mechanisms driving species interactions. Geographic information systems paired with unmanned aerial vehicles (i.e. UAVs or drones), may be used to accurately assess vegetation structure, habitat quality and connectivity (Yu *et al.*, 2012; Feng *et al.*, 2015), and crop damage from insect pests (Yue *et al.*, 2012). DNA-based techniques are providing new information on insect communities, their interactions and food-web complexity. DNA techniques can determine, for example, who eats whom and which insect species visit which plants, all of which can be difficult to study and observe in the field. Molecular methods (DNA fingerprinting techniques) have been used in rural agroecosystems to determine how farm and landscape management affect gut bacteria and fat content of insect natural enemies, which could potentially affect their effectiveness to provide ecosystem services (e.g. predation). In addition, environmental DNA has been used to determine the feeding preference of *Lutzomyia longipalpis* sand flies, an important vector of leishmaniasis disease in urban ecosystems (Lima *et al.*, 2016). It would be fruitful to use such techniques in urban agroecosystems to identify important pest control agents as well as to determine the factors that affect their health and their movement across the urban landscape. Further, these techniques could determine the essential plant–pollinator relationships and networks that lead to fruit production in UA systems.

Unexplored connections between humans and insects in urban agroecosystems

As described in this chapter, there are various pathways by which humans affect insects (e.g. by managing vegetation) and by which insects affect humans (through pollination, pest control services, etc.) in UA. Yet, there may be other less obvious connections that may be worth investigating. For example, recent investigations into the microbiome of bees in urban and in agricultural landscapes have shed light on indirect ways by which farm management affects insect health (Cohen *et al.*, 2020), but how insect health affects the delivery of ecosystem services is still unclear. Additionally, ecosystem disservices associated with urban agroecosystems, in relation to public health, have been largely unexplored. For example, standing water in community gardens can attract mosquitoes (Dongus *et al.*, 2009; Matthys *et al.*, 2010) and this may cause public health concerns, particularly in developing countries where diseases such as malaria are prevalent (Afrane *et al.*, 2004). Other indirect links, such as the potential spillover of insect pests from UA to urban parks or residential gardens, have not, to our knowledge, been thoroughly investigated.

Conclusions

Urban agroecosystems are multi-functional green spaces in urban areas that provide various environmental, ecological and social benefits (Fig. 12.1). Insects and other arthropods are important ecosystem service providers in the regulating services (e.g. pollination) and supporting services (e.g. soil formation) that ultimately contribute to food provision. Our understanding of the ecology of these important organisms in urban agroecosystems has spiked in recent decades, in parallel with urban agriculture popularity in the developed world. These developments are informing management practices, both at the garden/farm level and at the surrounding urban matrix level, to optimize ecosystem services and food production in a sustainable manner. However, much of the complexity behind the mechanisms driving species distributions, movement, interactions and ecosystem functions in these heterogeneous gardens and farms across heterogeneous urban landscapes remains to be unravelled. Because of the important link between insects and agricultural production, our understanding of these mechanisms will be of increasing importance for UA in future decades as urban areas become denser and climate change exacerbates urban weather extremes. Additionally, as relatively vegetatively complex urban habitats, UA could play an important role in the conservation of insect populations across regions. Given insect declines across the world due to anthropogenic forces, it is vital to understand how to recreate urban spaces, including urban agroecosystems, in order to complement local, regional and global conservation efforts.

References

Afrane, Y.A., Klinkenberg, E., Drechsel, P., Owusu-Daaku, K., Garms, R. *et al.* (2004) Does irrigated urban agriculture influence the transmission of malaria in the city of Kumasi, Ghana? *Acta Tropica* 89(2), 125–134.

Ahrné, K., Bengtsson, J. and Elmqvist, T. (2009) Bumble bees (*Bombus* spp) along a gradient of increasing urbanization. *PLOS ONE* 4(5), e5574. DOI: 10.1371/journal.pone.0005574.

Alig, R.J., Kline, J.D. and Lichtenstein, M. (2004) Urbanization on the US landscape: looking ahead in the 21st century. *Landscape and Urban Planning* 69(2-3), 219–234. DOI: 10.1016/j.landurbplan.2003.07.004.

Altieri, M.A. (1995) *Agroecology: The Science of Sustainable Agriculture*. Westview Press, Boulder, Colorado.

Altieri, M.A. (1999) The ecological role of biodiversity in agroecosystems. In: Paoletti, M.G. (ed.) *Invertebrate Biodiversity as Bioindicators of Sustainable Landscapes*. Elsevier, Amsterdam, pp.. 19–31

Altieri, M.A. and Nicholls, C. I. (2018) Urban agroecology: designing biodiverse, productive and resilient city farms. *Agro. Sur* 46, 49–60.

Andersson, E., Barthel, S. and Ahrné, K. (2007) Measuring social–ecological dynamics behind the generation of ecosystem services. *Ecological Applications* 17(5), 1267–1278. DOI: 10.1890/06-1116.1.

Baldock, K.C.R., Goddard, M.A., Hicks, D.M., Kunin, W.E., Mitschunas, N. *et al.* (2015) Where is the UK's pollinator biodiversity? The importance of urban areas for flower-visiting insects. *Proceedings of the Royal Society, London [Biology]* 282, 20142849

Banaszak-Cibicka, W. and Żmihorski, M. (2012) Wild bees along an urban gradient: winners and losers. *Journal of Insect Conservation* 16(3), 331–343. DOI: 10.1007/s10841-011-9419-2.

Bascompte, J., Jordano, P., Melián, C.J. and Olesen, J.M. (2003) The nested assembly of plant–animal mutualistic networks. *Proceedings of the Royal Society, London* 100, 9383–9387.

Bell, S., R., Fox-Kämper, N., Keshavarz, M., Benson, S., Caputo, S.N. *et al.* (2016) *Urban Allotment Gardens in Europe*. Routledge.

Beninde, J., Veith, M. and Hochkirch, A. (2015) Biodiversity in cities needs space: a meta-analysis of factors determining intra-urban biodiversity variation. *Ecology Letters* 18(6), 581–592. DOI: 10.1111/ele.12427.

Bennett, A.B. and Gratton, C. (2012) Local and landscape scale variables impact parasitoid assemblages across an urbanization gradient. *Landscape and Urban Planning* 104(1), 26–33. DOI: 10.1016/j.landurbplan.2011.09.007.

Bennett, A.B. and Lovell, S. (2019) Landscape and local site variables differentially influence pollinators and pollination services in urban agricultural sites. *PLOS ONE* 14(2), e0212034. DOI: 10.1371/journal.pone.0212034.

Blaine, T.W., Grewal, P.S., Dawes, A. and Snider, D. (2010) Profiling community gardeners. *Journal of Extension* 48.

Blüthgen, N. and Klein, A.-M. (2011) Functional complementarity and specialisation: the role of biodiversity in plant–pollinator interactions. *Basic and Applied Ecology* 12(4), 282–291. DOI: 10.1016/j.baae.2010.11.001.

Bohan, D.A., Boursault, A., Brooks, D.R. and Petit, S. (2011) National-scale regulation of the weed seedbank by carabid predators. *Journal of Applied Ecology* 48(4), 888–898. DOI: 10.1111/j.1365-2664.2011.02008.x.

Bolger, D.T., Suarez, A.V., Crooks, K.R., Morrison, S.A. and Case, T.J. (2000) Arthropods in urban habitat fragments in southern California: area, age, and edge effects. *Ecological Applications* 10(4), 1230–1248. DOI: 10.1890/1051-0761(2000)010[1230:AIUHFI]2.0.CO;2.

Bourioug, M., Gimbert, F., Alaoui-Sehmer, L., Benbrahim, M., Aleya, L. *et al.* (2015) Sewage sludge application in a plantation: effects on trace metal transfer in soil–plant–snail continuum. *Science of the Total Environment* 502, 309–314. DOI: 10.1016/j.scitotenv.2014.09.022.

Boyd, R.S., Wall, M.A. and Jaffré, T. (2006) Nickel levels in arthropods associated with Ni hyperaccumulator plants from an ultramafic site in New Caledonia. *Insect Science* 13(4), 271–277. DOI: 10.1111/j.1744-7917.2006.00094.x.

Brown, K.H. and Jameton, A.L. (2000) Public health implications of urban agriculture. *Journal of Public Health Policy* 21(1), 20–39. DOI: 10.2307/3343472.

Burkle, L.A. and Alarcón, R. (2011) The future of plant-pollinator diversity: understanding interaction networks across time, space, and global change. *American Journal of Botany* 98(3), 528–538. DOI: 10.3732/ajb.1000391.

Burkman, C.E. and Gardiner, M.M. (2014) Urban greenspace composition and landscape context influence natural enemy community composition and function. *Biological Control* 75, 58–67. DOI: 10.1016/j.biocontrol.2014.02.015.

Burks, J.M. and Philpott, S.M. (2017) Local and landscape drivers of parasitoid abundance, richness, and composition in urban gardens. *Environmental Entomology* 46(2), 201–209. DOI: 10.1093/ee/nvw175.

Cadenasso, M.L., Pickett, S.T.A. and Schwarz, K. (2007) Spatial heterogeneity in urban ecosystems: reconceptualizing land cover and a framework for classification. *Frontiers in Ecology and the Environment* 5(2), 80–88. DOI: 10.1890/1540-9295(2007)5[80:SHIUER]2.0.CO;2.

Cane, J.H., Minckley, R.L., Kervin, L.J., Roulston, T'ai H.. and Williams, N.M. (2006) Complex responses within a desert bee guild (Hymenoptera: Apiformes) to urban habitat fragmentation. *Ecological Applications* 16(2), 632–644. DOI: 10.1890/1051-0761(2006)016[0632:CRWADB]2.0.CO;2.

Clinton, N., Stuhlmacher, M., Miles, A., Uludere Aragon, N., Wagner, M. *et al.* (2018) A global geospatial ecosystem services estimate of urban agriculture. *Earth's Future* 6(1), 40–60. DOI: 10.1002/2017EF000536.

Clucas, B., Parker, I.D. and Feldpausch-Parker, A.M. (2018) A systematic review of the relationship between urban agriculture and biodiversity. *Urban Ecosystems* 21(4), 635–643. DOI: 10.1007/s11252-018-0748-8.

Cohen, H., Philpott, S., Liere, H., Lin, B. and Jha, S. (2020) The relationship between pollinator community and pollination services is mediated by floral abundance in urban landscapes. *Urban ecosystems*. DOI: 10.1007/s11252-020-01024-z.

Colding, J., Lundberg, J. and Folke, C. (2006) Incorporating green-area user groups in urban ecosystem management. *AMBIO: A Journal of the Human Environment* 35(5), 237–244. DOI: 10.1579/05-A-098R.1.

Daily, G. (1997) *Nature Services*. Island Press, Washington, DC.

Dale, A.G. and Frank, S.D. (2014) Urban warming trumps natural enemy regulation of herbivorous pests. *Ecological Applications* 24(7), 1596–1607. DOI: 10.1890/13-1961.1.

Delgado de la Flor, Y.A., Burkman, C.E., Eldredge, T.K. and Gardiner, M.M. (2017) Patch and landscape-scale variables influence the taxonomic and functional composition of beetles in urban greenspaces. *Ecosphere* 8(11), e02007. DOI: 10.1002/ecs2.2007.

Dongus, S., Nyika, D., Kannady, K., Mtasiwa, D., Mshinda, H. *et al.* (2009) Urban agriculture and Anopheles habitats in Dar es-Salaam, Tanzania. *Geospatial Health* 3(2), 189–210.

Draper, C. and Freedman, D. (2010) Review and analysis of the benefits, purposes, and motivations associated with community gardening in the United States. *Journal of Community Practice* 18(4), 458–492. DOI: 10.1080/10705422.2010.519682.

Drescher, A.W., Holmer, R.J. and Iaquinta, D.L. (2006) Urban homegardens and allotment gardens for sustainable livelihoods: management strategies and institutional environments. In: Kumar, B.M. and Nair, P.K.R. (eds) *Tropical Homegardens: A Time-Tested Example of Sustainable Agroforestry*. Dordrecht, Springer Netherlands, pp. 317–338.

Eatough Jones, M., Paine, T.D., Fenn, M.E. and Poth, M.A. (2004) Influence of ozone and nitrogen deposition on bark beetle activity under drought conditions. *Forest Ecology and Management* 200(1-3), 67–76. DOI: 10.1016/j.foreco.2004.06.003.

Egerer, M.H., Arel, C., Otoshi, M.D., Quistberg, R.D., Bichier, P. *et al.* (2017a) Urban arthropods respond variably to changes in landscape context and spatial scale. *Journal of Urban Ecology* 3(1). DOI: 10.1093/jue/jux001.

Egerer, M.H., Bichier, P. and Philpott, S.M. (2017b) Landscape and local habitat correlates of lady beetle abundance and species richness in urban agriculture. *Annals of the Entomological Society of America* 110(1), 97–103. DOI: 10.1093/aesa/saw063.

Egerer, M.H., Liere, H., Bichier, P. and Philpott, S.M. (2018a) Cityscape quality and resource manipulation affect natural enemy biodiversity in and fidelity to urban agroecosystems. *Landscape Ecology* 33(6), 985–998. DOI: 10.1007/s10980-018-0645-9.

Egerer, M.H., Liere, H., Lin, B.B., Jha, S., Bichier, P. *et al.* (2018b) Herbivore regulation in urban agroecosystems: direct and indirect effects. *Basic and Applied Ecology* 29, 44–54. DOI: 10.1016/j.baae.2018.02.006.

Egerer, M.H., Philpott, S.M., Liere, H., Jha, S., Bichier, P. *et al.* (2018c) People or place? neighborhood opportunity influences community garden soil properties and soil-based ecosystem services. *International Journal of Biodiversity Science, Ecosystem Services & Management* 14(1), 32–44. DOI: 10.1080/21513732.2017.1412355.

Egerer, M., Li, K. and Ong, T. (2018d) Context matters: contrasting ladybird beetle responses to urban environments across two US regions. *Sustainability* 10(6), 1829 DOI: 10.3390/su10061829.

Epstein, P.R. (1995) Emerging diseases and ecosystem instability: new threats to public health. *American Journal of Public Health* 85(2), 168–172. DOI: 10.2105/AJPH.85.2.168.

Faeth, S.H., Warren, P.S., Shochat, E. and Marussich, W.A. (2005) Trophic dynamics in urban communities. *BioScience* 55(5), 399–407. DOI: 10.1641/0006-3568(2005)055[0399:TDIUC]2.0.CO;2.

Faeth, S.H., Bang, C. and Saari, S. (2011) Urban biodiversity: patterns and mechanisms. *Annals of the New York Academy of Sciences* 1223(1), 69–81. DOI: 10.1111/j.1749-6632.2010.05925.x.

Feltham, H., Park, K. and Goulson, D. (2014) Field realistic doses of pesticide imidacloprid reduce bumblebee pollen foraging efficiency. *Ecotoxicology* 23(3), 317–323.

Feng, Q., Liu, J. and Gong, J. (2015) UAV remote sensing for urban vegetation mapping using random forest and texture analysis. *Remote Sensing* 7(1), 1074–1094. DOI: 10.3390/rs70101074.

Fetridge, E.D., Ascher, J.S. and Langellotto, G.A. (2008) The bee fauna of residential gardens in a suburb of New York City (Hymenoptera: Apoidea). *Annals of the Entomological Society of America* 101(6), 1067–1077. DOI: 10.1603/0013-8746-101.6.1067.

Frankie, G.W., Thorp, R.W., Schindler, M., Hernandez, J., Ertter, B. *et al.* (2005) Ecological patterns of bees and their host ornamental flowers in two Northern California cities. *Journal of the Kansas Entomological Society* 78(3), 227–246. DOI: 10.2317/0407.08.1.

Frumkin, H. (2001) Beyond toxicity: human health and the natural environment. *American Journal of Preventive Medicine* 20, 234–240.

Gagic, V., Tscharntke, T., Dormann, C.F., Gruber, B., Wilstermann, A. *et al.* (2011) Food web structure and biocontrol in a four-trophic level system across a landscape complexity gradient. *Proceedings of the Royal Society B: Biological Sciences* 278(1720), 2946–2953. DOI: 10.1098/rspb.2010.2645.

Gall, J.E., Boyd, R.S. and Rajakaruna, N. (2015) Transfer of heavy metals through terrestrial food webs: a review. *Environmental Monitoring and Assessment* 187(4), 201. DOI: 10.1007/s10661-015-4436-3.

Gardiner, M.M. and Harwood, J.D. (2017) Editorial overview: current investigations of environmental drivers and community interactions that influence biological control. *Current Opinion in Insect Science* 20, vii–0. DOI: 10.1016/j.cois.2017.04.002.

Gardiner, M.M., Prajzner, S.P., Burkman, C.E., Albro, S. and Grewal, P.S. (2014) Vacant land conversion to community gardens: influences on generalist arthropod predators and biocontrol services in urban greenspaces. *Urban Ecosystems* 17(1), 101–122. DOI: 10.1007/s11252-013-0303-6.

Geslin, B., Gauzens, B., Thébault, E. and Dajoz, I. (2013) Plant pollinator networks along a gradient of urbanisation. *PLOS ONE* 8(5), e63421. DOI: 10.1371/journal.pone.0063421.

Giller, K.E., Witter, E. and McGrath, S.P. (2009) Heavy metals and soil microbes. *Soil Biology and Biochemistry* 41(10), 2031–2037. DOI: 10.1016/j.soilbio.2009.04.026.

Goddard, M.A., Dougill, A.J. and Benton, T.G. (2010) Scaling up from gardens: biodiversity conservation in urban environments. *Trends in Ecology & Evolution* 25(2), 90–98. DOI: 10.1016/j.tree.2009.07.016.

Goulson, D., Whitehorn, P. and Fowley, M. (2012) Influence of urbanisation on the prevalence of protozoan parasites of bumblebees. *Ecological Entomology* 37(1), 83–89.

Gregory, M.M., Leslie, T.W. and Drinkwater, L.E. (2016) Agroecological and social characteristics of New York City community gardens: contributions to urban food security, ecosystem services, and environmental education. *Urban Ecosystems* 19(2), 763–794. DOI: 10.1007/s11252-015-0505-1.

Grewal, S.S., Cheng, Z., Masih, S., Wolboldt, M., Huda, N. *et al.* (2011) An assessment of soil nematode food webs and nutrient pools in community gardens and vacant lots in two post-industrial American cities. *Urban Ecosystems* 14(2), 181–194. DOI: 10.1007/s11252-010-0146-3.

Grimm, N.B., Faeth, S.H., Golubiewski, N.E., Redman, C.L., Wu, J. *et al.* (2008) Global change and the ecology of cities. *Science* 319(5864), 756–760. DOI: 10.1126/science.1150195.

Guitart, D., Pickering, C. and Byrne, J. (2012) Past results and future directions in urban community gardens research. *Urban Forestry & Urban Greening* 11(4), 364–373. DOI: 10.1016/j.ufug.2012.06.007.

Hain, F.P. (1987) Interactions of insects, trees and air pollutants. *Tree Physiology* 3(1), 93–102. DOI: 10.1093/treephys/3.1.93.

Heikens, A., Peijnenburg, W.J.G.M. and Hendriks, A.J. (2001) Bioaccumulation of heavy metals in terrestrial invertebrates. *Environmental Pollution* 113(3), 385–393. DOI: 10.1016/S0269-7491(00)00179-2.

Hernandez, J., Frankie, G. and Thorp, R. (2009) *Ecology of urban bees: a review of current knowledge and directions for future study*. Cities and the Environment (CATE) 2.

Hodgson, K., Campbell, M.C. and Bailkey, M. (2011) *Urban Agriculture: Growing Healthy, Sustainable Places*. American Planning Association, Chicago, Illinois.

Hunter, M.C.R. and Brown, D.G. (2012) Spatial contagion: gardening along the street in residential neighborhoods. *Landscape and Urban Planning* 105(4), 407–416. DOI: 10.1016/j.landurbplan.2012.01.013.

Hunter, M.R. and Hunter, M.D. (2008) Designing for conservation of insects in the built environment. *Insect Conservation and Diversity* 313(1), 189–196. DOI: 10.1111/j.1752-4598.2008.00024.x.

Imperato, M., Adamo, P., Naimo, D., Arienzo, M., Stanzione, D. *et al.* (2003) Spatial distribution of heavy metals in urban soils of Naples City (Italy). *Environmental Pollution* 124(2), 247–256. DOI: 10.1016/S0269-7491(02)00478-5.

Jha, S. and Kremen, C. (2013) Resource diversity and landscape-level homogeneity drive native bee foraging. *Proceedings of the National Academy of Sciences* 110(2), 555–558. DOI: 10.1073/pnas.1208682110.

Klein, A.-M., Vaissière, B.E., Cane, J.H., Steffan-Dewenter, I., Cunningham, S.A. *et al.* (2007) Importance of pollinators in changing landscapes for world crops. *Proceedings of the Royal Society B: Biological Sciences* 274(1608), 303–313. DOI: 10.1098/rspb.2006.3721.

Krupke, C.H., Hunt, G.J., Eitzer, B.D., Andino, G. and Given, K. (2012) Multiple routes of pesticide exposure for honey bees living near agricultural fields. *PLoS ONE* 7(1), e29268.

Kulak, M., Graves, A. and Chatterton, J. (2013) Reducing greenhouse gas emissions with urban agriculture: a life cycle assessment perspective. *Landscape and Urban Planning* 111, 68–78. DOI: 10.1016/j.landurbplan.2012.11.007.

Langellotto, G., Melathopoulos, A., Messer, I., Anderson, A., McClintock, N. *et al.* (2018) Garden pollinators and the potential for ecosystem service flow to urban and peri-urban agriculture. *Sustainability* 10(6), 2047. DOI: 10.3390/su10062047.

Letourneau, D.K., Jedlicka, J.A., Bothwell, S.G. and Moreno, C.R. (2009) Effects of natural enemy biodiversity on the suppression of arthropod herbivores in terrestrial ecosystems. *Annual Review of Ecology, Evolution, and Systematics* 40(1), 573–592. DOI: 10.1146/annurev.ecolsys.110308.120320.

Liere, H., Egerer, M.H. and Philpott, S.M. (2019) Environmental and spatial filtering of ladybird beetle community composition and functional traits in urban landscapes. *Journal of Urban Ecology* 5(1), juz014. DOI: 10.1093/jue/juz014.

Lima, L.H.G., Mesquita, M.R., Skrip, L., de Souza Freitas, M.T., Silva, V.C. *et al.* (2016) DNA barcode for the identification of the sand fly *Lutzomyia longipalpis* plant feeding preferences in a tropical urban environment. *Scientific Reports* 6(1), 29742. DOI: 10.1038/srep29742.

Lin, B.B. and Egerer, M.H. (2018) Urban agricultre. In: Ossola, A. and Niemelä, J. (eds) *Urban Biodiversity*. Routledge, pp. 71–86.

Lin, B.B. and Fuller, R.A. (2013) FORUM: Sharing or sparing? How should we grow the world's cities? *The Journal of Applied Ecology* 50, 1161–1168.

Lin, B.B., Philpott, S.M. and Jha, S. (2015) The future of urban agriculture and biodiversity-ecosystem services: challenges and next steps. *Basic and Applied Ecology* 16(3), 189–201. DOI: 10.1016/j.baae.2015.01.005.

Ljung, K., Selinus, O., Otabbong, E. and Berglund, M. (2006) Metal and arsenic distribution in soil particle sizes relevant to soil ingestion by children. *Applied Geochemistry* 21(9), 1613–1624. DOI: 10.1016/j.apgeochem.2006.05.005.

Longcore, T. and Rich, C. (2004) Ecological light pollution. *Frontiers in Ecology and the Environment* 2(4), 191–198. DOI: 10.1890/1540-9295(2004)002[0191:ELP]2.0.CO;2.

Loram, A., Thompson, K., Warren, P.H. and Gaston, K.J. (2008) Urban domestic gardens (XII): the richness and composition of the flora in five UK cities. *Journal of Vegetation Science* 19(3), 321–330. DOI: 10.3170/2008-8-18373.

Losey, J.E. and Vaughan, M. (2006) The economic value of ecological services provided by insects. *BioScience* 56(4), 311–323. DOI: 10.1641/0006-3568(2006)56[311:TEVOES]2.0.CO;2.

Lovell, S.T. (2010) Multifunctional urban agriculture for sustainable land use planning in the United States. *Sustainability* 2(8), 2499 DOI: 10.3390/su2082499.

Lowenstein, D.M. and Minor, E.S. (2018) Herbivores and natural enemies of Brassica crops in urban agriculture. *Urban Ecosystems* 21(3), 519–529. DOI: 10.1007/s11252-018-0738-x.

Lowenstein, D.M., Matteson, K.C., Xiao, I., Silva, A.M. and Minor, E.S. (2014) Humans, bees, and pollination services in the city: the case of Chicago, IL (USA). *Biodiversity and Conservation* 23(11), 2857–2874. DOI: 10.1007/s10531-014-0752-0.

Madrid, L., Díaz-Barrientos, E. and Madrid, F. (2002) Distribution of heavy metal contents of urban soils in parks of Seville. *Chemosphere* 49(10), 1301–1308. DOI: 10.1016/S0045-6535(02)00530-1.

Mathieu, R., Freeman, C. and Aryal, J. (2007) Mapping private gardens in urban areas using object-oriented techniques and very high-resolution satellite imagery. *Landscape and Urban Planning* 81(3), 179–192. DOI: 10.1016/j.landurbplan.2006.11.009.

Matteson, K.C. and Langellotto, G.A. (2010) Determinates of inner city butterfly and bee species richness. *Urban Ecosystems* 13(3), 333–347. DOI: 10.1007/s11252-010-0122-y.

Matteson, K.C. and Langellotto, G.A. (2011) Small scale additions of native plants fail to increase beneficial insect richness in urban gardens. *Insect Conservation and Diversity* 4(2), 89–98. DOI: 10.1111/j.1752-4598.2010.00103.x.

Matteson, K. and Langellotto, G. (2009) *Bumble bee abundance in New York City community gardens: implications for urban agriculture*. Cities and the Environment (CATE) 2.

Matteson, K.C., Ascher, J.S. and Langellotto, G.A. (2008) Bee richness and abundance in New York City urban gardens. *Annals of the Entomological Society of America* 101(1), 140–150. DOI: 10.1603/0013-8746(2008)101[140:BRAAIN]2.0.CO;2.

Matthys, B., Koudou, B.G., N'Goran, E.K., Vounatsou, P., Gosoniu, L. *et al.* (2010) Spatial dispersion and characterisation of mosquito breeding habitats in urban vegetable-production areas of Abidjan, Côte d'Ivoire. *Annals of Tropical Medicine & Parasitology* 104(8), 649–666.

McIntyre, N.E. (2000) Ecology of urban arthropods: a review and a call to action. *Annals of the Entomological Society of America* 93(4), 825–835. DOI: 10.1603/0013-8746(2000)093[0825:EOUAAR]2.0.CO;2.

McKinney, M.L. (2002) Urbanization, biodiversity, and conservation: the impacts of urbanization on native species are poorly studied, but educating a highly urbanized human population about these impacts can greatly improve species conservation in all ecosystems. *BioScience* 52, 883–890.

McKinney, M.L. (2008) Effects of urbanization on species richness: a review of plants and animals. *Urban Ecosystems* 11(2), 161–176. DOI: 10.1007/s11252-007-0045-4.

Meindl, G.A. and Ashman, T.-L. (2013) The effects of aluminum and nickel in nectar on the foraging behavior of bumblebees. *Environmental Pollution* 177, 78–81. DOI: 10.1016/j.envpol.2013.02.017.

Morales, H., Ferguson, B., Marín, L., Gutiérrez, D., Bichier, P. *et al.* (2018) Agroecological pest management in the city: experiences from California and Chiapas. *Sustainability* 10(6), 2068. DOI: 10.3390/su10062068.

Neilson, S. and Rajakaruna, N. (2015) Phytoremediation of agricultural soils: using plants to clean metal-contaminated arable land. In: Ansari, A.A., Gill, S.S., Gill, R., Lanza, G.R. and Newman, L. (eds) *Phytoremediation: Management of Environmental Contaminants*. Springer, pp. 159–168.

Normandin, É.., Vereecken, N.J., Buddle, C.M. and Fournier, V. (2017) Taxonomic and functional trait diversity of wild bees in different urban settings. *PeerJ* 5(1748), e3051. DOI: 10.7717/peerj.3051.

Oberholtzer, L., Dimitri, C. and Pressman, A. (2014) Urban agriculture in the United States: characteristics, challenges, and technical assistance needs. *Journal of Extension* 52.

Ordóñez, A., Loredo, J., De Miguel, E. and Charlesworth, S. (2003) Distribution of heavy metals in the street dusts and soils of an industrial city in northern Spain. *Archives of Environmental Contamination and Toxicology* 44(2), 160–170. DOI: 10.1007/s00244-002-2005-6.

Otoshi, M.D., Bichier, P. and Philpott, S.M. (2015) Local and landscape correlates of spider activity density and species richness in urban gardens. *Environmental Entomology* 44(4), 1043–1051. DOI: 10.1093/ee/nvv098.

Pardee, G.L. and Philpott, S.M. (2014) Native plants are the bee's knees: local and landscape predictors of bee richness and abundance in backyard gardens. *Urban Ecosystems* 17(3), 641–659. DOI: 10.1007/s11252-014-0349-0.

Pawelek, J.C., Frankie, G.W., Thorp, R.W. and Przybylski, M. (2009) Modification of a community garden to attract native bee pollinators in urban San Luis Obispo, California. *Cities and the Environment* 2(1), 1–20. DOI: 10.15365/cate.2172009.

Peisley, R.K., Saunders, M.E. and Luck, G.W. (2015) A systematic review of the benefits and costs of bird and insect activity in agroecosystems. *Springer Science Reviews* 3(2), 113–125. DOI: 10.1007/s40362-015-0035-5.

Peralta-Videa, J.R., Lopez, M.L., Narayan, M., Saupe, G. and Gardea-Torresdey, J. (2009) The biochemistry of environmental heavy metal uptake by plants: implications for the food chain. *The International Journal of Biochemistry & Cell Biology* 41(8-9), 1665–1677. DOI: 10.1016/j.biocel.2009.03.005.

Pereira-Peixoto, M.H., Pufal, G., Martins, C.F. and Klein, A.-M. (2014) Spillover of trap-nesting bees and wasps in an urban–rural interface. *Journal of Insect Conservation* 18(5), 815–826.

Pereira-Peixoto, M.H., Pufal, G., Staab, M., Feitosa Martins,C. and Klein, A.M. (2016) Diversity and specificity of host-natural enemy interactions in an urban-rural interface. *Ecological Entomology* 41(3), 241–252. DOI: 10.1111/een.12291.

Perugini, M., Manera, M., Grotta, L., Abete, M.C., Tarasco, R. *et al.* (2011) Heavy metal (Hg, Cr, CD, and Pb) contamination in urban areas and wildlife reserves: honeybees as bioindicators. *Biological Trace Element Research* 140(2), 170–176. DOI: 10.1007/s12011-010-8688-z.

Peterson, L.R., Trivett, V., Baker, A.J.M., Aguiar, C. and Pollard, A.J. (2003) Spread of metals through an invertebrate food chain as influenced by a plant that hyperaccumulates nickel. *Chemoecology* 13(2), 103–108. DOI: 10.1007/s00049-003-0234-4.

Philpott, S.M. and Bichier, P. (2017) Local and landscape drivers of predation services in urban gardens. *Ecological Applications* 27(3), 966–976. DOI: 10.1002/eap.1500.

Philpott, S.M., Cotton, J., Bichier, P., Friedrich, R.L., Moorhead, L.C. *et al.* (2014) Local and landscape drivers of arthropod abundance, richness, and trophic composition in urban habitats. *Urban Ecosystems* 17(2), 513–532. DOI: 10.1007/s11252-013-0333-0.

Philpott, S.M., Albuquerque, S., Bichier, P., Cohen, H., Egerer, M.H. *et al.* (2019) Local and landscape drivers of Carabid activity, species richness, and traits in urban gardens in coastal California. *Insects* 10(4), 112–125. DOI: 10.3390/insects10040112.

Pickett, S.T.A. and Cadenasso, M.L. (2009) Altered resources, disturbance, and heterogeneity: a framework for comparing urban and non-urban soils. *Urban Ecosystems* 12(1), 23–44. DOI: 10.1007/s11252-008-0047-x.

Pocock, M.J.O., Evans, D.M. and Memmott, J. (2012) The robustness and restoration of a network of ecological networks. *Science* 335(6071), 973–977. DOI: 10.1126/science.1214915.

Politi Bertoncini, A., Machon, N., Pavoine, S. and Muratet, A. (2012) Local gardening practices shape urban lawn floristic communities. *Landscape and Urban Planning* 105(1-2), 53–61.

Ponge, J.-F. (2003) Humus forms in terrestrial ecosystems: a framework to biodiversity. *Soil Biology and Biochemistry* 35(7), 935–945. DOI: 10.1016/S0038-0717(03)00149-4.

Prince, S.P., Senthilkumar, P. and Subburam, V. (2001) Mulberry-silkworm food chain – a templet to assess heavy metal mobility in terrestrial ecosystems. *Environmental Monitoring and Assessment* 69(3), 231–238. DOI: 10.1023/A:1010715606097.

Quistberg, R.D., Bichier, P. and Philpott, S.M. (2016) Landscape and local correlates of bee abundance and species richness in urban gardens. *Environmental Entomology* 45(3), 592–601. DOI: 10.1093/ee/nvw025.

Raupp, M.J., Shrewsbury, P.M. and Herms, D.A. (2010) Ecology of herbivorous arthropods in urban landscapes. *Annual Review of Entomology* 55, 19–38. DOI: 10.1146/annurev-ento-112408-085351.

Ricketts, T.H., Regetz, J., Steffan-Dewenter, I., Cunningham, S.A., Kremen, C. *et al.* (2008) Landscape effects on crop pollination services: are there general patterns? *Ecology Letters* 11(5), 499–515. DOI: 10.1111/j.1461-0248.2008.01157.x.

Rudd, H., Vala, J. and Schaefer, V. (2002) Importance of backyard habitat in a comprehensive biodiversity conservation strategy: a connectivity analysis of urban green spaces. *Restoration Ecology* 10(2), 368–375. DOI: 10.1046/j.1526-100X.2002.02041.x.

Sánchez-Bayo, F. and Wyckhuys, K.A.G. (2019) Worldwide decline of the entomofauna: a review of its drivers. *Biological Conservation* 232, 8–27. DOI: 10.1016/j.biocon.2019.01.020.

Sattler, T., Duelli, P., Obrist, M.K., Arlettaz, R. and Moretti, M. (2010) Response of arthropod species richness and functional groups to urban habitat structure and management. *Landscape Ecology* 25(6), 941–954. DOI: 10.1007/s10980-010-9473-2.

Seto, K.C., Güneralp, B. and Hutyra, L.R. (2012) Global forecasts of urban expansion to 2030 and direct impacts on biodiversity and carbon pools. *Proceedings of the National Academy of Sciences* 109(40), 16083–16088. DOI: 10.1073/pnas.1211658109.

Shochat, E., Warren, P.S., Faeth, S.H., McIntyre, N.E. and Hope, D. (2006) From patterns to emerging processes in mechanistic urban ecology. *Trends in Ecology & Evolution* 21(4), 186–191. DOI: 10.1016/j.tree.2005.11.019.

Siegner, A., Sowerwine, J. and Acey, C. (2018) Does urban agriculture improve food security? Examining the nexus of food access and distribution of urban produced foods in the United States: a systematic review. *Sustainability* 10(9), 2988 DOI: 10.3390/su10092988.

Smith, R.M., Thompson, K., Hodgson, J.G., Warren, P.H. and Gaston, K.J. (2006) Urban domestic gardens (IX): composition and richness of the vascular plant flora, and implications for native biodiversity. *Biological Conservation* 129(3), 312–322. DOI: 10.1016/j.biocon.2005.10.045.

Speak, A.F., Mizgajski, A. and Borysiak, J. (2015) Allotment gardens and parks: provision of ecosystem services with an emphasis on biodiversity. *Urban Forestry & Urban Greening* 14(4), 772–781. DOI: 10.1016/j.ufug.2015.07.007.

Sridhara Chary, N., Kamala, C.T. and Samuel Suman Raj, D. (2008) Assessing risk of heavy metals from consuming food grown on sewage irrigated soils and food chain transfer. *Ecotoxicology and Environmental Safety* 69(3), 513–524. DOI: 10.1016/j.ecoenv.2007.04.013.

Theodorou, P., Radzevičiūtė, R., Settele, J., Schweiger, O., Murray, T.E. *et al.* (2016) Pollination services enhanced with urbanization despite increasing pollinator parasitism. *Proceedings of the Royal Society B: Biological Sciences* 283(1833), 20160561. DOI: 10.1098/rspb.2016.0561.

Tilman, D. (2001) Functional diversity. In: Levin, S.A. (ed.) *Encyclopedia of Biodiversity*. Elsevier, New York, pp. 109–120.

Tommasi, D., Miro, A., Higo, H.A. and Winston, M.L. (2004) Bee diversity and abundance in an urban setting. *The Canadian Entomologist* 136(6), 851–869. DOI: 10.4039/n04-010.

Tscharntke, T., Klein, A.M., Kruess, A., Steffan-Dewenter, I. and Thies, C. (2005) Landscape perspectives on agricultural intensification and biodiversity – ecosystem service management. *Ecology Letters* 8, 857–874. DOI: 10.1111/j.1461-0248.2005.00782.x.

Tylianakis, J.M. and Binzer, A. (2014) Effects of global environmental changes on parasitoid–host food webs and biological control. *Biological Control* 75, 77–86. DOI: 10.1016/j.biocontrol.2013.10.003.

Tzoulas, K., Korpela, K., Venn, S., Yli-Pelkonen, V., Kaźmierczak, A. *et al.* (2007) Promoting ecosystem and human health in urban areas using green infrastructure: a literature review. *Landscape and Urban Planning* 81(3), 167–178. DOI: 10.1016/j.landurbplan.2007.02.001.

United Nations (2010) *World Urbanization Prospects: 2009 Revisions*. Department of Economic and Social Affairs of the United Nations Secretariat, New York, NY.

United States Census Bureau (2020). Available at: https://www.census.gov/en.html (accessed 4 August 2020).

Uyttenbroeck,R., Hatt,S., Paul,A., Boeraeve,F., Piqueray,J. *et al.* (2016) Pros and cons of flowers strips for farmers: a review. *Biotechnology, Agronomy, Society and Environment* 20, 225–235.

Vergnes, A., Viol, I.L. and Clergeau, P. (2012) Green corridors in urban landscapes affect the arthropod communities of domestic gardens. *Biological Conservation* 145(1), 171–178. DOI: 10.1016/j.biocon.2011.11.002.

Walker, B., Kinzig, A. and Langridge, J. (1999) Original articles: plant attribute diversity, resilience, and ecosystem function: the nature and significance of dominant and minor species. *Ecosystems* 2, 95–113. DOI: 10.1007/s100219900062.

Werrell, P.A., Langellotto, G.A., Morath, S.U. and Matteson, K.C. (2009) The influence of garden size and floral cover on pollen deposition in urban community gardens. *Cities and the Environment* 2(1), 1–16. Available at: https://digitalcommons.lmu.edu/cate/vol2/iss1/ DOI: 10.15365/cate.2162009.

Whittinghill, L.J. and Rowe, D.B. (2012) The role of green roof technology in urban agriculture. *Renewable Agriculture and Food Systems* 27(4), 314–322. DOI: 10.1017/S174217051100038X.

Williams, M. (2009) Butterflies and day-flying moths in a fragmented urban landscape, south-west Western Australia: patterns of species richness. *Pacific Conservation Biology* 15(1), 32–46. DOI: 10.1071/PC090032.

Williams, D.R., Brule, H., Kelley, S.S. and Skinner, E.A. (2018) Science in the learning gardens: a study of students' motivation, achievement, and science identity in low-income middle schools. *International Journal of STEM Education* 5(1), 8. DOI: 10.1186/s40594-018-0104-9.

Woodward, G. and Bohan, D. (eds) (2013) *Ecological Networks in an Agricultural World*. Academic Press.

Yang, E.-C., Chang, H.-C., Wu, W.-Y. and Chen, Y.-W. (2012) Impaired olfactory associative behavior of honeybee workers due to contamination of imidacloprid in the larval stage. *PLoS ONE* 7(11), e49472.

Yu, D., Xun, B., Shi, P., Shao, H. and Liu, Y. (2012) Ecological restoration planning based on connectivity in an urban area. *Ecological Engineering* 46, 24–33. DOI: 10.1016/j.ecoleng.2012.04.033.

Yue, J., Lei, T., Li, C. and Zhu, J. (2012) The application of unmanned aerial vehicle remote sensing in quickly monitoring crop pests. *Intelligent Automation & Soft Computing* 18(8), 1043–1052. DOI: 10.1080/10798587.2008.10643309.

Zasada, I. (2011) Multifunctional peri-urban agriculture—a review of societal demands and the provision of goods and services by farming. *Land Use Policy* 28(4), 639–648. DOI: 10.1016/j.landusepol.2011.01.008.

Zeeuw, H., Veenhuizen, R. and Dubbeling, M. (2011) The role of urban agriculture in building resilient cities in developing countries. *The Journal of Agricultural Science* 149(S1), 153–163. DOI: 10.1017/S0021859610001279.

13 Future Prospects and Challenges in Urban Ecology

Gail A. Langellotto*
Department of Horticulture, Oregon State University, Corvallis, Oregon, USA

Early and influential ecologists have long studied the patterns and associations of living organisms within and near cities. For example, Frank Lutz, first chair of the entomology department at the American Museum of Natural History, meticulously documented the 1402 insect species collected from his small suburban lot just outside of New York City (Lutz, 1941). Eugene Odum and Thomas Burleigh noted the geographic expansion of the breeding ranges of robins, song-sparrows, and warblers, and linked these changes to the abundance of early seral vegetation 'or their artificial equivalent' in human-altered landscapes (Odum and Burleigh, 1946). Botanists studied successional patterns in Chicago-area plant communities in the late 19th and early 20th century (Cowles, 1901). Despite these and other early contributions, the ecological sciences more often focused on study organisms and field sites with limited to no direct human influence (McDonnell, 2011). For example, less than one half of one per cent of all papers published in nine leading ecological journals between 1995 and 2000 focused on urban systems or urban species (Collins *et al.*, 2000). Many of these studies focused on single species, in small-scale patches, using a single disciplinary approach (McDonnell, 2011). Although such studies are important for understanding basic patterns of urban species distributions, they fail to capture the complexity of cities as social-ecological systems, where human beings are central to patterns and processes (Grimm *et al.*, 2000; McDonnell, 2011; Pickett *et al.*, 2016). Interdisciplinary, multi-scalar and long-term studies are needed to generate knowledge and advance sustainability in, of and for urban systems (Pickett *et al.*, 2016).

Today, the importance and influence of urban ecology in modern science is impossible to ignore. While there were 1.4 billion people living on earth when ecology emerged as a discipline in the 1940s (Demeny, 1990; McDonnell, 2011), there are now more than 7.7 billion people sharing this planet (United Nations *et al.*, 2019), 55% of whom live in urban areas (United Nations *et al.*, 2018a). Global increases in urbanization are expected to continue throughout this century, with urban areas expected to absorb virtually all future population growth. Urbanization has been identified as one of four globally important trends in population demography by the United Nations *et al.* (2019). In order to meet the United Nations Sustainable Development Goal of making urban areas safe, inclusive, resilient and sustainable (United Nations *et al.*, 2018b),

*Gail.Langellotto@oregonstate.edu

continued and increased investment in the science of urban ecology is needed.

In this chapter, I suggest key areas of investigation to guide the urban ecological sciences over the next 30 years. Specifically, I examine the state of knowledge at the population, community, landscape and ecosystem scales, and identify key gaps that are ripe for future study. I look at potential applications of urban ecological knowledge to advance urban sustainability, before concluding with key challenges that threaten progress. Throughout this chapter, I assume the importance of adopting a research paradigm that endeavours to conduct ecology *for* the city (Pickett *et al.*, 2016), where scientific advances that yield sustainable urban systems depend upon long-term and collaborative interactions between multi-disciplinary scientists, urban residents and decision makers.

Designing Resilient Cities Requires an Understanding of Urban Evolution

Humans are exerting strong selective pressures on a multitude of wild populations via anthropogenic influence on climate, land cover, hydrology, nutrient cycling, resource abundance and pollutants. Even Darwin's iconic Galapagos finches have been affected, as access to human foods have shifted dietary preferences of several species towards processed foods (e.g. chips and sweets) and away from species-specific diets of seeds, insects or fruits that drove adaptive radiation and maintained niche separation (De León *et al.*, 2018). Preliminary analyses suggest that phenotypic changes in urban systems exceed those in natural systems, and that phenotypic changes in urban systems are abrupt rather than gradual (Hendry *et al.*, 2008). Much of the work on evolutionary change in urban systems has focused on non-adaptive change due to restricted gene flow in fragmented urban landscapes, as well as to genetic drift within an environment where stochastic events are common (Johnson and Munshi-South, 2017; Santangelo *et al.*, 2018). However, there are documented cases of adaptive evolution as well. For example, birds in urban environments sing at a higher pitch than their rural counterparts, presumably so as to be heard by prospective mates above city noise (Slabbekoorn and Peet, 2003). Weedy plants reduce seed dispersal in as few as five generations in urban environments, because a majority of dispersed seeds settle on asphalt or concrete (Cheptou *et al.*, 2008). Urban killifish exhibit rapid evolution of tolerance to pollutants (Reid *et al.*, 2016). Acorn ants in urban environments display rapid evolution of tolerance to heat, and greater sensitivity to cooler temperatures (Diamond *et al.*, 2017).

Research in urban evolution is expected to proliferate in the coming decades, as scientists try to better understand the impact of urbanization on non-adaptive evolution (via restricted gene flow or drift) and adaptive evolution (via natural selection). Key questions include whether or not informed landscape design can ameliorate the impact of habitat fragmentation on gene flow or genetic drift contingent on the availability of habitat corridors. In addition, evolutionary biologists are asking whether or not adaptive evolution is capable of keeping pace with the rapid and dramatic selection pressures associated with urbanization, and identifying the precursors to evolutionary resilience in the face of urban change (Hendry *et al.*, 2008; Johnson and Munshi-South, 2017; Santangelo *et al.*, 2018).

Current evidence suggests that phenotypic plasticity and genetic variation play key roles in species' capacity to respond to urbanization (Hendry *et al.*, 2008; Reid *et al.*, 2016; Diamond *et al.*, 2017). The importance of genetic diversity to sustainable urban systems should be a key concern to urban planners and landscape architects (Paap *et al.*, 2017), where traditional designs of urban streetscapes lean towards plantings of a few genetically monotonous cultivars, i.e. cultivars that have and transfer the same genetic pattern from one generation to the next (MacFarlane and Meyer, 2005). The lack of genetic variation in horticulturally important plantings is often maintained via asexual propagation to retain morphological traits that are unique, aesthetically pleasing and/or that conform to the compact planting spaces that are typical of urban environments. However, such plantings are prone to decline via invasive pests or other stochastic stresses.

Many professionals within the horticultural trade fail to recognize the importance of species diversity, much less genetic diversity, to the long-term persistence of plantings in the face of urban stresses (Polakowski *et al.*, 2011). As scientists develop a better understanding of the mechanisms that promote or constrain adaptive evolution in urban systems, they have an opportunity to affect the sustainable design of cities, particularly at the first trophic level. Specifically, urban evolutionary biologists could collaborate with plant breeders to develop plant materials that meet the needs of urban arborists, landscape architects, landscapers and gardeners, while retaining the genetic variation needed to respond to the multitude of selection pressures in an urban environment. Ideally, there should be a way to retain, or at least balance, genetic variation with favoured plant traits and practical propagation methods in horticultural production and distribution systems.

Diverse Urban Communities Contribute to Urban Sustainability

Biodiversity is central to the idea of ecosystem function (Cardinale *et al.*, 2012), and, by extension, ecosystem resilience and sustainability. The conservation and restoration of biodiversity should thus be integral to urban planning, where increasing focus is placed on the importance of resilient urban systems to human well-being (McPhearson *et al.*, 2015). For many taxa, conservation and restoration goals are challenging. Urbanization can contribute to biodiversity loss via habitat loss and fragmentation, the introduction of exotic and invasive species, and novel environmental stresses. These factors can promote biotic filtering and homogenization (Lizée *et al.*, 2011), ultimately degrading multiple measures of biodiversity, including species richness and community evenness (Shochat *et al.*, 2010).

The diverse plantings that are typical of urban gardens could benefit organisms at higher trophic levels. Specifically, urban gardens can have higher levels of plant α-diversity (within gardens), β-diversity (between gardens), and γ-diversity (in aggregate across the city) than nearby agricultural or natural areas (Hope *et al.*, 2003; Grove *et al.*, 2006; Grimm *et al.*, 2008). Whether or not the diversity of plants within gardens has biodiversity benefits at higher trophic levels seems to depend, at least in part, on the decisions that gardeners make when planting and maintaining their gardens.

For bees, research suggests that gardeners can increase local bee richness via fairly simple actions related to plant selection and care, such as planting more flowering plants (Matteson and Langellotto, 2010), tolerating lawn weeds (Larson *et al.*, 2014) and decreasing the frequency of lawn mowing (Lerman *et al.*, 2018). Provisioning native plants in urban landscape designs is also important to urban biodiversity. Insectivorous birds benefit from yards with a high proportion of native plants, because these plants support larger and more diverse populations of caterpillars (Burghardt *et al.*, 2009; Narango *et al.*, 2018). Specifically, Burghardt *et al.* (2009) and Narango *et al.* (2018) found that gardeners that choose to plant or protect native grasses, forbs, shrubs and trees in their yards and gardens were able to positively impact the diversity of breeding birds via increases in caterpillar diversity and abundance. These studies show the potential influence that gardeners can have at higher trophic levels (herbivores and insectivorous birds) via decisions that determine the plant community on the first trophic level. In field comparisons of native plants with cultivated varieties of these same natives, bees, butterflies and moths prefer to forage from the native plants (White, 2016).

Even though the provisioning and selection of plants can benefit birds and pollinating insects, it does not guarantee that these or other components of a biological community will assemble and persist in urban systems (Faeth *et al.*, 2011). Despite the documented benefits of urban systems to bees (Hall *et al.*, 2017), ground-nesting bees are disproportionately absent from urban gardens (Langellotto, 2017). The bees that do persist in urban habitats tend to be phylogenetically and functionally homogenous (Harrison *et al.*, 2018), which limits their collective contribution to urban ecosystem function. Further, despite the documented benefits of native plants to diversity at higher trophic levels, exotic plants typically comprise 70% of the plant palette in urban gardens and other ornamental landscapes (Loram *et al.*, 2008). Small-scale installations of native plants are insufficient to provide a net benefit to pollinators (Matteson and Langellotto, 2011) or birds (Narango *et al.*, 2018). Unfortunately, many barriers exist that limit the production and sale of native plants, which ultimately limits gardeners' exposure to and acceptance of native plants in urban garden and landscape design (Anderson, 2019).

The challenges associated with increasing the availability and acceptance of native plants to benefit urban biodiversity is one example of how building more sustainable ecosystems relies on developing a better understanding of urban yards and gardens as social-ecological systems. Social norms and expectations are important drivers of yard and garden management (Locke *et al.*, 2018). Even if individuals want to change their garden to include nesting sites for bees, an abundance of native plants or a less tidy lawn, neighbours or neighborhood associations often push back on aesthetics that do not conform to the rest of the neighbourhood. These pressures are even more pronounced for front-facing gardens, which has given rise to the landscape mullet concept (Locke *et al.*, 2018). A landscape mullet results when homeowners manage front-facing spaces according to social norms, while using rear spaces to express personal preferences (Locke *et al.*, 2018). Challenging and changing these social norms will take concerted and collaborative efforts by urban residents, professionals within the nursery and landscape industries, homeowners, neighbourhood associations, and social and ecological scientists. However, successfully shifting social norms related to gardening and landscaping could have an outsized impact on urban sustainability. While residential gardens cover only 2% of land area in continental USA (derived from Milesi *et al.*, 2005), they can comprise 25–35% of urban land area and more than 50% of urban green space (Loram *et al.*, 2007; Mathieu *et al.*, 2007).

Questions of Scale, Linkages and Equity in Urban Landscapes

Ecological studies across urban to rural gradients have generally found that the lowest levels of species richness occur towards the urban core and that the highest levels of species richness occur in more suburban sites (reviewed in McKinney, 2008). Urbanization encompasses a gradient of disturbance, including major disturbance, such as development, and more frequent, lower-level disturbance associated with landscape maintenance (Faeth *et al.*, 2011). Intermediate levels of disturbance are thought to select for high species richness via the coexistence of colonizers and late seral species. In fact, the intermediate disturbance hypothesis (sensu Connell, 1978) has been used to explain patterns of species richness along rural-to-urban gradients (Blair, 1996; Blair and Launer, 1997, but see Collins *et al.*, 1995 and Fox, 2013). Suburban sites tend to have more green space than the urban core, and these green spaces are generally managed in ways that approximate 'intermediate' levels of disturbance (e.g. composting, less intensive weeding, reduced pesticide use and cultivation of diverse plant species). This combination offers greater resource abundance but less resource stability, which might explain why peak species richness can often be found at intermediate levels of urbanization, such as in suburban sites (Faeth *et al.*, 2011).

Despite the potential of suburban gardens to foster biodiversity in urban systems, these same sites tend to exhibit ecologically homogenized plant communities and microclimates (Groffman *et al.*, 2014). This homogenization results from the common and convergent decisions that individuals make related to plant selection and irrigation regimes across cities in different ecoregions (Groffman *et al.*, 2014). These individual decisions are limited at the point of purchase by the plant selection offered by retail nurseries (Avolio *et al.*, 2018). This winnowing of choice, plant palettes and management regimes to a few narrowly defined examples of urban and suburban gardens has measurable and mostly negative impacts on ecosystem processes and services, which likely scale up across broad spatial scales (Groffman *et al.*, 2014).

The question of scale is an important one in urban ecology. Studies have shown local, within-site characteristics are more important to the biodiversity of bees and butterflies in urban gardens than are landscape-level characteristics (Matteson and Langellotto, 2010; Pardee and Philpott, 2014). However, there is evidence that neighbourhood-level characteristics are also important (Lowenstein *et al.*, 2014). For birds, the characteristics of groups of gardens, in aggregate, were more important for species richness than characteristics at the neighbourhood or landscape scale (Belaire *et al.*, 2014). This begs the question, at what scale do the characteristics of individual urban parcels combine to promote urban sustainability at the neighbourhood and landscape scales? In other words, how many urban or suburban lots within a neighbourhood would have to adopt sustainable landscape design and management practices before the benefits are both measurable at broader spatial scales and persistent across long-term temporal scales?

The proliferation of urbanization is also creating landscapes where urban settlements are immediately adjacent to agricultural land (Harvey and Works, 2002). This creates an opportunity to study how these systems are linked socially (via the production and consumption of agricultural products) and ecologically (via pollinator movement between urban gardens and farms). In the USA, 90% of fruits, nuts and berries and 78% of vegetables and melons (all pollinator-dependent crops) are now grown in countries with significant urban encroachment (Francis *et al.*, 2012). If pollinators readily cross the boundary between urban and agricultural landscapes, urban gardens could provision pollinators to farms at the peri-urban boundary, or may provide a refuge to bees from agricultural stresses (Langellotto *et al.*, 2018).

Reducing Vulnerability to Urban Heat and Flooding

The average mean temperature within the urban landscape can be between 1–3 degrees Celsius warmer than adjacent non-urban areas. Increased urban temperature in the midst of global climate change drives energy demand, which contributes to greenhouse gas emissions, air and water pollution, and heat-related illness and mortality. Building resilient cities requires a thorough accounting of the areas and residents that are most vulnerable to urban heat and climate-induced hazards (Shandas *et al.*, 2018), as well as data-informed design and management responses. Perhaps not surprisingly, the most vulnerable groups tend to have the least economic, political and/or social capital (Huang *et al.*, 2011; Voelkel *et al.*, 2018). Urban heat and urban particulate pollution can be tempered via tree-planting programmes (Livesley *et al.*, 2016). Unfortunately, the most vulnerable areas are often the least likely recipients of municipal tree-planting programmes (Watkins *et al.*, 2017).

Combating urban heat requires an understanding of how the built environment and green environment interact. For example, although morning temperatures are affected by the presence of low-lying vegetation, afternoon and evening temperatures are most affected by building height and variation in building height (Voelkel *et al.*, 2018). This suggests a path forward in planning future cities, but also an opportunity to retrofit existing cities by integrating herbaceous plants, shrubs and street trees throughout the urban landscape.

Urban Waste as a Component of Ecosystems Ecology in the Anthropocene

Urban systems exhibit biogeochemical and hydrological cycles that are distinct from agricultural or unmanaged counterparts (Kaye *et al.*, 2006). Scientists are developing a better understanding of how human actions at the household, neighbourhood and city scales influence fluxes in urban nutrient and hydrological cycles. The built landscape, the influence of demographic trends on ecological stoichiometry and household-scale actions have been identified as the three primary drivers of biogeochemical fluxes in urban systems (Kaye *et al.*, 2006). In addition to these drivers, urban green space also plays a role, and has been associated with reductions in urban greenhouse gasses, particulate pollution and stormwater run-off. To date, scientists have not been able to model how robust these benefits are to variable anthropogenic demands and inputs (Dijst *et al.*, 2018).

The concept of urban metabolism has been developed to provide a framework that expands urban ecosystem ecology beyond nutrient and water cycles, and to include anthropogenic materials and waste (Dijst *et al.*, 2018). This is an important development, since the creation, consumption and disposal of materials is a defining characteristic of cities. In fact, in her book entitled *Cities: The First 6,000 Years*, anthropologist Monica Smith notes that rubbish is perhaps the most telling characteristic of urban activity (Smith, 2019). Although the impact of solid waste, particularly plastic waste, has been well studied in marine systems, there has been relatively little attention to this issue in terrestrial ecosystems. Within terrestrial ecosystems, 19 species of birds and 13 species of mammals have been observed to forage from urban rubbish dumps in only 25 hours of observation time (Katlam *et al.*, 2018). Bees (MacIvor and Moore, 2013), ants (Michlewicz and Tryjanowski, 2018), crows (Townsend and Barker, 2014) and rodents (Cavia *et al.*, 2009) use anthropogenic waste for nest-site materials. The interaction of the biotic components of an ecosystem with anthropogenic waste can influence organisms' fitness and evolution (Suárez-Rodríguez *et al.*, 2017) but may also be important to the transformation and degradation of rubbish. Given the dominance of rubbish as a characteristic of urban systems, understanding how organisms interact with and transform (or are transformed by) anthropogenic waste will be essential to modelling and planning sustainable cities.

Key Challenges in Urban Ecology

The characteristics of urban landscapes are shaped by social norms and interactions, the consumption of resources, the production of refuse, the built environment and urban green space, all of which occur amidst global climate change. The complexity of urban ecosystems is unlike any other, with multiple land managers making a series of independent decisions about how to build, plant and manage a site according to current policy and social norms. It is no coincidence that the term 'wicked problems' was coined by urban planners (Rittel and Webber, 1973) to describe complex problems that are characterized by multiple interacting systems and social and institutional uncertainty. Rittel and Webber (1973) argue that perfect solutions do not exist for wicked problems, but that solutions can only be judged as relatively better or worse, depending upon how the problem is framed. Thus, one of the biggest challenges to progress in the urban ecological sciences is the complex nature of urban ecosystems, where a singular approach towards sustainable cities will not suffice.

Meeting this challenge requires a convener who can bring together multi-disciplinary scientists, urban residents and decision makers to collaborate to find the best solutions to the wicked problems inherent in urban systems. Although a great deal of progress has been made towards advancing the science of urban ecology, there has also been movement away from transdisciplinary studies, with ecological sciences dominating other disciplines (Young and Wolf, 2006).

The National Science Foundation's funding of urban long-term ecological research efforts in Phoenix, Arizona, and Baltimore, Maryland, arguably did more to advance the transdisciplinary study of urban systems than any other singular effort. When the creation of these research programmes was announced, it was characterized as 'a quantum leap in studying the way the urban environment works' (National Science Foundation, 1997). Unfortunately, National Science Foundation funding for the Baltimore long-term research effort came to an end in 2018 (Wheeler, 2019), although a call for a new urban long-term research site was posted in 2019. This highlights the second major challenge facing the urban ecological sciences – funding for long-term, transdisciplinary and collaborative studies in, of and for urban systems (*sensu* Pickett *et al.*, 2016; see also Chapter 7, 'Urban ecology as an integrative science and practice'). Given that urbanization is a globally important driver of ecological properties and processes across spatial and temporal scales, continued and increased investments in the science of urban ecology are needed now, more than ever, in order to make cities safe, inclusive, resilient and sustainable.

References

Anderson, A. (2019) Native plant production and marketing. *Digger* 63, 33–35.

Avolio, M.L., Pataki, D.E., Trammell, T.L.E. and Endter-Wada, J. (2018) Biodiverse cities: the nursery industry, homeowners, and neighborhood differences drive urban tree composition. *Ecological Monographs* 88(2), 259–276. DOI: 10.1002/ecm.1290.

Belaire, J.A., Whelan, C.J. and Minor, E.S. (2014) Having our yards and sharing them too: the collective effects of yards on native bird species in an urban landscape. *Ecological Applications* 24(8), 2132–2143. DOI: 10.1890/13-2259.1.

Blair, R.B. (1996) Land use and avian species diversity along an urban gradient. *Ecological Applications* 6(2), 506–519. DOI: 10.2307/2269387.

Blair, R.B. and Launer, A.E. (1997) Butterfly diversity and human land use: species assemblages along an urban grandient. *Biological Conservation* 80(1), 113–125. DOI: 10.1016/S0006-3207(96)00056-0.

Burghardt, K.T., Tallamy, D.W. and Gregory Shriver, W. (2009) Impact of native plants on bird and butterfly biodiversity in suburban landscapes. *Conservation Biology* 23(1), 219–224. DOI: 10.1111/j.1523-1739.2008.01076.x.

Cardinale, B.J., Duffy, J.E., Gonzalez, A., Hooper, D.U., Perrings, C. *et al.* (2012) Biodiversity loss and its impact on humanity. *Nature* 486(7401), 59–67. DOI: 10.1038/nature11148.

Cavia, R., Cueto, G.R. and Suárez, O.V. (2009) Changes in rodent communities according to the landscape structure in an urban ecosystem. *Landscape and Urban Planning* 90(1-2), 11–19. DOI: 10.1016/j.landurbplan.2008.10.017.

Cheptou, P.-O., Carrue, O., Rouifed, S. and Cantarel, A. (2008) Rapid evolution of seed dispersal in an urban environment in the weed *Crepis sancta*. *Proceedings of the National Academy of Sciences* 105(10), 3796–3799. DOI: 10.1073/pnas.0708446105.

Collins, S.L., Glenn, S.M. and Gibson, D.J. (1995) Experimental analysis of intermediate disturbance and initial floristic composition: decoupling cause and effect. *Ecology* 76(2), 486–492. DOI: 10.2307/1941207.

Collins, J.P., Kinzig, A., Grimm, N.B., Fagan, W.F., Hope, D. *et al.* (2000) A new urban ecology: modeling human communities as integral parts of ecosystems poses special problems for the development and testing of ecological theory. *American Scientist* 88, 416–425.

Connell, J.H. (1978) Diversity in tropical rain forests and coral reefs. *Science* 199(4335), 1302–1310. DOI: 10.1126/science.199.4335.1302.

Cowles, H.C. (1901) The physiographic ecology of Chicago and vicinity; a study of the origin, development, and classification of plant societies. *Botanical Gazette* 31(2), 73–108. DOI: 10.1086/328080.

De León, L.F., Sharpe, D.M.T., Gotanda, K.M., Raeymaekers, J.A.M., Chaves, J.A. *et al.* (2018) Urbanization erodes niche segregation in Darwin's finches. *Evolutionary Applications* 12(7).

Demeny, P. (1990) Populations. In: Turner II, B.L., Clark, W.C., Kates, R.W., Richards, J.F., Mathews, J.T., *et al.* (eds) *The Earth as Transformed by Human Action: Global and Regional Changes in the Biosphere over the Past 300 Years*. Cambridge University Press and Clark University, New York, pp. 41–54.

Diamond, S.E., Chick, L., Perez, A., Strickler, S.A. and Martin, R.A. (2017) Rapid evolution of ant thermal tolerance across an urban-rural temperature cline. *Biological Journal of the Linnean Society* 121(2), 248–257. DOI: 10.1093/biolinnean/blw047.

Dijst, M., Worrell, E., Böcker, L., Brunner, P., Davoudi, S. *et al.* (2018) Exploring urban metabolism—towards an interdisciplinary perspective. *Resources, Conservation and Recycling* 132, 190–203. DOI: 10.1016/j.resconrec.2017.09.014.

Faeth, S.H., Bang, C. and Saari, S. (2011) Urban biodiversity: patterns and mechanisms. *Annals of the New York Academy of Sciences* 1223, 69–81. DOI: 10.1111/j.1749-6632.2010.05925.x.

Fox, J.W. (2013) The intermediate disturbance hypothesis should be abandoned. *Trends in Ecology & Evolution* 28(2), 86–92. DOI: 10.1016/j.tree.2012.08.014.

Francis, C.A., Hansen, T.E., Fox, A.A., Hesje, P.J., Nelson, H.E. *et al.* (2012) Farmland conversion to non-agricultural uses in the US and Canada: current impacts and concerns for the future. *International Journal of Agricultural Sustainability* 10(1), 8–24. DOI: 10.1080/14735903.2012.649588.

Grimm, N.B., Morgan Grove, J., Pickett, S.T.A. and Redman, C.L. (2000) Integrated approaches to long-term studies of urban ecological systems. *BioScience* 50(7), 571–584. DOI: 10.1641/0006-3568(2000)050[0571:IATLTO]2.0.CO;2.

Grimm, N.B., Faeth, S.H., Golubiewski, N.E., Redman, C.L., Wu, J. *et al.* (2008) Global change and the ecology of cities. *Science* 319(5864), 756–760. DOI: 10.1126/science.1150195.

Groffman, P.M., Cavender-Bares, J., Bettez, N.D., Grove, J.M., Hall, S.J. *et al.* (2014) Ecological homogenization of urban USA. *Frontiers in Ecology and the Environment* 12(1), 74–81. DOI: 10.1890/120374.

Grove, J.M., Troy, A.R., O'Neil-Dunne, J.P.M., Burch, W.R., Cadenasso, M.L. *et al.* (2006) Characterization of households and its implications for the vegetation of urban ecosystems. *Ecosystems* 9(4), 578–597. DOI: 10.1007/s10021-006-0116-z.

Hall, D.M., Camilo, G.R., Tonietto, R.K., Ollerton, J., Ahrné, K. *et al.* (2017) The city as a refuge for insect pollinators. *Conservation Biology* 31(1), 24–29. DOI: 10.1111/cobi.12840.
Harrison, T., Gibbs, J. and Winfree, R. (2018) Phylogenetic homogenization of bee communities across ecoregions. *Global Ecology and Biogeography* 27(12), 1457–1466. DOI: 10.1111/geb.12822.
Harvey, T. and Works, M.A. (2002) Urban sprawl and rural landscapes: perceptions of landscape as amenity in Portland, Oregon. *Local Environment* 7(4), 381–396. DOI: 10.1080/1354983022000027509.
Hendry, A.P., Farrugia, T.J. and Kinnison, M.T. (2008) Human influences on rates of phenotypic change in wild animal populations. *Molecular Ecology* 17(1), 20–29. DOI: 10.1111/j.1365-294X.2007.03428.x.
Hope, D., Gries, C., Zhu, W., Fagan, W.F., Redman, C.L. *et al.* (2003) Socioeconomics drive urban plant diversity. *Proceedings of the National Academy of Sciences* 100(15), 8788–8792. DOI: 10.1073/pnas.1537557100.
Huang, G., Zhou, W. and Cadenasso, M.L. (2011) Is everyone hot in the city? Spatial pattern of land surface temperatures, land cover and neighborhood socioeconomic characteristics in Baltimore, MD. *Journal of Environmental Management* 92(7), 1753–1759. DOI: 10.1016/j.jenvman.2011.02.006.
Johnson, M.T.J. and Munshi-South, J. (2017) Evolution of life in urban environments. *Science* 358(6363), eaam8327. DOI: 10.1126/science.aam8327.
Katlam, G., Prasad, S., Aggarwal, M. and Kumar, R. (2018) Trash on the menu:patterns of animal visitation and foraging behaviour at garbage dumps. *Current Science* 115(12), 2322–2326. DOI: 10.18520/cs/v115/i12/2322-2326.
Kaye, J.P., Groffman, P.M., Grimm, N.B., Baker, L.A. and Pouyat, R.V. (2006) A distinct urban biogeochemistry? *Trends in Ecology & Evolution* 21(4), 192–199. DOI: 10.1016/j.tree.2005.12.006.
Langellotto, G.A. (2017) An analysis of bee communities in home and community gardens. *Acta Horticulturae* 1189, 491–496. DOI: 10.17660/ActaHortic.2017.1189.98.
Langellotto, G., Melathopoulos, A., Messer, I., Anderson, A., McClintock, N. *et al.* (2018) Garden pollinators and the potential for ecosystem service flow to urban and peri-urban agriculture. *Sustainability* 10(6), 2047–2062. DOI: 10.3390/su10062047.
Larson, J.L., Kesheimer, A.J. and Potter, D.A. (2014) Pollinator assemblages on dandelions and white clover in urban and suburban lawns. *Journal of Insect Conservation* 18(5), 863–873. DOI: 10.1007/s10841-014-9694-9.
Lerman, S.B., Contosta, A.R., Milam, J. and Bang, C. (2018) To mow or to mow less: lawn mowing frequency affects bee abundance and diversity in suburban yards. *Biological Conservation* 221, 160–174. DOI: 10.1016/j.biocon.2018.01.025.
Livesley, S.J., McPherson, G.M. and Calfapietra, C. (2016) The urban forest and ecosystem services: impacts on urban water, heat, and pollution cycles at the tree, street, and city scale. *Journal of Environmental Quality* 45(1), 119–124. DOI: 10.2134/jeq2015.11.0567.
Lizée, M.-H., Mauffrey, J.-F., Tatoni, T. and Deschamps-Cottin, M. (2011) Monitoring urban environments on the basis of biological traits. *Ecological Indicators* 11(2), 353–361. DOI: 10.1016/j.ecolind.2010.06.003.
Locke, D.H., Roy Chowdhury, R., Grove, J.M., Martin, D.G., Goldman, E. *et al.* (2018) Social norms, yard care, and the difference between front and back yard management: examining the landscape mullets concept on urban residential lands. *Society & Natural Resources* 31(10), 1169–1188. DOI: 10.1080/08941920.2018.1481549.
Loram, A., Tratalos, J., Warren, P.H. and Gaston, K.J. (2007) Urban domestic gardens (X): the extent & structure of the resource in five major cities. *Landscape Ecology* 22(4), 601–615. DOI: 10.1007/s10980-006-9051-9.
Loram, A., Thompson, K., Warren, P.H. and Gaston, K.J. (2008) Urban domestic gardens (XII): the richness and composition of the flora in five UK cities. *Journal of Vegetation Science* 19(3), 321–330. DOI: 10.3170/2008-8-18373.
Lowenstein, D.M., Matteson, K.C., Xiao, I., Silva, A.M. and Minor, E.S. (2014) Humans, bees, and pollination services in the city: the case of Chicago, IL (USA). *Biodiversity and Conservation* 23(11), 2857–2874. DOI: 10.1007/s10531-014-0752-0.
Lutz, F.E. (1941) *A Lot of Insects: Entomology in a Suburban Garden*. G. P. Putnam's Sons, New York, New York.
MacFarlane, D.W. and Meyer, S.P. (2005) Characteristics and distribution of potential ash tree hosts for emerald ash borer. *Forest Ecology and Management* 213(1-3), 15–24. DOI: 10.1016/j.foreco.2005.03.013.

MacIvor, J.S. and Moore, A.E. (2013) Bees collect polyurethane and polyethylene plastics as novel nest materials. *Ecosphere* 4(12), art155–156. DOI: 10.1890/ES13-00308.1.

Mathieu, R., Freeman, C. and Aryal, J. (2007) Mapping private gardens in urban areas using object-oriented techniques and very high-resolution satellite imagery. *Landscape and Urban Planning* 81(3), 179–192. DOI: 10.1016/j.landurbplan.2006.11.009.

Matteson, K.C. and Langellotto, G.A. (2010) Determinates of inner city butterfly and bee species richness. *Urban Ecosystems* 13(3), 333–347. DOI: 10.1007/s11252-010-0122-y.

Matteson, K.C. and Langellotto, G.A. (2011) Small scale additions of native plants fail to increase beneficial insect richness in urban gardens. *Insect Conservation and Diversity* 4(2), 89–98. DOI: 10.1111/j.1752-4598.2010.00103.x.

McDonnell, M.J. (2011) The history of urban ecology: an ecologist's perspective. In: Niemelä, J., Breuste, J.H., Elmqvist, T., Guntenspergen, G., James, P., *et al.* (eds) *Urban Ecology: Patterns, Processes and Applications*. Oxford University Press, Oxford, UK, pp. 5–13.

McKinney, M.L. (2008) Effects of urbanization on species richness: a review of plants and animals. *Urban Ecosystems* 11(2), 161–176. DOI: 10.1007/s11252-007-0045-4.

McPhearson, T., Andersson, E., Elmqvist, T. and Frantzeskaki, N. (2015) Resilience of and through urban ecosystem services. *Ecosystem Services* 12, 152–156. DOI: 10.1016/j.ecoser.2014.07.012.

Michlewicz, M. and Tryjanowski, P. (2018) Anthropogenic waste products as preferred nest sites for *Myrmica rubra* (L.) (Hymenoptera, Formicidae). *Journal of Hymenoptera Research* 57, 103–114. DOI: 10.3897/jhr.57.12491.

Milesi, C., Running, S.W., Elvidge, C.D., Dietz, J.B., Tuttle, B.T. *et al.* (2005) Mapping and modeling the biogeochemical cycling of turf grasses in the United States. *Environmental Management* 36(3), 426–438. DOI: 10.1007/s00267-004-0316-2.

Narango, D.L., Tallamy, D.W. and Marra, P.P. (2018) Nonnative plants reduce population growth of an insectivorous bird. *Proceedings of the National Academy of Sciences* 115(45), 11549–11554. DOI: 10.1073/pnas.1809259115.

National Science Foundation (1997) NSF funds first long-term studies of urban ecology. Available at: https://www.nsf.gov/news/news_summ.jsp?cntn_id=102837 (accessed June 2019).

Odum, E.P. and Burleigh, T.D. (1946) Southward invasion in Georgia. *The Auk* 63(3), 388–401. DOI: 10.2307/4080124.

Paap, T., Burgess, T.I. and Wingfield, M.J. (2017) Urban trees: bridge-heads for forest pest invasions and sentinels for early detection. *Biological Invasions* 19(12), 3515–3526. DOI: 10.1007/s10530-017-1595-x.

Pardee, G.L. and Philpott, S.M. (2014) Native plants are the bee's knees: local and landscape predictors of bee richness and abundance in backyard gardens. *Urban Ecosystems* 17(3), 641–659. DOI: 10.1007/s11252-014-0349-0.

Pickett, S.T.A., Cadenasso, M.L., Childers, D.L., Mcdonnell, M.J. and Zhou, W. (2016) Evolution and future of urban ecological science: ecology *in*, *of*, and *for* the city. *Ecosystem Health and Sustainability* 2(7), e01229. DOI: 10.1002/ehs2.1229.

Polakowski, N.R., Lohr, V. I. and Cerny-Koenig, T. (2011) Survey of wholesale production nurseries indicates need for more education on the importance of plant species diversity. *Arboriculture & Urban Forestry* 37, 259–264.

Reid, N.M., Proestou, D.A., Clark, B.W., Warren, W.C., Colbourne, J.K. *et al.* (2016) The genomic landscape of rapid repeated evolutionary adaptation to toxic pollution in wild fish. *Science* 354(6317), 1305–1308. DOI: 10.1126/science.aah4993.

Rittel, H.W.J. and Webber, M.M. (1973) Dilemmas in a general theory of planning. *Policy Sciences* 4(2), 155–169. DOI: 10.1007/BF01405730.

Santangelo, J.S., Rivkin, L.R. and Johnson, M.T.J. (2018) The evolution of city life. *Proceedings of the Royal Society B:* 285, 20180230.

Shandas, V., van Diepen, A., Voelkel, J. and Rao, M. (2018) Coproducing resilience through understanding vulnerability. In: Inan, K.N. and Inoue, M. (eds) *Building a Climate Resilient Economy and Society: Challenges and Opportunities*. Edward Elgar, Cheltenham, UK and Northamptom, Maine, pp. 78–95.

Shochat, E., Lerman, S.B., Anderies, J.M., Warren, P.S., Faeth, S.H. *et al.* (2010) Invasion, competition, and biodiversity loss in urban ecosystems. *BioScience* 60(3), 199–208. DOI: 10.1525/bio.2010.60.3.6.

Slabbekoorn, H. and Peet, M. (2003) Ecology: birds sing at a higher pitch in urban noise. *Nature* 424(6946), 267–267. DOI: 10.1038/424267a.

Smith, M. (2019) *Cities: The First 6,000 Years*. Viking Books (Penguin Random House), New York.

Suárez-Rodríguez, M., Montero-Montoya, R.D. Macías Garcia, C, Garcia, C.M. (2017) Anthropogenic nest materials may increase breeding costs for urban birds. *Frontiers in Ecology and Evolution* 5, 1–10. DOI: 10.3389/fevo.2017.00004.

Townsend, A.K. and Barker, C.M. (2014) Plastic and the nest entanglement of urban and agricultural crows. *PLOS ONE* 9(1), e88oo6. DOI: 10.1371/journal.pone.0088006.

United Nations, Department of Economic and Social Affairs, Population Division (2018a) World urbanization prospects: the 2018 revision: key facts. Available at: https://population.un.org/wup/Publications/Files/WUP2018-KeyFacts.pdf (accessed June 2019).

United Nations, Department of Economic and Social Affairs, Population Division (2018b) The sustainable development goals report. Available at: https://www.un-ilibrary.org/economic-and-social-development/the-sustainable-development-goals-report-2018_7d014b41-en (accessed June 2019).

United Nations, Department of Economic and Social Affairs, Population Division (2019) World population prospects 2019: highlights (ST/ESA/SER.A/423). Available at: https://population.un.org/wpp/Publications/Files/WPP2019_Highlights.pdf (accessed June 2019).

Voelkel, J., Hellman, D., Sakuma, R. and Shandas, V. (2018) Assessing vulnerability to urban heat: a study of disproportionate heat exposure and access to refuge by socio-demographic status in Portland, Oregon. *International Journal of Environmental Research and Public Health* 15(4), E640. DOI: 10.3390/ijerph15040640.

Watkins, S.L., Mincey, S.K., Vogt, J. and Sweeney, S.P. (2017) Is planting equitable? an examination of the spatial distribution of nonprofit urban tree-planting programs by canopy cover, income, race, and ethnicity. *Environment and Behavior* 49(4), 452–482. DOI: 10.1177/0013916516636423.

Wheeler, T.B. (2019) Science foundation cuts 20-year-old Baltimore ecological study's funds. *Bay Journal*.

White, A. (2016) From nursery to nature: evaluating native herbaceous flowering plants versus native cultivars for pollinator habitat restoration. PhD dissertation. University of Vermont, Burlington, Vermont.

Young, R.F. and Wolf, S.A. (2006) Goal attainment in urban ecology research: a bibliometric review 1975–2004. *Urban Ecosystems* 9(3), 179–193. DOI: 10.1007/s11252-006-8589-2.

Index

Note: bold page numbers indicate table; italic page numbers indicate figures.

Abies fraseri 26
Abuja (Nigeria) 165
Acanthococcus lagerstroemiae 27
Acer see maple
Adelaide (Australia) 21
Adelges tsugae 26
Aedes albopictus 186–187
affluent communities 5, 6–7, 47–48, 130, 160
Africa 1, 2, 90, 162
agriculture 2, 5, 15, 50, 78, 88, 100, 109, 110, 112, 163, 193–213
agriculture, urban 5, 50, 109, 193–194
    benefits of 194, *196*
    and biodiversity 193, 194, 196–198, *196*
    and ecosystem services/disservices 193, 194, 205–206
    and gardens 109, 194, **195**, 198, 201, 202
    and pollinators 193, 194, 197, 200, 202, 204, 218
    types of **195**, 196
    see also arthropods and urban agriculture; pest species
Agrilus anxius 25
A. planipennis 17, 26, 42, 52
Ailanthus altissima 27, 73
air conditioning 4
air quality/pollution 6, 8, 51, **70**, 88, 105, 123, 166, 174
ALAN (artificial light at night) *see* light pollution
algae 152, 178
alligators 48
allotments 41, 194, **195**, 198
amphibians 43, 52
Anatidae 52
Anderson, J.R. 129
Anisota senatoria 23
anoles
    brown (*Anolis sagrei*) 154
    crested (*Anolis cristatellus*) 154
Anoplophora glabripennis 17
Anthropocene 1, 2, 4, 218–219
ants 42, 88–89, 150, 196, 197, 215
aphids 20, 23, 27, 179, 180, 197, 201
Apis mellifera 101, 102–103, 106, 204
aquatic systems 129, 135, 174–175, 178–179
Arabidopsis thaliana 74
architects 173
Argentine ant (*Linepithema humile*) 42
arid regions 4, 16, 47, 69, 73, 74
Arionidae 86, 178
Aristotle 162
Arizona (US) 5
arthropods 38, 45, 48, 49, 51, 52, 89
arthropods and urban agriculture 193–206
    and abiotic factors 202, 203–205
    and biodiversity/richness/abundance patterns 196–198
    and ecosystem functions/services 198, 199–200, 200–201, 202, 206
    and functional trait diversity 198–199

arthropods and urban agriculture (*continued*)
 future research directions for 202–205
 and herbivory/plant damage/pest control 201–202
 and landscape management approach 194–196
 and movement/connectivity/spillover 202–203
 and new research technologies 205
 and pest control, biological 194, 196, 198
 and pesticides 203, 204
 and pollination/fruit set 202
 and pollution 203, 204–205
 and rural agriculture, compared 194, 197, 202, 203
 and species interactions/food webs/networks 199–200
arthropods and urban forests 15, 16, 18–27, 28, 52
 butterflies/moths 19–20
 exotic, damage to native trees by 25–26
 and exotic plants 22–25, *23*, 27
 and habitat fragmentation 19–20
 and herbivory patterns 20–21
 and novel interactions 22–23
 and plant density/diversity 18–19
 and resistance/susceptibility 23–25
 and urban heat island 18
arthropod communities 15–36, 198, 202, 203, 204
ash (*Fraxinus* spp.) 26, 42, 52
 F. mandshurica 26
Asia 1, 2, 27, 106, 162
Asian citrus psyllid 20
Asian long-horned beetle (*Anoplophora glabripennis*) 17
Attenborough, David 174–175
attitudinal factors 48–49, 51, 85, 126
Aurora (US) 19
Australia 20, 21, 39, 162, 167
azalea/azalea lace bug 27

Babylon 162
Bakersfield, California (US) 49
balsam woolly adelgid 26
Baltimore Ecosystem Study (BES) 3, 5–6, 7, 9, 122, 124–134, 219
 annual stakeholders' meeting 127–128
 as community of practice 128
 and ecological urban design 133–134
 and economics 131
 and field trips 133, 134, 137
 and historical geography 131–132, 137
 and human ecosystem 127, *128*
 integration/synthesis in 126–128
 and land use/riparian function 128–129
 and patch dynamics/watershed theory 124–126, *125*, 127, 129, 137
 research questions in 127
 social-ecological integration in 134–137, *136*
 and urban heat island 132–133
Baltimore, Maryland (US) 71, *75*, 110, 123, 124, 146
 Department of Public Works (DPW) 134, 135
Baltimore Mosquito Study (BMS) 184, 186–189
 and agency of participants 187, 189
 and improvement of environment 187, 188, 189
 and Photovoice 187–188
 and science–community alignment 189
Basel (Switzerland) 16
bats 42, 44, 100, 178
bear, black (*Ursus americanus*) 52
beech bark scale (*Cryptococcus fagisuga*) 26
beech (*Fagus grandifolia*) 26
beekeeping, urban 109–112
 and bee conservation 111–112
 benefits of 109–110
 challenges of 110–111
 future of 111–112
 growth in 109
 regulations/guidelines for 111
bees (Hymenoptera) 5, 38, 41, 42, 52, 100
 books on 114–115
 bumble (*Bombus*) 101, 102, 103, 104, 105, 107, 204
 cavity-nesting 107–108, 198
 conservation of 101, 106, 108, 216
 dispersal ability/foraging range of 102
 and floral diversity 101–102
 honey (*Apis mellifera*) 101, 102–103, 106, 204
 interactions with plants *see* plant–pollinator interactions
 and nesting sites 102–103
 and pollen/nectar profiles 101–102, 104
 and rural-urban gradients 106–109
 solitary 102, 103, 107, 197
 species/families 101
 sweat (*Lasioglossum*) 102
 and toxic defences of flowers 102
 and urban agriculture 194, 196, 197, 198, 199, 200, 203, 204
 and urban beekeeping *see* beekeeping, urban
 and urban heat island 105
 and urbanization 107–108
beetles 17, 26, 42, 199
 see also ladybird beetles
Beijing (China) 21
below-ground processes 5, 9, 73, 76
Berlin (Germany) 89, 147, 167
BES *see* Baltimore Ecosystem Study
biodiversity, global 87, 88, 90, 180
biodiversity, urban 5, 6, 37–38, 46, 85–93, *86*, 109, 137, 216–217
 in arid/tropical cities 48, 52
 and attitudes 48–49, 85, 130
 and connectedness/interaction with nature 92–93
 conservation 37, 40, 44, 45, 50–51, 52–53, 85, 87, 90–93
 and ecosystem services 50–52

and elsewhere-nature/post-wild world 85, 86–88, *87*, 90, 91
and global biodiversity 88, 90
and habitat fragmentation 89
and local uniqueness/global homogenization 90
loss 7, 87, 88, 90, 100, 216
measurement of 85–86
and possibility-focused management 90, 91–92
and rigorous curiosity 90–91
and social/demographic variations 47–48
and spatial/temporal scales 86, 90, 91
and species richness 89
and urban agriculture 193, 194, 196–198, *196*
and urban forests 18, 22
variations in 88
biogeochemical cycles 6, 7, 9, 218
Biological Diversity, Convention on 167–168
biologists 124, 161, 170–171, 173, 174, 215, 216
biophysical urban ecology 151–154, 155
and modelling 153
and predictors of body temperature 152–154
and thermal performance curves 151–152, *151*, 153
birch 25
birds 3, 5, 7, 20–21, 24, 38, 41, 42, 86, 87, 89, 214, 216
and attitudes/health/well-being 48, 49, 51, 172
and ecosystem services 50, 51
feeding 46–47, 51, 172
negative impacts of 52
and phenotypic changes 215
and pollution 43, 44, 204
and species interactions 44, 45
and vegetation composition 45–46, 53
Blattodea 38, 51, 89
BMS *see* Baltimore Mosquito Study
boar, wild 89
Bombus 101, 102, 103, 104, 105, 107, 204
Boston, Massachusetts (US) 76–77
Bournville (UK) 165
boxtree (*Buxus* spp.) 27
Brasilia (Brazil) 165
Brassica crops 197, 201
Brazil 21, 48, 160, 165, 167
Britain (UK) 41, 47–48, 106, 113, 114, 200
Green Cities Movement in 163, 164, 165, 168, 170
bronze birch borer (*Agrilus anxius*) 25
brown marmorated stink bug (*Halyomorpha halys*) 26
Burch, William 124, 127
Burleigh, Thomas 214
butterflies/moths (Lepidoptera) 19–20, 24, 38, 46, 52, 178, 179, 196, 197, 216
butterfly bush (*Buddleja davidii*) *86*

California (US) 19, 49, 72, 76, 109, 197, 198, 201, 202
Canada 38, 43, 106, 111, 162, 199
Canis latrans 40–41, 43, 48, 52
cankerworms 20, 22
canopy UHI 145, 147–148
and biome 147
and city size 148
and diurnal cycle 147
and latitude 148
measurement of 147
and season 147
and surface UHI, compared 147
within-city variation in 148
carbon dioxide (CO_2) 2, 6, *68*, 69–71, **70**, 76
carbon sequestration 51, 76, 194, *196*
Caretta caretta 179
Castanea dentata 26
Cato the Elder 163
cats, domestic (*Felis catus*) 40, 45, 89
Ceaușescu, Nicolae 167
cemeteries 40, 41
Central Arizona-Phoenix LTER 3, 9, 122, 133, 219
Charlotte, North Carolina (US) 22–23
Chesapeake Bay (US) 126, 127
chestnut, American (*Castanea dentata*) 26
chestnut blight (*Cryphonectria parasitica*) 26
Chicago, Illinois (US) 40–41, 45–46, 48, 88, 110, 113, 132–133, 201, 202, 214
children 52, 87, 92, 113–114, 186
China 15, 167, 171, 186
Chionaspis pinifoliae 25
cicadas 18
Cincinnati, Ohio (US) 20, 24
cities
in ancient world 162–163
ecological footprint of 6
physical environment of 4–5, 160
range of population sizes/densities in 2
citizen science 92, 114, 184–189
benefits of 184, 185, 186
co-created projects 185, 187, 188
contributory projects 185, 187
and data collection 184, 185, 186, 187–188
and diversity 186
and environmental justice 188
history of 184
three levels of engagement in 184–185
see also Baltimore Mosquito Study
city size 21, 146–147, 148
civic engagement/awareness 93, 184, 185, 189
Cleveland, Ohio (US) 17, 21, 150, 199
climate 4–6, 7
climate change 2, 7, 8, 73, 77, 105, 144, 219
and resilience 7, 8
and species interactions 150
see also global warming
climate, urban 4, 9
see also urban heat island
clover 106, 109, 150
coastal cities 7, 8

cockroaches (Blattodea) 38, 51, 89
coevolved relationships 22, *23*, 24, 25, 26, 27, 28, 89, 90, 101
cognitive psychology 166, 168, 171, 173, 174
Columba livia 49, *86*, 172
commercial/industrial districts 38
communities of colour 131–132, 137, 184, 186, 188
community ecology 67, 68, 108–109, 110, *110*, 124
community engagement 109, 135, 189
community forests 123, 124
community gardens 103, 106, 109, **195**, 197, 201, 205
community of practice 128
connectivity 41, 89, 91, 92, 197, 198, 199, 203
conservation 37, 40, 44, 45, 50–51, 52–53, 85, 87, 185
    and Green Cities Movement 161, 166, 168, 169, 172–173
consumption 130
Corvus spp. *see* crows
coyote (*Canis latrans*) 40–41, 43, 48, 52
crape myrtle aphid (*Tinocallis kahawaluokalani*) 27
crape myrtle bark scale (*Acanthococcus lagerstroemiae*) 27
creation–destruction–creation cycle 163
Crithidia bombi 200
crows (*Corvus* spp.) 51
    C. brachyrhynchos 52
Cryphonectria parasitica 26
Cryptococcus fagisuga 26
cucumber 202
cues to care 108, 112
culture 127–128, 169, 173
Cynthia moth (*Samia cynthia*) 27

dandelion (*Taraxacum officinale*) 89, 106, 109
de-icing salts 69, 73, 134–135
deer 69, **71**
    white-tailed (*Odocoileus virginianus*) 41, 52, *86*
democracy 164, 165, 168, 185
determinants of diversity 19, 39, 41, 45, 48, 73–74
determinants of the biology and ecology of plants and animals in urban environments 43, 193–213, 199, 216
Detroit, Michigan (US) 112
developing countries, urbanization in 1, 2
disease 51, 52, 53, **71**, 150, 160, 164, 186, 205
Douglas fir (*Pseudotsuga menziesii*) 24
drought 7, 18, 28, 52, 69
dumping, illegal 108, 130, 131, 188
Dutch elm disease (*Ophiostoma ulmi/O. novo-ulmi*) 17, 26

EAB *see* emerald ash borer
early-successional habitat 92
ecological footprint 6, 100
ecological knowledge 133, 135–136, 137, 215
ecological restoration 50, 53, 61, 91, 216
ecological urban design 133–136
ecology of prestige 130, 137
economics 122, 128, 131, 133, 137, 169
ecosystem concept 3–4
ecosystem services/disservices *3*, 6, 8, 15, 16, 17–18, 23, 37, 38, 39, 40, 44, 45, 47, 48, 50–52, 53, 67, 77, 78, 91, 100, 144, 150, 180, 193, 194, 198, 200, 201, 202, 203, 204, 205
    social/cultural 51
    and urban agriculture 193, 194, 205–206
    and urban forests 15, 17
    and wildlife 37, 45, 50–52, 53
    see also pollinators
ecosystem, urban 1, 4, 5–7
    functions 6–7, 42, 198, 199, 200–201, 202, 206
    future of research on 171–172
    imports to 6
ectotherms and urban warming 105, 144, 150–151
    and adaptation 150
    and biophysical ecology *see* biophysical urban ecology
    and body temperature 151
    and species interactions 150
    and validation 151
    variations in vulnerability 151
Edmonton (Canada) 43
education 113–114, 130, 171, *196*
Edward I 163
elm (*Ulmus* sp.) 17
    American (*U. americana*) 26
    English (*U. procera*) 26
elsewhere-nature 85, 86–87, *87*, 88, 90, 91
emerald ash borer (EAB, *Agrilus planipennis*) 17, 26, 42, 52
energy budget 132, 133, 149
energy flow 6
Engels, Friedrich 164–165
Engystomops pustulosus 180
environmental justice 137, 188
Environmentalism–A Global History 162
ESA (Ecological Society of America) 1, *2*
eucalyptus 27
Eugenia see myrtle
euonymous/euonymous scale 27
Europe 1, 15, 17, 21, 24, 27, 38, 39, 90, 106, 133, 146
    GCMs in *see* Green Cities Movement
evapotranspiration 43, 72, 76, 77
exotic plants 9, 40, 41, 45–46, 50, 68, *68*, 172, 216
    and pollinators 103, 104, 108
    see also invasive species; ornamental plants
exotic plants and urban forests 15, 16, 17, 21–27
    and arthropods 22, *23*
    increase in proportion of 21
Exotic plants, native plants and urbanization 17, 20, 21–23, 40, 74, 78
extreme events 2, 7

Fagus grandifolia 26
Fallopia japonica 172
fauna *see* wildlife, urban
fertilizers 6, **70**, 71, 74, 75, 130, 194, 199
fir, Fraser (*Abies fraseri*) 26
Fisher, Frank 168
Flint, Michigan (US) 185, 189
flood control 50
flooding 5, 7
floral resources 46, 48, 50
Florida (US) 24, 48
flux towers 76
food webs 6, 44, 179, 199–200, 204
Forest(s) 2, 5, 15–36, 37, 38, 41, 43, 50, 51, 52, 69, 71, 74, 75, 88, 122, 123, 124, 126, 131, 145, 146, 147, 154, 198, 201
Forest remnants 15, 16, 20
Fortaleza (Brazil) 21
fossil fuels 6
foxes (*Vulpes* spp.) 44, 49, 89
  San Joaquin kit (*V. macrotis mutica*) 49
France 41
Fraxinus spp. *see* ash
frogs 47, 180
functional diversity 16, 21, 41, 50, 198–199
functional traits 9, 77, 198–199
funding of urban ecology 8, 87, 113, 123, 124, 127, 171, 172, 189, 219
fungi 26, 172
  mycorrhizal **71**, 77–78
future for urban ecology 214–219
  and scale/linkages/equity 217–218
  and urban biodiversity 216–217
  and urban evolution 215–216
  and urban heat/flooding 218

Galapagos finches 215
garden cities 164, 165
Garden Festival Movement (Germany) 165
gardens 15, 20, 24, 25, 38, 40, 41, 45, 46, 67, 69, 76–77, 92
  in ancient world 162, 163
  community *see* community gardens
  management of 48, 53
  in Middle Ages 163–164
  and pollinators 103, 104, 106, 216–217
  and prestige 130
  and urban agriculture 109, 194, **195**, 198, 201, 202
  see also lawns
GCM *see* Green Cities Movement
Geddes, Patrick 164, 165, 170
geese (*Anatidae*) 52
geography 122, 123, 128, 171
  historical 131–132
Germany 17, 49, 151, 165, 166, 200
  see also Berlin
ghost crab 179
Gilbert, Oliver 172
Ginkgo biloba 27, 73
global environmental change 7–8
  and scale 7
global warming 4, 145, 149–154
  and urban heat island, compared 149–150, 153
gloomy scale (*Melanaspis tenebricosa*) 18, 20, 51, 201
golf courses 41
grasslands 92
Greece, ancient 162
Green Cities Movement (GCM) 160–175
  in 19th/20th centuries 164–165
  abiotic aspects (AGCM) 160, 161
  in ancient world 162–163
  and architects 173
  in Australia 167
  biotic aspects (bGCM) 160, 161, 162, 164, 165, 166, 168
  in China 167
  and cognitive psychology 166, 168, 171, 173, 174
  and creation–destruction–creation cycle 163
  future for 169–173
  and garden cities 164, 165
  and international agencies 167–168
  and land/housing shortage 169–170
  and language 160–161
  literature on 161–162
  in Middle Ages 163–164
  and New Towns Movement 165
  in Poland 166
  and postwar redevelopment 165
  and research, pure/applied 171–172
  in Romania 167
  and specialization 170–171, 172
  and tension between professions 161
  and three conflicting approaches 166
  and 'two cultures' debate 160, 169
  and wildlife conservation 161, 166, 168, 169
green infrastructure 8, 107, 134
  and phytoremediation systems 78
  see also green roofs; street trees
green roofs 77–78, 105, 107, 109, **195**
green spaces 5, 6, 15, 39, 40, 67
  in ancient world 163
  enlargement/enhancement of 105, 131
  management of 50, 53, 71–72, 100–101
  standards for 168
gulls (*Larridae* spp.) 52, 172
Gwynns Falls (Baltimore) 131
gypsy moth 26

habitat analogues 71, 88, 122
habitat degradation 38, 100, 184, 186

habitat fragmentation 19–20, 27, 37, 92, 100, 103–104, 197, 215
and genetic diversity 89
and urban heat island 42
and wildlife crossings 44
habitat loss 44, 88, 100, 103
Halyomorpha halys 26
Hausmann, Georges-Eugene 164
health/well-being 4, 8, 15, 46–47, 51, 67, 77, 78, 92, 186–187, 188
heat archipelagoes 132, 133
heat island 4, 6, 7, 18, 42, 43, 50, 68, 70, 76, 77, 105, 132–133, 144, 152, 176, 194
heat loading 132
heatwaves 7, 132–133
heavy metals 69, **70**, 78, 123, 180, 203, 204
hemlock (*Tsuga canadensis*/*T. caroliniana*) 26
hemlock woolly adelgid (*Adelges tsugae*) 26
herbicides 46, **70**
herbivory in urban Areas 20–21, 22–28
herbivory, exotic versus native plants 16, 17, 20, 21, 22–23, 24, 25–28, 38, 74
herbivory, native and exotic arthropods 22–23, 24, 25–28
HERCULES (land cover classification) 129, 134, 137
herpetofauna 43, 48
high-resolution hyperspectral imagery 76
Hippodamus 162
homogenization, biotic 17–18, 25, 26, 37, 38, 41, 53, 90
Hong Kong 167
house prices 130, 131, 132, 168
house sparrow (*Passer domesticus*) 42, 49, 89
housing 2, 3, 5, 165, 168
shortage 169–170
slum 2, 170, 173, 174
hoverflies 27, 41
Howard, Ebenezer 164, 165, 175
Hubbard Brook LTER, New Hampshire (US) 126
Human Ecosystem Framework 3, *125*, 127, *128*
human settlement 2, 7, 67, 167
hydrology 4–5, 6, 7, 9, 126, 215, 218

impervious surfaces 5, 71, 73, 88, 103, 197, 198, 201
and heat island effect 6
and urban forests 16, 18
and wildlife 38, 41, 42
India 90, 164
Indiana (US) 52
infrastructure
ageing/inadequate 7
below-ground 9
built 3, 4, 5
and ecosystem processes 6
green 8, 77–78, 105
insect pollinators, *see also* bees
insecticides 105, 106
see also pesticides
insects 38, 43, 45, 51–52, 69, **71**, 152
decline in 203, 206
see also ants; bees; butterflies/moths
Integrated Social-Ecological System Model 3
intermediate disturbance hypothesis 217
introduced species 67, **70**, 73, 88, 89, 90, 91, 103, 173
invasive species 24, 38, 40, 41–42, 45, 50, 51–52, **70**, 90, 103, 216
irrigation 69, 73, 74, 75, 104, 130, 194, 199, 217

jalapeno pepper 202
Japan 26, 38
Japanese knotweed (*Fallopia japonica*) 172
Jekyll, Gertrude 164
Jellicoe, Geoffrey/Jellicoe, Susan 161
Jones Falls (Baltimore) 131

Korea 18
Kumasi (Ghana) 21

ladybird beetles 180, 196, 197, 198, 199, 202
land cover 6, 7, 21, 131, 202, 215
classification of 129, 137
land management 37, 40, 45
land use 40–41, *68*, *72*, 129, 197
classifications 129, 137
history 67, 71, 74
land-sparing/land-sharing approaches to 92
see also LULCC
land values 169
landscape architecture 90, 108, 133, 165, 166, 167, 172, 215, 216
Landscape of Man, The (Jellicoe/Jellicoe) 161
landscaping 45–46, 47, *68*, 72
Larridae spp. 52, 172
Lasioglossum 102
Latin America 2, 8, 90, 106, 164
lawns 38, 40, 67, 72–73, 75, 217
and insecticides 106
mowing 46, 53, **70**, 106, 113
Le Corbusier 164
leaf-miners 19, 20, 201
leafhoppers (Hemiptera: Auchenorrhyncha) 24
Leavis, F.R. 160, 169
Lees, Audrey 168
LiDAR data 76
light pollution 2, 4, 43, 44, 105, 106, 176–183, 203, 204, 205
and food webs 179
global extent of 176, *177*, 180
and host–parasite interactions 179, 180
impacts of 176–177
management of 180
and plant–herbivore interactions 178
and pollination 179

and predator–prey interactions 176, 177–179
sources of 177
lime leaf miner 27
Lincoln, Nebraska (US) 18
Linepithema humile 42
lizards 47, 153–154
London (UK) 163, 165, 168
Long-Term Ecological Research Program 3
Los Angeles, California (US) 19, 72, 76
low-income communities 1, 39, 40, 48, 49, 130, 193–194
LTER (Long-Term Ecological Research) programme 3, 9, 122, 124, 126
see also Baltimore Ecosystem Study; Central Arizona-Phoenix LTER
LULCC (land use/land-cover change) 7
Luniak, Maciej 166, 171
Lutyens, Edward 164
Lutz, Frank 214
Lutzomyia longipalpis 205
Lyme disease 52, 53

McDonnell, Mark 122, 123
McHarg, Ian 133, 166
mammals 38, 48, 50, 52, 69, 174
maple (*Acer*) 24, 25
Norway (*A. platanoides*) 24
red 73, 151
Marcus Terentius Varro 163
Marx, Karl 164–165
meadows 92
Mediterranean cities 16, 26
Medley, Kimberley 123
Melanaspis tenebricosa 18, 20, 51, 201
Melbourne (Australia) 167
Mephitis mephitis 44, 89
Mesopotamia 162
metabolism of urban systems 6, 9
Mexico City (Mexico) 17, 21
Miami, Florida (US) 154
Michigan (US) 52
microbes 5, 6, **71**, 178
micronutrients 101
midges (*Corethrella* spp.) 180
Mies van der Rohe, Ludwig 164
migratory species 43, 44, 46, 48, 50, 92
milkweed (*Asclepias*) 46
Milvus milvus 47
mimosa webworm 18
Minnesota (US) 16
mites 18, 20, 23, 201
Modern Environmentalism (Pepper) 162
MODIS (Moderate Resolution Imaging Spectroradiometer) 76
molecular techniques 91, 203, 205
monarch butterfly (*Danaus plexippus*) 46, *86*
More, Thomas 163–164, 175
Morris, William 164, 165
mosquitoes 48, 50, 51, 87, 130, 150, 205
Asian tiger (*Aedes albopictus*) 186–187
see also Baltimore Mosquito Study
moths *see* butterflies/moths
mountain lion (*Puma concolor*) 52
mouse, house (*Mus musculus*) 89
mulch cover 103, 197, 199, 201
Mumbai (India) 160
Mumford, Lewis 165
Munich (Germany) 49
mussels 153
mycorrhizal fungi **71**, 77–78, 150
myrtle (*Eugenia*) 24
crape 27

Napoleon III 164, 173
National Science Foundation 122, 124, 219
Natural England 168
natural history 162, 163, 170, 214
natural selection 68–69, 71, 73, 215
nature 86–87, 173
connectedness with 51, 85, 92–93
nature deficit disorder 85, 92
nature-based solutions *see* green infrastructure
NDVI (Normalized Difference Vegetation Index) 76
neighbourhood associations 126, 218
neighbourhoods 130, 131, 135–136, 218
see also affluent communities; low-income communities
neonicotinoids 106
Nephila pilipes 178
Nero 163, 173
network analysis 200
New Towns Movement 165
New York City (US) 8, 17, 21, 71, 75, 89, 109, 123, 124, 189, 214
Central Park 164
New York State (US) 16
New Zealand 39, 162
NGOs (nongovernmental organizations) 126, 137
nitrate 105, 129
nitrogen 46, 50, 69, 73, 105–106, 204
leaf 16, 18, 71, 76
soil 123
noise pollution 4, 6, 43, 174, 176, 179, 180, 205, 215
non-native species *see* exotic plants; introduced species; invasive species
Normalized Difference Vegetation Index (NDVI) 76
North America 1, 16, 17, 24, 39, 154
nutrient availability 18, 71, 73, 104
nutrient cycling *68*, 123, 215, 218
nutrients/nutrient flows in urban habitats 6, 18, 20, 67, 79, 215, 218
access to nutrients 70, 71, 73, 75, 104, 106
micronutrients 101
nutrient transport/uptake 51, 74, 75, 218

oak (*Quercus*) 18, 19, 20, 21, 24
  pin (*Q. palustris*) 23
  willow (*Q. phellos*) 21, 22–23, *86*
oak spider mite (*Oligonychus bicolor*) 18
Odocoileus virginianus 41, 52, *86*
Odum, Eugene 165–166, 170, 214
Ohio (US) 52
Oligonychus bicolor 18
Olmsted, Frederick 164
Oncorhnchus spp. 179
Ophiostoma ulmi/O. novo-ulmi 17, 26
orange-striped oak worm (*Anisota senatoria*) 23
orb weaver spider (*Nephila pilipes*) 178
Orléans (France) 21
ornamental plants 20, 22, 25, 26, 45, 46, 73, 104, 197
Oslo (Norway) 112
ozone 69, **70**, 71, 76, 204

Palo Alto, California (US) 19
Panthera tigris 49, 174
parasites 20, 24, 27, 42, 44, 47
Paris (France) 89, 164, 200
parks 5, 15, 16–17, 38, 40, 41, 51, 69, 165
  in ancient world 163
  and pollinators 103, 104, 107, 113
Parks & People Foundation (PPF, Baltimore) 135
Parthenolecanium spp. 18, 20
Passer domesticus 42, 49, 89
patch dynamics 124–126, *125*, 137
pattern of urbanization 16-17, 19-20, 21, 27
pavements *see* impervious surfaces
Pearce, Fred 172–173
Pepper, David 162
pest species 20, 23, 26, 27, 28, 49, 50, 51, 69, **71**
  biological control of 193, 194, 196, 198, 200, 201–202
pesticides 45, 46, 74, 130, 194, 203, 204
  see also insecticides
Peter the Great 164
pharmaceuticals 6, 129
Philadelphia, Pennsylvania (US) 108
Phoca vitulina 179
Phoenix, Arizona (US) 45–46, 49, 199–200
  see also Central Arizona-Phoenix LTER
photosynthesis rates 73
Photovoice 187–188
phytoremediation systems 78
Pickett, Steward 123, 127, 128
pigeon paradox *87*, 92
pigeons 38
  rock (*Columba livia*) 49, *86*, 172
pine (*Pinus*)
  North American red (*P. resinosa*) 25
  Scots (*P. sylvestris*) 25
pine processionary moth 26
place, power of 124
plant diversity 5, 16, 17, 21, 22, 101, 108, 200
plant ecophysiology 67
  urban *see* urban plant ecophysiology
plant phenology, alterations to 104–105
plant physiology *68*, 69, 71, 74–77
plant productivity 71, 73, 76, 78
plant–pollinator interactions 101, 103–106, 107, 205
  and environmental contaminants 105–106
  and habitat loss/fragmentation 103–104
  and impervious surfaces 103
  and non-native/managed plant species 104–105
  and urban heat island 105
Plato 162
Pliny the Elder 163
poison ivy 71
Poland 166
Pole Mokotowski Park (Warsaw) 166
pollen 50, 88, 100, 101, 104, 105, 111
  and community conservation 112–114
pollinators 5, 7, 38, 41, 46, 47–48, 50, 53, 105, 106, 176
  books on 114–115
  economic value of 100
  interaction with plants *see* plant–pollinator interactions
  and light pollution 179
  and urban agriculture 193, 194, 197, 200, 202, 204, 218
  variety of 100
  see also bees
pollinator conservation 100–101, 108, 112–114
  and collaborative framework *110*
  and ecology in/for the city 108–109
  and education 113–114
  and light pollution mitigation 105, 106
  and local communities 101, 108–109, *110*
  online resources for 114
  and urban aesthetics/function 100
  and wildflower planting 108, 109, 113, 114
pollution 5, 6, 7, 8, 9, 15, 22, 69, **70**, 76, 88, 103, 105, 106, 123, 127, 129, 130, 134, 147, 150, 161, 164, 166, 168, 173, 174, 203, 204–205, 215, 218
  and plants as remediators 78
  see also light pollution
Population density 2, 5, 18, 47, 88, 130, 214
Population dynamics 5, 9
Population growth 1–2, 42, 152, 153, 165, 201, 214
Population size/density 2, 15, 16, 148, 177
Populus spp. 78
Port Sunlight (UK) 165
Portland, Oregon (US) 109, 110, *111*
possibility-focused management 90, 91–92
post-wild world 85, 87–88, *87*, 90, 91
Pouyat, Richard 122, 123
poverty 5, 132–133, 160, 164, 166
PPF (Parks & People Foundation, Baltimore) 135
predation paradox 44
predator–prey dynamics *39*, 44, 45, 88, 176, 177–178

Press-Pulse Dynamics Model 3, *3*
Procyon lotor 38, 44, 52, 89
pruning 69, **70**
Pseudotsuga menziesii 24
Puma concolor 52
Pyrrhalta viburni 26

Quercus see oak

racial segregation 131–132, 137
racoon (*Procyon lotor*) 38, 44, 52, 89
ragweed 71
Raleigh, North Carolina (US) 21, 151
rats 38, 89, 130
  brown (*Rattus norvegicus*) 89
Reading (UK) 47
red kite (*Milvus milvus*) 47
remnant habitats 40, 41, 88, 90, 107
  forest 15, 16, 20, 22
remote sensing methods 76–77, 91, 145, 205
reptiles 47, 48, 52
research, pure/applied 171–172
residential areas 5, 38, 92
  and urban forests 16, 17, 20
resilience 2, 7–8, 199, 215–216, 218, 219
rigorous curiosity 90–91
Rio de Janeiro (Brazil) 160, 167
roads 3, 4, 38, 42–43, 123, 173
  and noise 43
roadside verges 41, 104
rodents 51, **71**
  see also rats
Rogers, Richard 167, 173
Romania 167
Rome, ancient 162–163
rural–urban gradient 17, 106, 107, 200

St Petersburg (Russia) 164
Salicaceae 78
salinity **70**, 134–135
Salix spp. 78
salmon 178
  Pacific (*Oncorhnchus* spp.) 179
Salt Lake City, Nevada (US) 21
Samia cynthia 27
sampling transect 123
San Francisco, California (US) 19, 109
sand fly (*Lutzomyia longipalpis*) 205
sanitation 2
scale insects 23, 24, 27, 151, 197
  gloomy scale (*Melanaspis tenebricosa*) 18, 20, 51, 201
  pine needle (*Chionaspis pinifoliae*) 25
scale, spatial 7, 16, 21, 37, 38, 39, 40, 43, 46, 68, 74, 86, 91, 129, 215, 217–218
Scandinavian cities 21
scavengers 51, 52
scenario co-development 8–9
Schmid, Calvin 165–166
Schoenauer, Norbert 161–162
Sciurus carolinensis *86*, 172
seal, harbour (*Phoca vitulina*) 179
Seddon, George 167
seed dispersal 50, 88, 215
selection, natural/domestic 68–69, 71, 73
SES (social-ecological systems) 6, 7, 68, *68*, 71, 91, 214
  models 2–3, *4*
SETS (social-ecological-technological) framework *3*, 4, 5, 7, 8–9, 122
sewers 9, 134, 174
shade 15, 16, 73, 74
Shanghai (China) 2
Sheffield (UK) 47, 172
6000 Years of Housing (Schoenauer) 161–162
Skaryszewski Park (Warsaw) 166
skunk (*Mephitis mephitis*) 44, 89
slugs (*Arionidae*) *86*, 178
slum housing 2, 170, 173, 174
Smith, Monica 218
snakes 48
Snow, C.P. 160, 169
social capital 124, 127, 130, 132–133, 218
social cohesion 131, 133, 135
social dimensions 1, 3, 8
  see also SES models; SETS framework
social ecology 124, 133, 135
social equity/inequity 7, 133, 137
social science 9, 78, 123–124, 125, 129–130, 132
social-ecological systems 2, 3, 7, 71, 91, 125, 127, 130, 214, 217
Socio-cultural factors 37, 38, 39, 40, 49, 168, 184–192
Socio-cultural income and urbanization 1, 39, 40, 47, 48, 49, 72, 130, 194
Socio-cultural issues and urbanization 37, 38, 39, 40, 45, 47–52, 78
Socio-cultural systems 3, 6, 7, 68, 71, 74, 77, 78, 79, 91, 122, 124, 125, 127, 130
socio-spatial heterogeneity 5–6
soft-scale complex (*Parthenolecanium* spp.) 18, 20
soil compaction 69, **70**, **71**, 73, 103, 145
soils 5, 6, 15, 69, **70**, 71, 88, 108, 125
South Africa 21
spatial heterogeneity 125, 126, 129, 132, 133
specialization 170–171, 172
species diversity 21, 38, 41, 78, 87, 194, 196, 203, 215–216
species extinction 44, 87, 88, 100
species interactions 5, 9, 22–23, 40, 44–45, 88, 91, 199–200
  and climate change 150
  and light pollution 177–180
  see also predator–prey dynamics

species richness 16, 17, 18–20, 21, 24, 41, 42, 48–49, 51, 89, 92, 196, 217
spiders 18, 42, 178, 196, 197, 198, 204
Spirn, Anne Whiston 133
springtails 198, 204
squirrels 47, 88
 grey (*Sciurus carolinensis*) *86*, 172
Sri Lanka 90
starling (*Sturnus vulgaris*) 52
Staten Island (US) 19
stewardship 47, 53, 85, 91, 92–93, 108, 114
Stockholm (Sweden) 196
stormwater 4, 8, 52, **70**, 129, 131, 134, 135, 137, *196*, 218
street trees 15, 16, 17, 20, 21, 25, 26, 41, 69, 71–72, 78, 189
 as nature-based solution 77
 and pollinators 105, 107
stresses 4, 8, 9, 77
 gradients 69, 123
 soil compaction 69, **70**, **71**, 73, 103, 145
 tolerance of *68*
Sturnus vulgaris 52
suburban areas 17, 41, 130, 217
Sukopp, Herbert 166, 171
surface UHI 145–147
 and biome 145–146
 and canopy UHI, compared 147
 and city size 146
 and diurnal cycle 145
 and latitude 146
 satellite measurement of 145, 146–147
 and season 146
 within-city variations in 146
sustainability 2, 7, 8, 23, 77, 113, 173, 202
 and urban design 67, 134, 214, 215, 216, 217, 219
 see also Green Cities Movement
Switzerland 16, 19
Sydney (Australia) 20
synanthropic species 41–42, 52, 89

tall buildings 4, 5, **70**, 74
Tansley, Arthur 3
Taraxacum officinale 89, 106, 109
technological dimension 3, 8
temperature 69–71, 73, 102–103, 201, 215
 and organism size 42
 and urban forests 20, 22, 27, 28, 73
 Atkinson's temperature-size rule 42
 see also global warming; urban heat island
Thatcher, Margaret 168
Theoka, Andrew 165
Theophrastus 162
ticks 51, 53
tiger (*Panthera tigris*) 49, 174
Tilia spp. 27
timber, salvaging 135–136
Tinocallis kahawaluokalani 27
Toronto (Canada) 111
trade-offs 9, 52, 77, 78, 152, 203
traffic 6, 123, 173, 174
transdisciplinary approach 7, 9, 91, 122–123, 127, 128–129, 133–134, 214, 215, 219
 and over-specialization 170–171, 172
transportation corridors 38, 104
transportation systems 3, *3*, 42
trees 5, 71–72, 104–105
 ecophysiology of 67, 69, 71, 72, 73, 75, 76, 77
 and evapotranspiration 43, 72
 and impervious surfaces 16, 18, 73
 planting 50, 77, 108, 113, 131, 135, 137, 162, 218
 street *see* street trees
 as wind breaks 74
 see also urban forests
Tree diversity 16–17
TRI (Toxic Release Inventory) 131, 132, 137
tropical regions 52, 151
trust 124, 136, *136*, 184, 189
 see also social capital
trypanosomatids 200
Tsuga canadensis/*T. caroliniana* 26
turf grasses 72, 73, 104, 106
turtle, loggerhead (*Caretta caretta*) 179
turtles 48
'two cultures' debate 160, 169

UHI *see* urban heat island
Ulmus sp. *see* elm
United Nations 167–168
 Sustainable Development Goals 7, 214
United States (USA) 1, 3, 45, 52, 69, 73, 162
 citizen science in 184, 186–188
 gardens/lawns in 73, 76–77, 104, 106, 217
 Green Cities Movement in 164, 165–166
 Long-Term Ecological Research programme in *see* LTER research
 pollinators in 104, 106, 108, 109, 110, *111*, 113
 resilience to climate change in 8
 social sciences in 129–130
 socio-spatial heterogeneity in 5–6
 urban agriculture in 194, 197, 198, 199–200, 201
 urban forests in 16, 17, 18, 19, 20, 21, 22–23, 24, 26, 27
 urban heat island in 145, 146, 150, 151, 154
 urban wildlife in 40–41, 45–46, 48, 49
 see also California; New York City; Baltimore; Chicago
urban agriculture 50, 77–78, 109, 193–213
urban bee abundance and richness 105
urban biodiversity 37, 38, 47, 52, 72, 85–99, 196, 216, 217
urban climate 4, 9, 73

see also urban heat island
urban cooling 4, 15, 109, 144
urban design 108, 113, 122, 123, 128, 133–134, 137
sustainable 67, 134
urban development 15, 37, 38, 39, 40, 41, 112
see also Green Cities Movement
urban ecology
funding for *see* funding of urban ecology
future for *see* future for urban ecology
and global environmental change 7–8
growth of 1, *2*, 100
history of 122–124, 214
key challenges in 219
knowledge gaps/barriers in 9, 214, 215
new directions for 9
resistance to 124
social/ecological/technological dimensions of 3, *3*
urban ecosystem fluxes 68, 69, 74
urban ecosystem services 8, 77
Urban effect 18, 20, 27–28, 70
see also global warming;
urban heat island
urban forests 4, 5, 15–36, 40, 41, 50, 52, 123
and arthropods *see* arthropods and urban forests
composition of 15–18
and exotic species 15, 16, 17, 21–27, 28, 172
historical development of 15–16
and impervious surfaces 16
simplification/homogenization of 17, 25, 26, 37, 41, 53, 90, 107, 216, 217
and street trees 15, 16, 17, 20, 21, 25, 26, 41, 69
and tree density/diversity 16–17, 27
and water stress 18
and watershed theory 126
urban form 6, 40, 41, 90
Urban Green Space Standards 168
urban greening initiatives 78, 202
urban heat island (UHI) 4, 6, 7, 42, 50, 132–133, 144–155, 176, *196*, 205, 218
and air temperature 147, 148, 149, 151, 152, 153–154
and arthropods 18
canopy *see* canopy UHI
complexity of 145
and ectotherms *see* ectotherms and urban warming
four types of 145
and global warming, compared 149–150, 153
and habitat fragmentation 42
and pollinators 103, 105
subsurface 145, 148–149
surface *see* surface UHI
and urban plant ecophysiology 68, *68*, **70**, 76, 77
urban hydrology 4, 6, 9, 215, 218
urban infrastructure/physical environment of cities 4–5, 7, 8, 9, 40, 42, 43, 44, 70–71, 77, 102, 105, 107, 108, 122, 126, 127, 129, 130, 131, 134, 135, 145, 161, 187, 194
urban metabolism 218–219
urban planning 47, **74**, 92, 145, 162, 165, 216
and bees 102
and green cities *see* Green Cities Movement
see also urban design
urban plant ecophysiology 67–79
approaches to 68–74
and below-ground processes 73, 74, 76
factors affecting plant growth 69, **70–71**
field research methods 74–77, *75*
future research 78–79
and horticultural practices 73–74
in humid/arid cities 74
and management 67, 71–72, 77
and migration/species range 73
and natural/domestic selection 68–69, 71, 73
and nature-based solutions 77–78
and related disciplines 68, *68*
and stresses 69
and urban-to-rural gradients 69–71
and variations in site types 71, *72*, 78
urban riparian zones 5, 129, 178
urban sites 5, 16, 17, 20, 69, 73, 75, 76, 154
urban-to-rural gradient 16, 41, 69–71, 73, 197, 200
and bee trends 106–109
and intermediate disturbance hypothesis 217
urbanization 1–2, 4, 16, 71, 144, 153, 176, 193, 200, 214
and biodiversity 37–38, 40, 41, 42, 85, 88, 90
and body size 19, 20, 39, 42, 102, 199
and exotic species 21
gradient of *see* urban-to-rural gradient
and habitat fragmentation 19–20
and homogenization 37, 38, 41, 90
and light pollution 176, 179
and pollinators 100, 103, 107–108
URGE (Urban-Rural Gradient Ecology) project 123–124
Ursus americanus 52
Utopia (More) 163–164

vacant lots 16–17, 21, 40, 89, 109, 110, 131
and pollinators 103, 104, 107
and salvaging timber 135–136
vegetation 3, 5
and urban heat island 4
viburnum leaf beetle (*Pyrrhalta viburni*) 26
Virgil 163
Vitruvius 163
Vulpes spp. *see* foxes

Warsaw (Poland) 166
Washington, DC (US) 48
wasps 111, 179, 198
waste 6, 187, 188, 218–219

waste disposal 51, 173, 174
water flows 6, 88, 129, 135
water quality 2, 5, 8, 126, 174, 185
water supply/removal systems 3, *3*, 4
water stress in urban habitats 18, 23
water use by plants 71–73, 74, 75, 77
watersheds 124, *125*, 126, 129, 137
weeds 87, *87*, 89, 93, 104, 200, 216
well-being *see* health/well-being
West Nile virus 48, 51
White, Carl 123
wicked problems 7, 219
wildflower plantings 108, 109, 113, 114
wildlands 37, 38, 41, 45
wildlife crossings 44
wildlife, urban 37–53
  and anthropogenic resource provision 45, 46–47, 48
  and attitudinal factors 48–50, 51
  and body size 20, *39*, 42
  and built structures 42–44
  conservation/promotion of 37, 40, 44, 45, 50–51, 52–53, 85
  and ecosystem services/disservices 37, 45, 50–52, 53
  and environmental features 37, 38, *39*, 40–41
  and filters 9, 38, 40–42
  future research on 52–53
  generalist/synanthropic species 41–42, 52, 89
  and habitat changes 37, 38
  and health/well-being 46–47, 51
  and human drivers 39–40
  and human variation 47–48
  and light pollution 43–44
  and local level 37, 40, 41, 43, 45–46
  and management 37, 38, 39–40, 44–45, 48, 49, 52–53
  mesopredators 44, 45
  and noise pollution 43
  and socio-cultural factors 37, 38, *39*, 40
  and spatial scales 37, 38, 39, 40, 43
  specialist species 37, 38, 42
  and species interactions 5, 9, 40, 44–45
  species pool, interacting factors in 38, *39*
  and species traits 37, 38, *39*, 40
  and tolerance threshold 49
  and urban heat island 42, 43
  and urban-to-rural gradient 41
  and urbanization 37–38, 40
  see also biodiversity, urban
wind 4, 15, **70**, 133
  breaks, trees as 74
Wright, Frank Lloyd 164

Printed and bound by CPI Group (UK) Ltd, Croydon, CR0 4YY

13/09/2026

14961109-0001